ANTARCTICA

ANTARCTICA

ARNOLD ZAGERIS

Fitzhenry & Whiteside

Published in Canada by Fitzhenry & Whiteside, 195 Allstate Parkway, Markham, ON L3R 4T8
Published in the United States in 2016 by Fitzhenry & Whiteside, 311 Washington Street, Brighton, Massachusetts 02135

Fitzhenry & Whiteside acknowledges with thanks the Canada Council for the Arts, and the Ontario Arts Council for their support of our publishing program. We acknowledge the financial support of the Government of Canada through the Canada Book Fund (CBF) for our publishing activities.

Canada Council for the Arts Conseil des Arts du Canada

ONTARIO ARTS COUNCIL
CONSEIL DES ARTS DE L'ONTARIO
an Ontario government agency
un organisme du gouvernement de l'Ontario

Library and Archives Canada Cataloguing in Publication
Zageris, Arnold, 1948-, author
Antarctica : Arnold Zageris.
First edition.
ISBN 978-1-55455-367-9 (bound)
1. Antarctica—Pictorial works. 2. Zageris, Arnold, 1948—Travel—Antarctica. 3. Zageris, Arnold, 1948—Travel—Antarctica—Pictorial works. I. Title.
G860.Z33 2016 919.89′00222 C2016-906736-7

Publisher Cataloging-in-Publication Data (U.S)
Zageris, Arnold, 1948-
Antarctica : Arnold Zageris.
[ ] pages : color photographs ; cm.
Summary: "Antarctica is a photographic journey compiling the images of trips that the author took to Antarctica, and also shared in the book his emotions, discoveries and feelings about this sublime land" – Provided by publisher.
ISBN-13: 978-1-55455-367-9
1. Antarctica – Description and travel. 2. Antarctica – Pictorial works. I. Title.
919.89 dc23 G860.Z343 2016

Text and cover design by Kerry Designs
Printed in China
10 9 8 7 6 5 4 3 2 1

To all my former students. Listen up. No talking.

Here's the deal.
Study the images, read it all, respect our planet,
do your homework and learn from our mistakes
and consider this book dedicated to you.

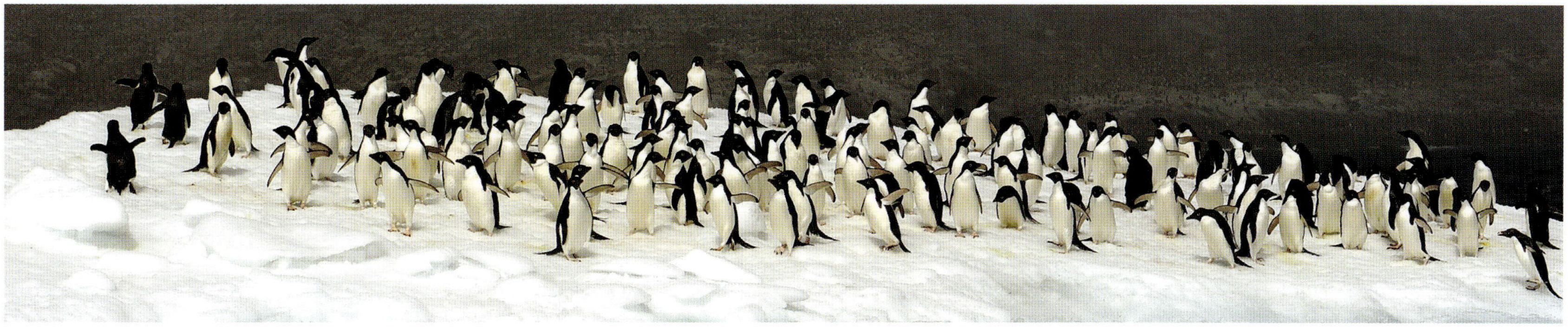

"Even seen for the first time, the Adelie penguin gives you the impression of a very smart man in an evening dress suit, so absolutely immaculate is he, with his shimmering white front and black back and shoulders. He stands about two feet five inches in height, walking very upright on his little legs."

Dr. G. Murray Levick

staff- surgeon on the Terra Nova

(British Antarctic Expedition of 1910-1913)

the ship that carried Captain Scott to his rendezvous with death.

Adelie display

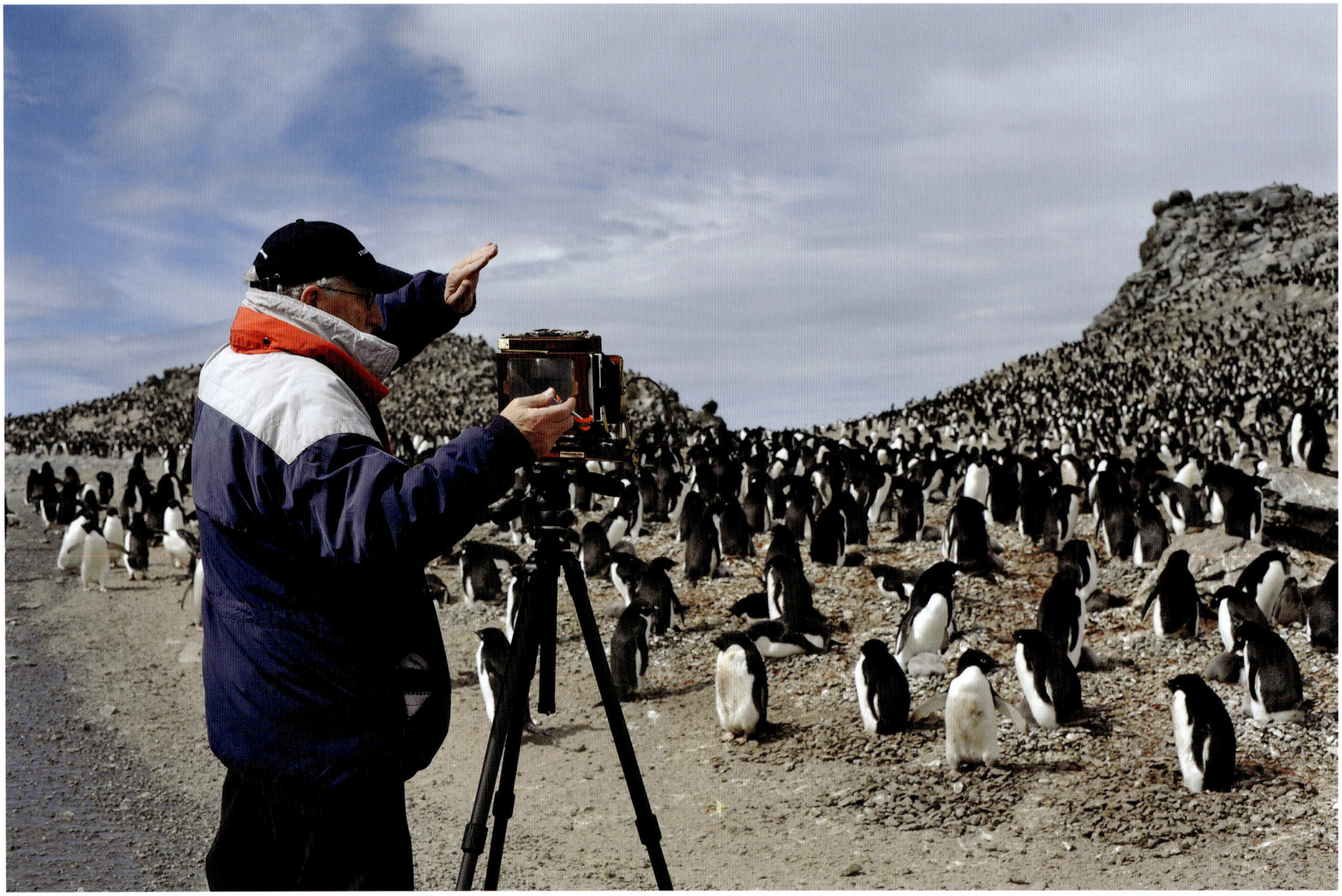

Blocking the sun

WHEN I WAS about ten years old I came across a sketchy drawing of a musk ox. I asked my father if this kind of animal really does exist. He explained that it does and some live in Canada but very far north where it was very cold and mostly covered in ice and snow. To prove it he showed me a picture in one of his books. I was fascinated. Here was a black and white close up picture of a primeval looking animal with a huge head and an enormous bare forehead elongated into two drooping and sharp upward curving horns. Its unusual long hair was streaming out sideways in a windstorm and from beneath this mass of long hair were what seemed to me, at the time, legs that were too short and small. I wanted to see more. I vaguely remember turning the page and immediately becoming disappointed because there was no musk ox, only penguins. Lots of them. I do not remember what my father had told me but it must have been fascinating because I had forgotten the musk ox and my attention turned elsewhere. This was when I first learned of Antarctica.

One has to make a special effort to reach Antarctica and in 1995 I did just that. With the seasons reversed in the southern hemisphere the only time I could get there was during our Canadian winter. I was granted a sabbatical, packed my bags and a new 4 x 5 inch view camera. The two week trip was a revelation. It was wild in the truest sense of the word, untouched, without people, without culture. Nothing I saw or experienced was restricted. Size, temperature, colour, and light all expressed their attributes and the Antarctic. Included on the trip would be a staff of scientists, explorers, researchers, historians, and artists who contributed to the expedition with their inspiring lectures fostering an awareness of the importance of the poles to the planet. I felt this organization made things count by directing its energy to developing young minds. It had a worthy cause and I wanted to contribute and be part of it.

Soon after I contacted Geoff and asked if he might consider including me on his next trip south. After a few introductory and preliminary questions he said that he and his team would consider the request. After a couple days of anxious waiting he called to advise me that I had indeed been accepted as photographer on the next trip. All I had to do now was to get all my papers in order and I would be welcomed. The trip was such a success that I requested to return another time and I think that it was my visible enthusiasm and exuberance that were the motivating factors that allowed Geoff, with much good will, to squeeze into, what I know, was an already complete roster.

Even though I returned another two times with a different crew it was this initial exposure to his extraordinary staff and also his tangible commitment to the accompanying youth that set in motion my resolution to help portray the importance and sublime beauty of this fragile continent.

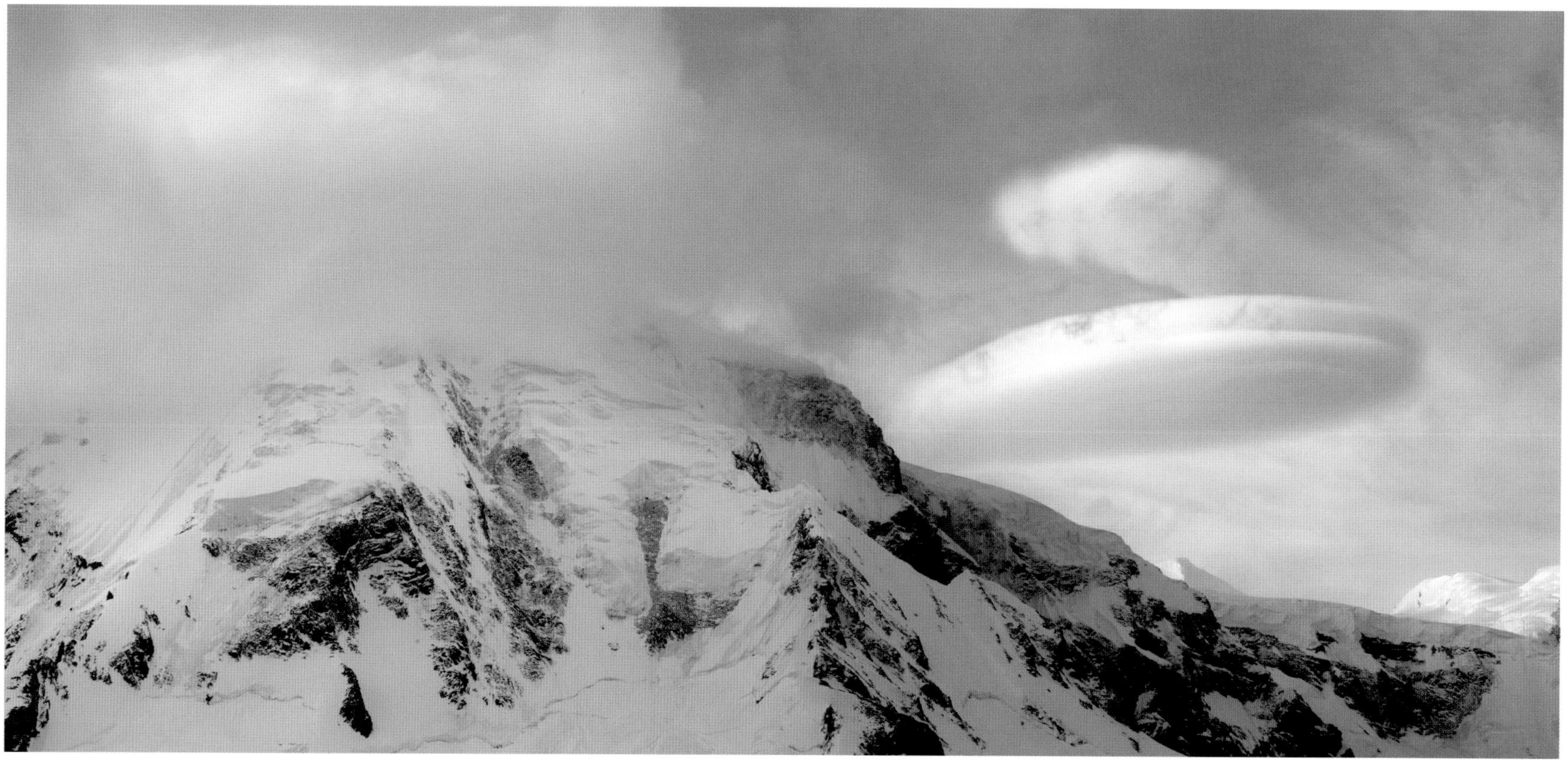

Antarctic sky

MOST ANTARCTIC CRUISES depart from either Puenta Arenas Chile, Port Stanley Falkland Islands, or Ushuaia Argentina. My five visits have always been from Ushuaia and when I climb the shifting gangplank and step on board ship I can already feel Antarctica's presence. This is where I consider my trips to begin.

With a population of nearly sixty thousand Ushuaia is the world's most southerly city and the capital of the province of Tierra del Fuego. Snugly nestled along and against the base of the Le Martial mountain range this city has grown quickly over the years and has become evermore a destination for countrymen and tourists alike. Duty free shops, new restaurants and hotels catering to more expensive tastes as well as the more modest have all claimed prime locations closer to the port or higher up the mountains. Across the water white capped mountains peer from over the horizon and below the city of Ushuaia wakes to the rumbling ships arriving and departing. Since 1995 when I first arrived the town has evolved dramatically into a bustling city strongly focusing on the needs of adventurers heading into the Patagonia highlands or the shores of Antarctica.

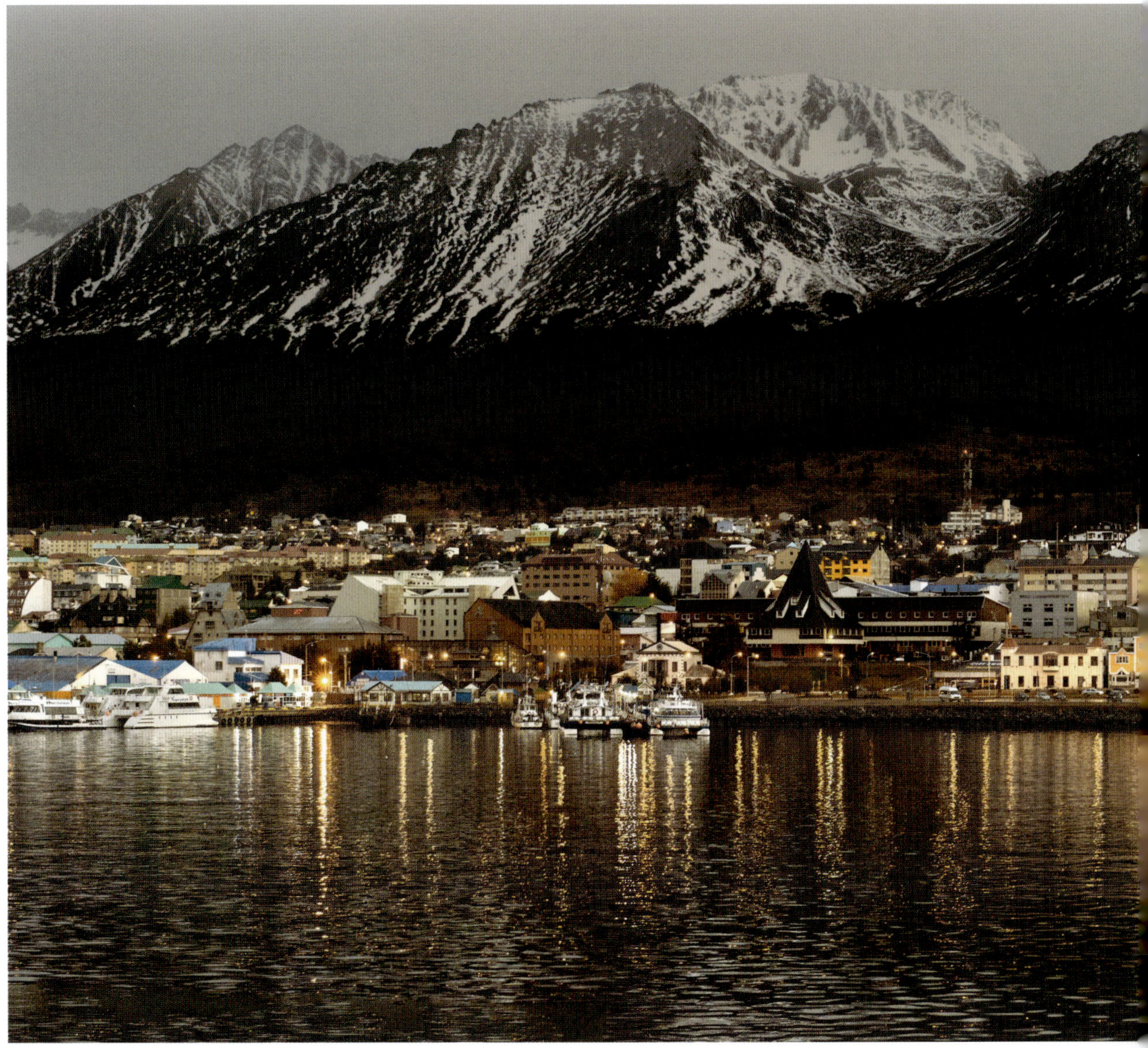

Morning before the lights go out in Ushuaia 2014

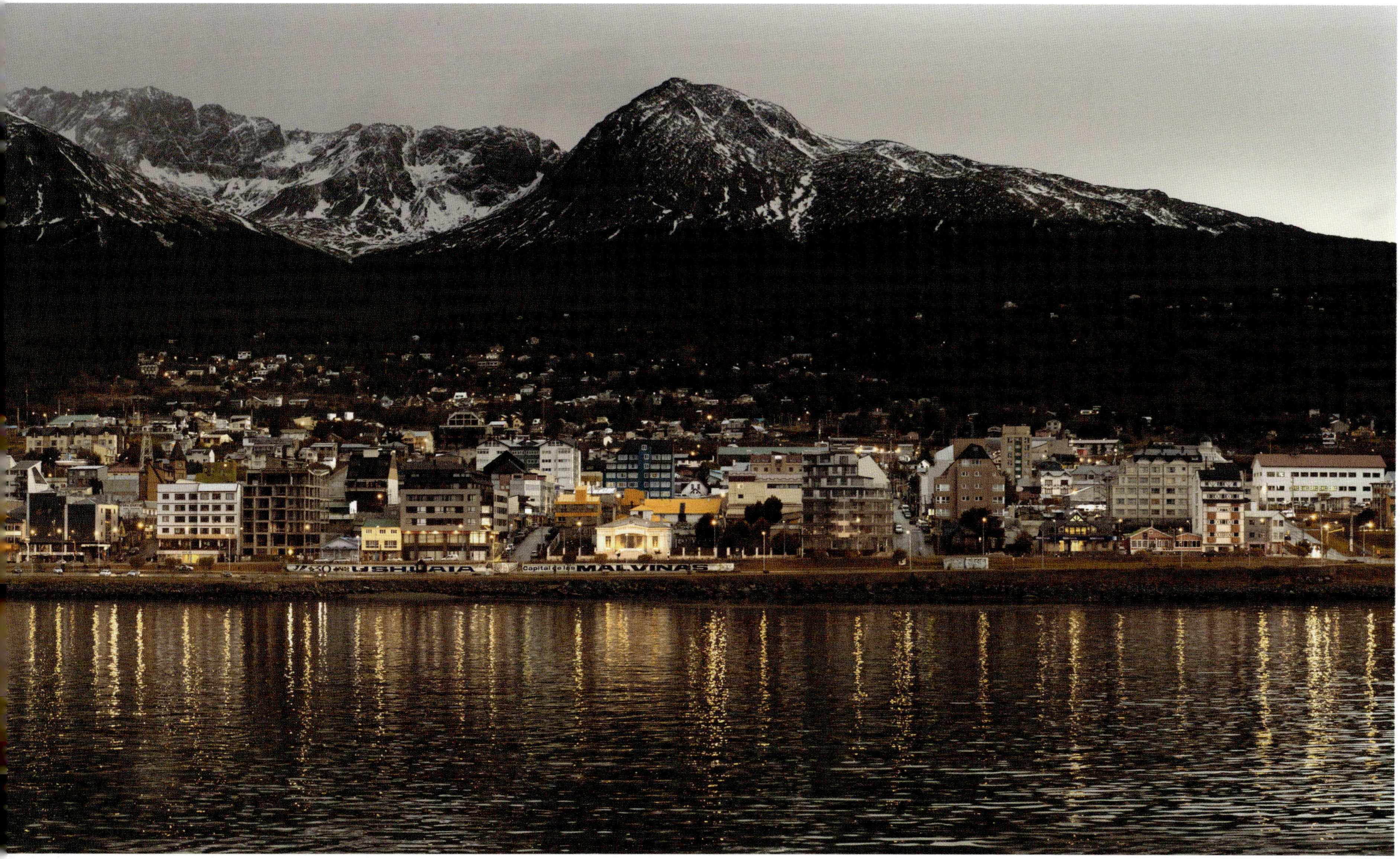
USHUAIA
Capital de las MALVINAS

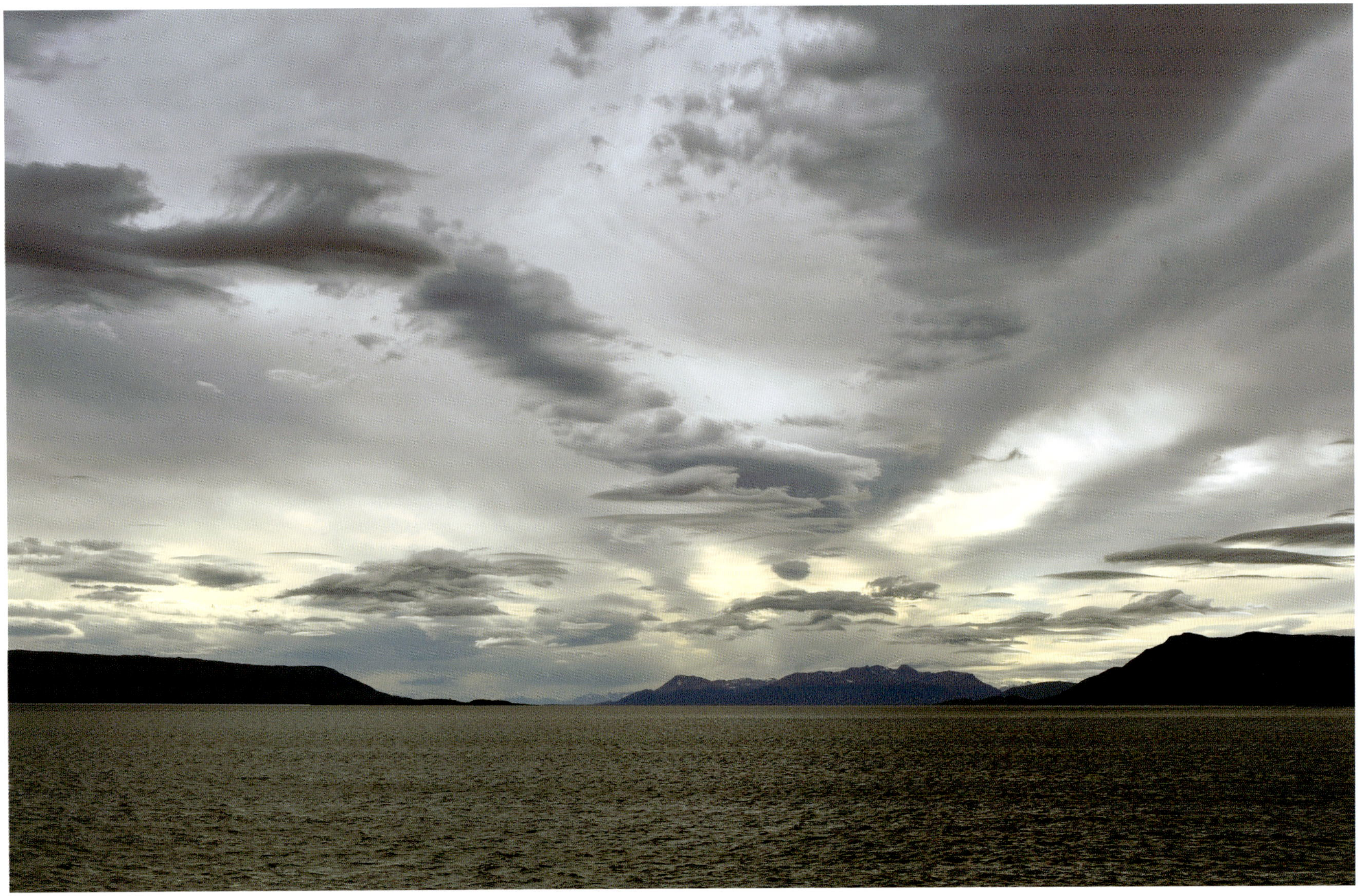

Beagle Channel

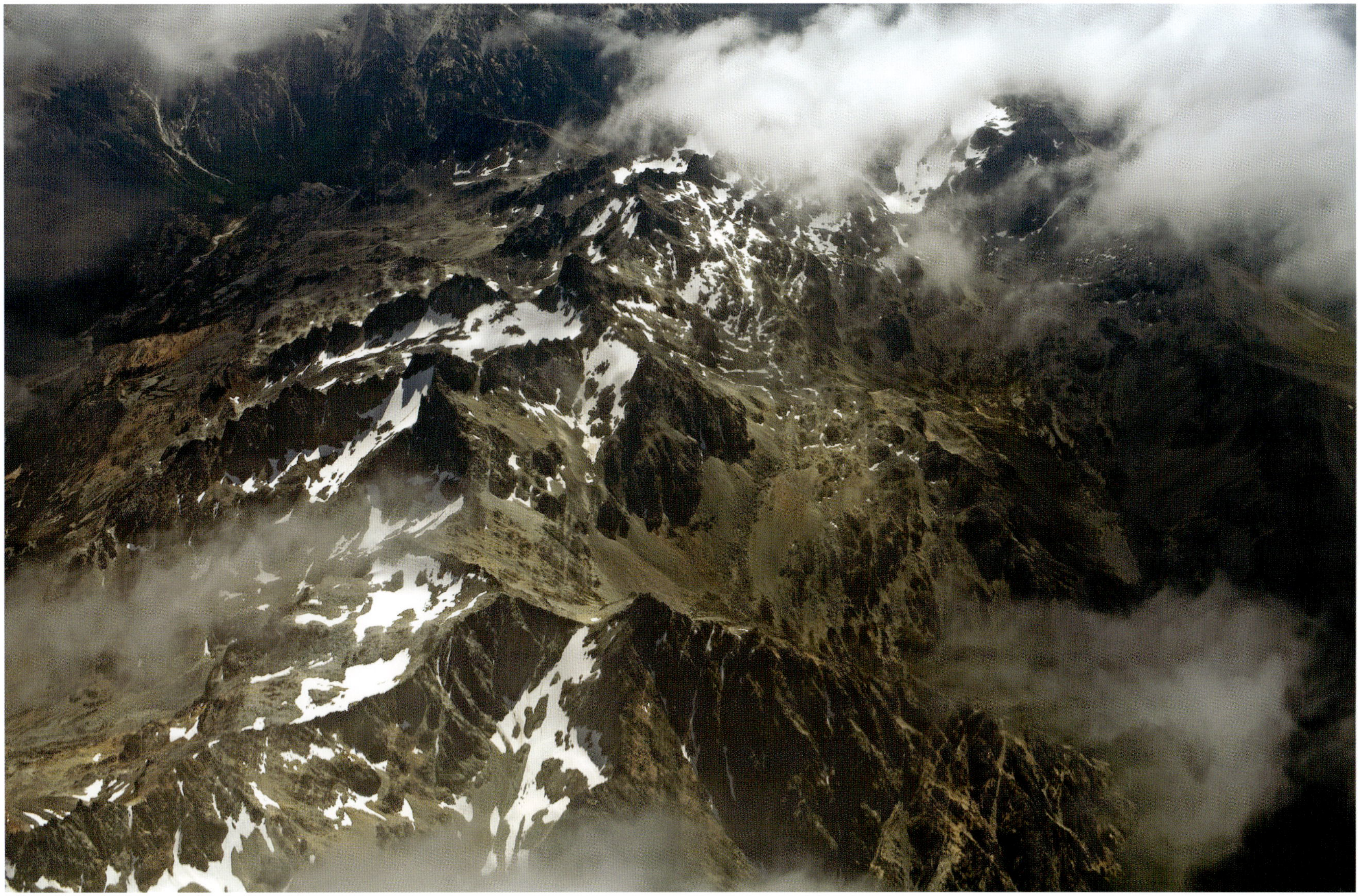

Andes Patagonia

WE WERE ABOUT to untie at the dock when the port pilot came aboard to safely lead us out of Ushuaia's harbour and through the Beagle Channel into the Drake Passage. This is a very official event so even though the ship has an open bridge policy, for the moment, everyone but the crew was ordered off the bridge to reduce all possible distractions as the ship was skilfully steered and guided out into more predictable waters. When the pilot deemed it safe he once again handed over the duties to the ship's captain and boarded a smaller pick-up vessel home. Upon our return home into the Beagle Channel the process was again repeated and we were reliably escorted to port. It's the law even though our experienced captain passed this way more than forty times. You could hear the banter and laughs between captain and harbour pilot as they no doubt exchanged the latest news and gossip. I doubt there was one necessary word of caution or advice shared between them.

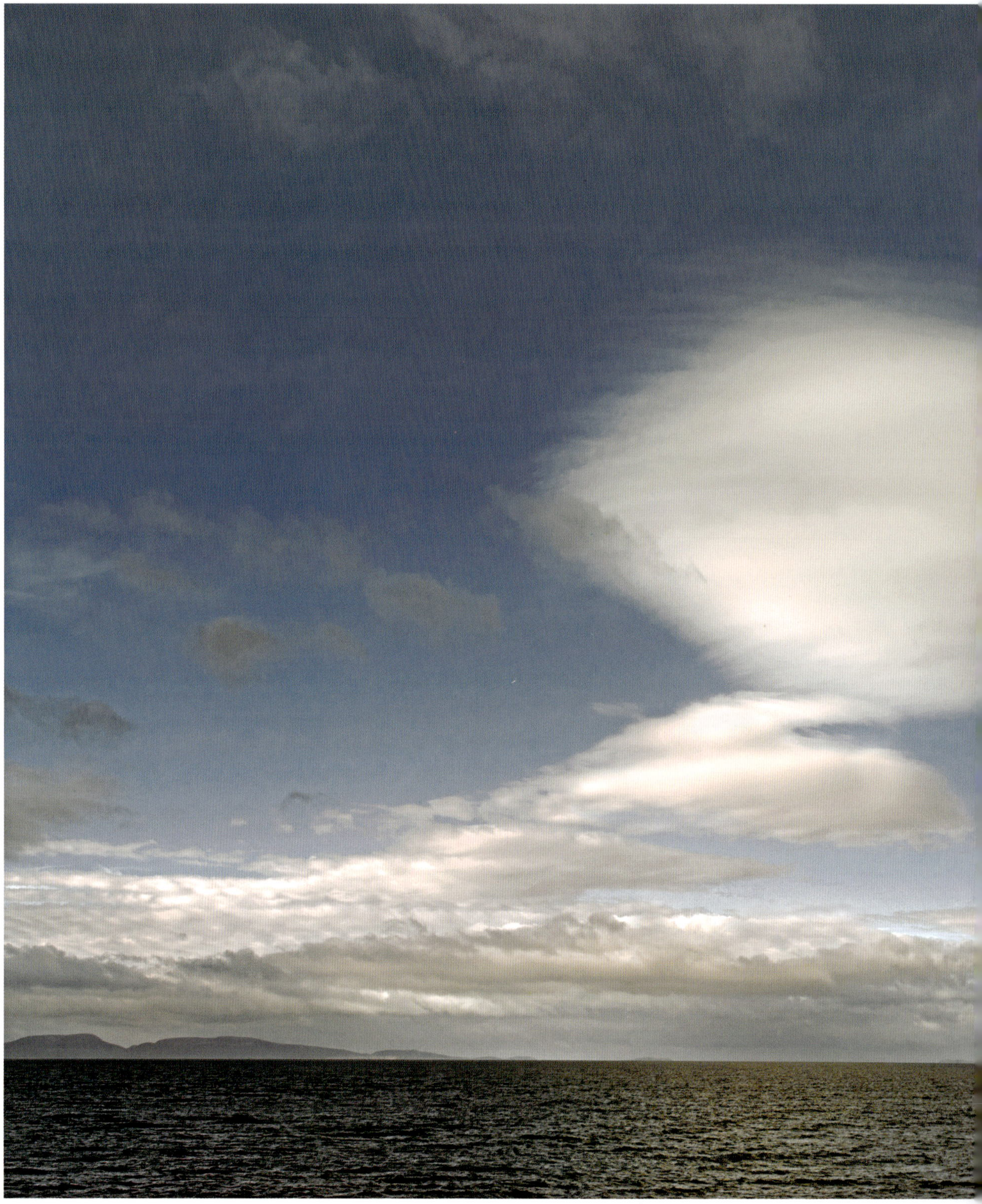

Opening to the Drake Passage and Antarctica

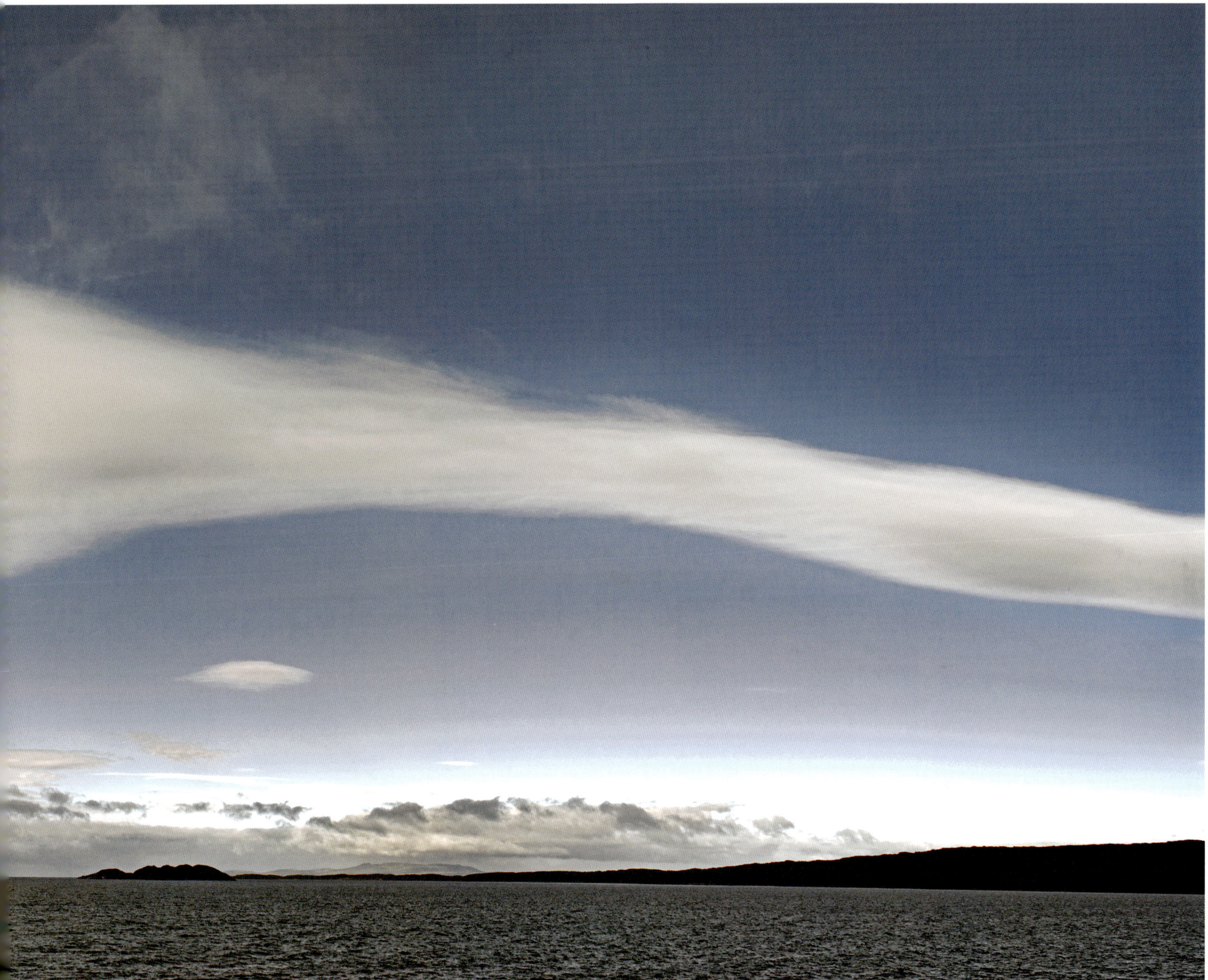

Nevertheless this departure is still new and fresh to me so I manage my excitement and with mounting anticipation at the unpredictable events to come I watch the ship cautiously slide out of the Beagle Channel into the notorious Drake Passage, the Antarctic convergence and the Antarctic continent.

This channel was named after the HMS *Beagle*, a 200 ton ship that for five years was home to Charles Darwin as he sailed around the world. I move to the ship's port side and stare at famous Mount Olivia, a landmark mountain that reveals its pronounced beauty in form and size. Snow capped and solitary with a sharp peak it is Ushuaia's Matterhorn. Its presence moves me and I reflect on what Darwin nearly two hundred years ago might have been thinking when, I am sure, he and many others also studied and admired its appeal.

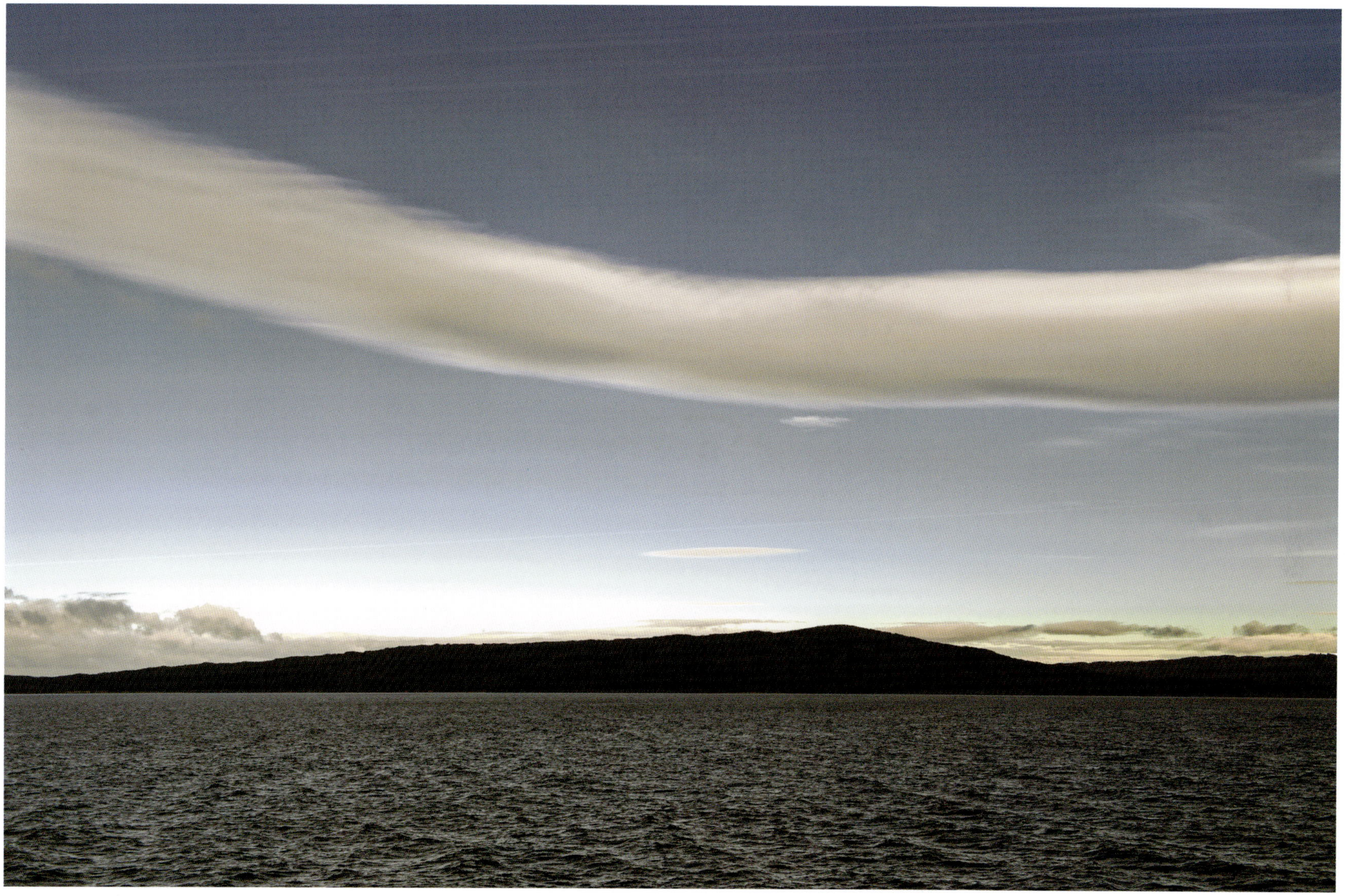

Nacreous cloud Beagle Channel

WHEN YOU LEAVE Argentina from the port of Ushuaia to cross the sometimes unruly nine hundred kilometre Drake Passage, the first encounter with Antarctica are the South Shetland Islands. Cut off from the continent by the Bransfield Strait, this archipelago runs parallel to the Antarctic peninsula and extends four hundred kilometres north-east to Elephant Island. They are a good indicator of things to come and after two and a half days at sea in one of the stormiest waters in the world they are a most compelling and welcome sight. Rising abruptly out of the ocean the high snow and ice-covered mountains do not disappoint. Their ice caps gleam brilliantly in the morning sun or in a fog loom like alarming apparitions. The air becomes cooler and the winds more disruptive. One easily senses much is about to change. You are entering a dream and sailing into a landscape that so far you have only experienced in pictures but never witnessed. Eyes strain to understand dimension and distances. Without any familiar reference points your mind floats unanchored through the captivating landscape.

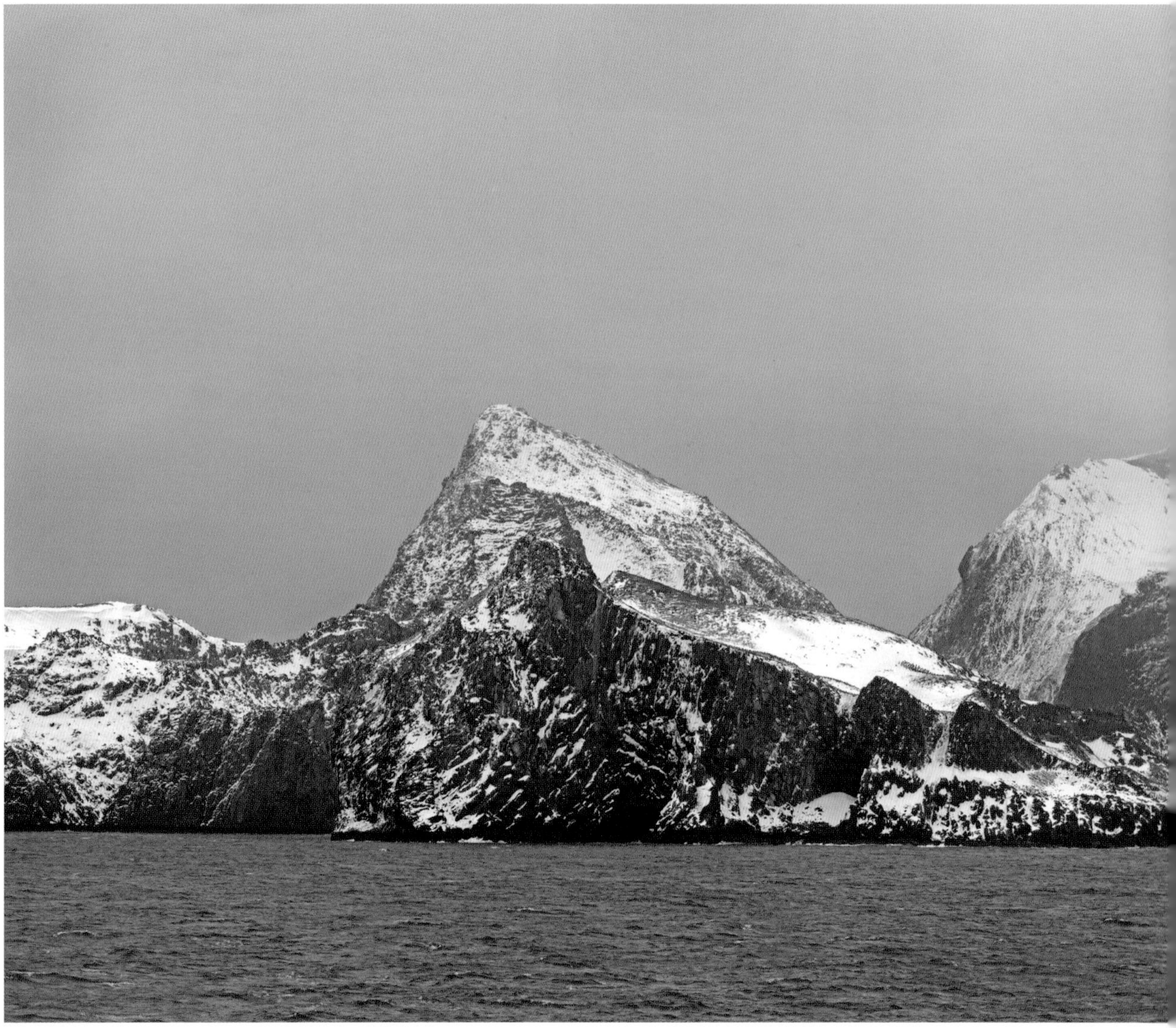

South Shetland archipelago

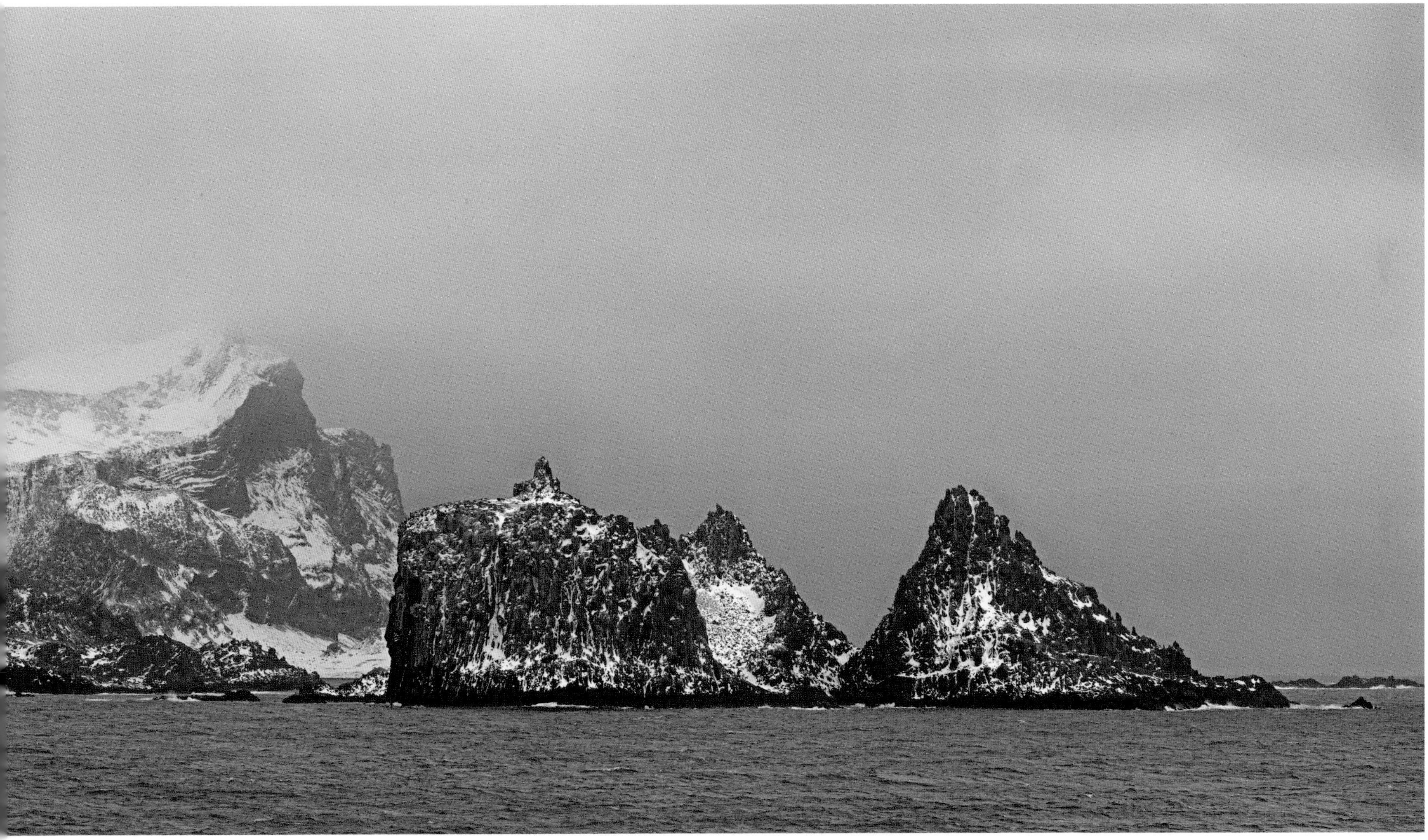

WHEN I ARRIVED home from Antarctica and began reviewing my images I felt that this picture of a bursting sky filled with lenticular clouds had not lost its impact.

It was one of the first images that I studied and immediately felt it had some merit. I also felt that just witnessing this unique spectacle cleared any doubts about the success of my trip.

I had seen similar cloud development before and I knew that if there was a design I wanted to record I would have to act quickly. I would scan the sky, study it for a few seconds, look back, and already find the roiling clouds emerging into something new. Subject, composition and exposure, the essentials of a well-designed photograph, must be quickly determined. Not only must you consider where to point the camera but also when.

These billowing masses of condensation are caused by very strong winds that rise rapidly up a mountain side to burst over the peaks and mix in a mass confusion with an accelerated wind. At these high altitudes the moisture laden air is quickly cooled and spun into smooth compressed clusters. Curious shapes seem to momentarily come alive and briskly stretch upwards and outwards into another configuration. Just when you think it is all over another whirlwind of activity develops and everything is reworked. These lively clouds are entertaining not only in shape but also in their swift animated tempo.

I was on the uppermost deck of the ship watching the fading light change the colours of the distant mountains when the kitchen personnel announced that supper was served. The passengers on board watching with me, who

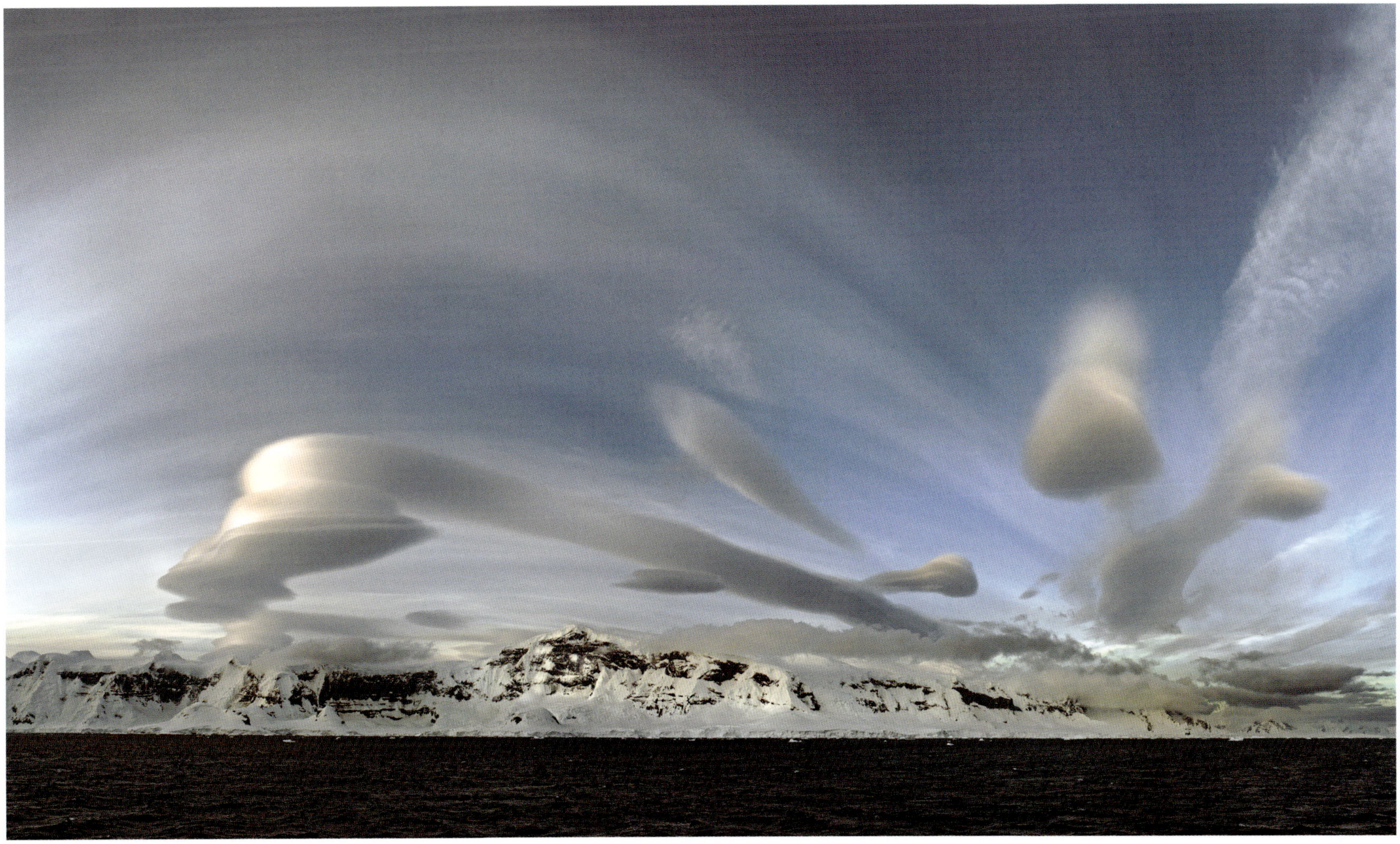

Lenticular clouds, spinning winds

were maybe a little colder and more hungry than I, quickly gathered their things and headed inside. I however did not. I had spotted a small smooth cloud far to the horizon and from experience hoped that this was a sign of more to come. I decided to forego supper and wait.

Within minutes the sky began to change. Clouds that only moments before seemed fixed were now moulded into a smooth silky look. All the shapes began to reconfigure and I feverishly searched for a merger of contours and profiles. None was evident at first but things were always in motion and transforming enough that I just had to be patient and wait. I now settled down ever intently studying the scene hoping for a composition. I especially wished for something higher in the sky to strengthen the composition and fill the entire picture frame.

The brisk wind at sea level blew directly from my left but another wind much higher up came from straight behind me. In this strong wind the bulbous clouds began to stretch and elongate seemingly to the breaking point when suddenly even higher up a nebulous fog began to materialize. Immediately this hovering mass was ripped apart by a powerful invisible force and exploded into a high curving veil. This, now a gauze-like curtain, began to bend sharply from the top left and severely curve down to the right towards the far horizon. Here at the distant vanishing point the warm yellow of the setting sun mixed with cold glacier blues turning the evening glow into a shade of light celery green. A visual phenomenon not unusual for Antarctica.

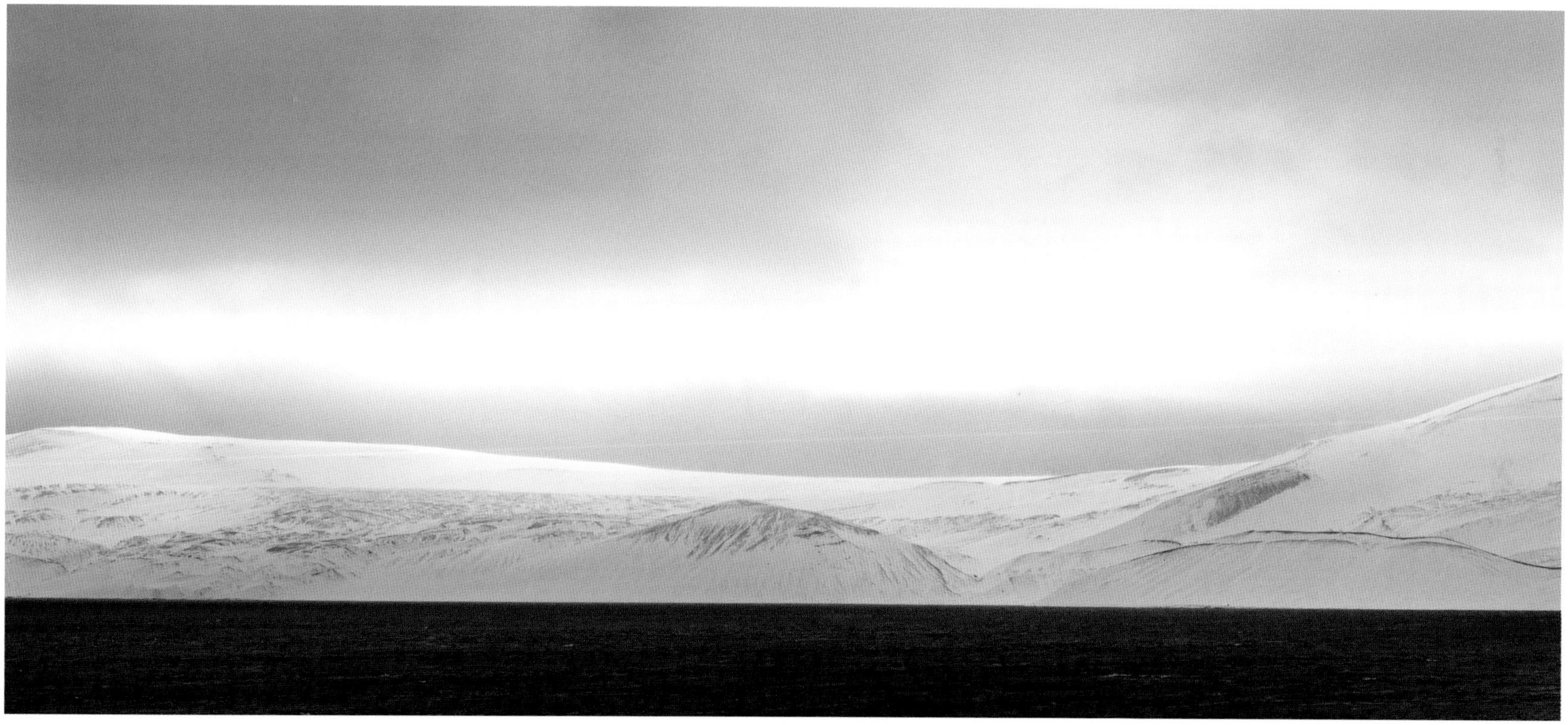

Port Foster Deception Island

BEING SO CLOSE to the Antarctic Circle the long evenings with the lingering sunsets must surely give rise to some of the most beautiful colour spectacles anywhere.

We had just come through a very windy and squally day where all traces of colour were absent. Whites and blacks had disappeared and were now visible only in neutral shades of grey. All contrast was gone and whatever one could see looked flat.

The light was visually mundane with no relief.

For most of the day wind beaten clouds blended with a bone chilling fog that swept incessantly through the gaps between the mountains forcing me to tie down my cap and raise my hood. I had my gloves in my pockets but hesitated to use them. I could never get used to working the camera with them and sacrificed comfort for efficiency. Years ago I had frozen my fingers badly enough by too often handling the brass knobs on a 4 x 5 inch view camera in minus 35 degree Celcius weather. That made my fingers very sensitive to colder temperatures and very quickly they would sting and lose feeling. I have to work swiftly.

Over the years on my different journeys on the same ship I had found a small sheltered corner on deck that would give me the best vantage point and also the best weather protection. Today I stationed myself here and waited for

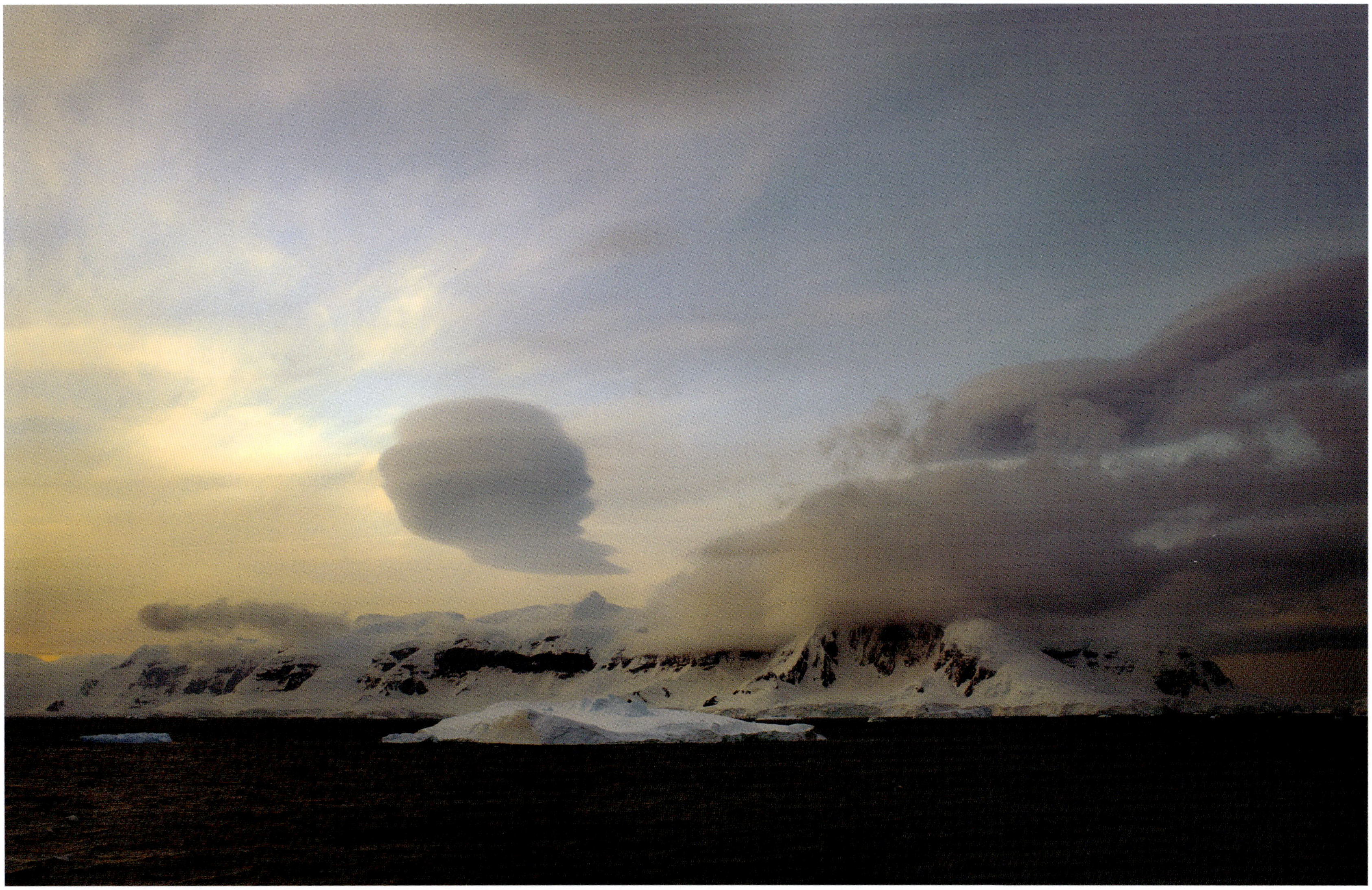

Evening light Antarctica

change and then periodically slipped behind a sheltering door to warm up. Given enough time and patience the ice fog would give way to the wind and wear off.

Finally in a flurry that lasted only a few minutes a different wind reworked the sky and light penetrated the landscape. An overabundance of colours materialized. Where once nothing was happening now everything seemed to happen at once.

Countless hues transformed the remaining clouds and their luminance spread everywhere.

This was when I quickly realized my inefficient description of colour. With only a few colours off the primary I understood how inept I was trying to put into words the variation of hues that kept changing and disappearing into intense shades of unexpected colours. Snow became a soft pink and sometimes a cloud took on a light greenish hue. Horizons were lost in an inky black and in a moment the whole world turned into a warm pastel of reddish beige.

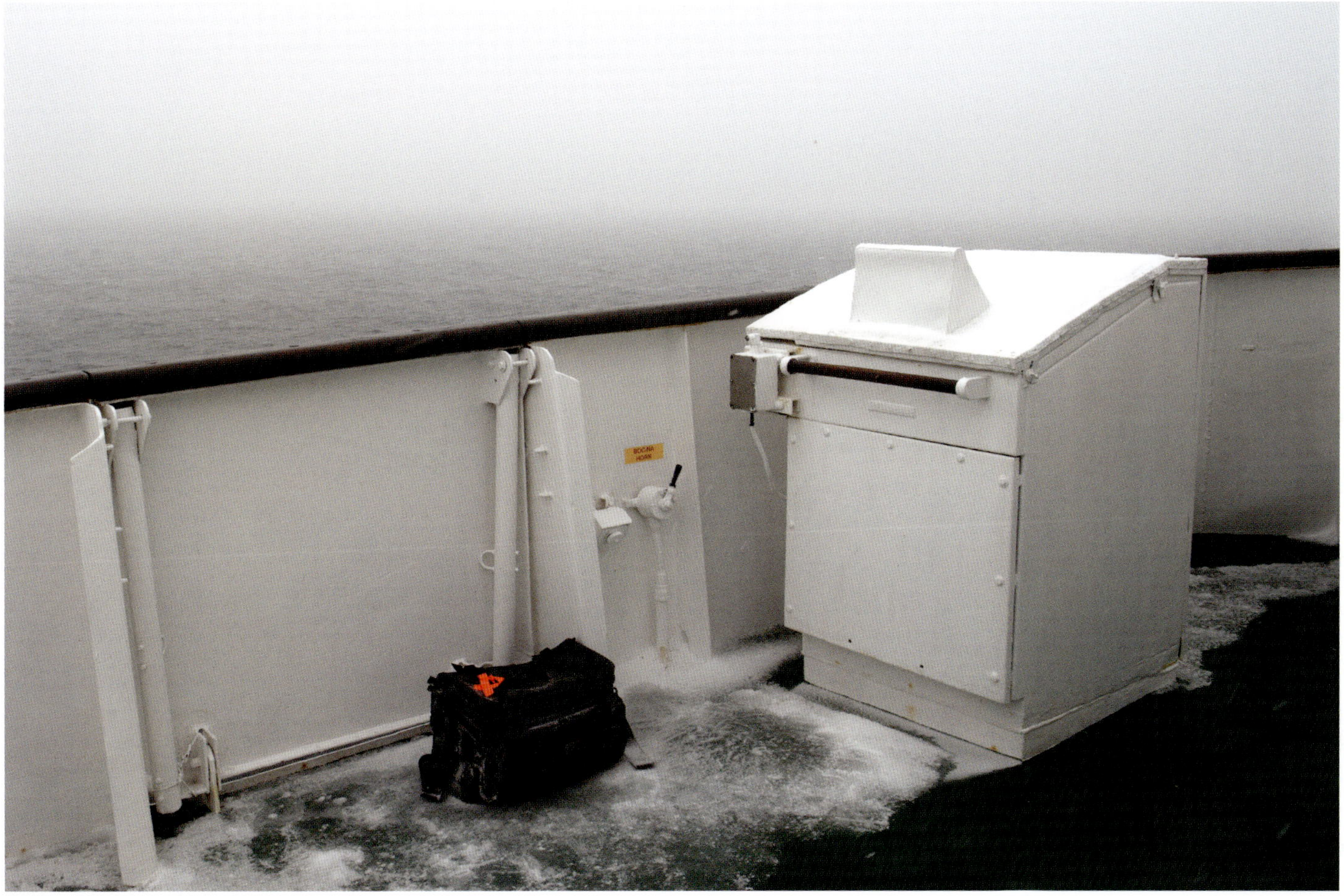

Scant shelter

FROM A DISTANCE there was nothing conspicuous about this cloud. However, as we approached its lower layer became darker as it slowly tunneled and sank towards the sea. It became apparent that there was a great weight descending. Swirling above its upper surface a strong wind was picking up moisture as it streamed upward spreading higher into the clear blue sky. The dense descending column now widened and filled with a piercing light into a haystack of opulent yellows mixed with delicate greens.

Like a bouquet of wavering light this broad column transformed steadily as it wandered across a placid sea. After it dissolved I realized I had witnessed something special when our experienced captain commented that he had seen many cloudburst but never one so distinct and exceptionally beautiful.

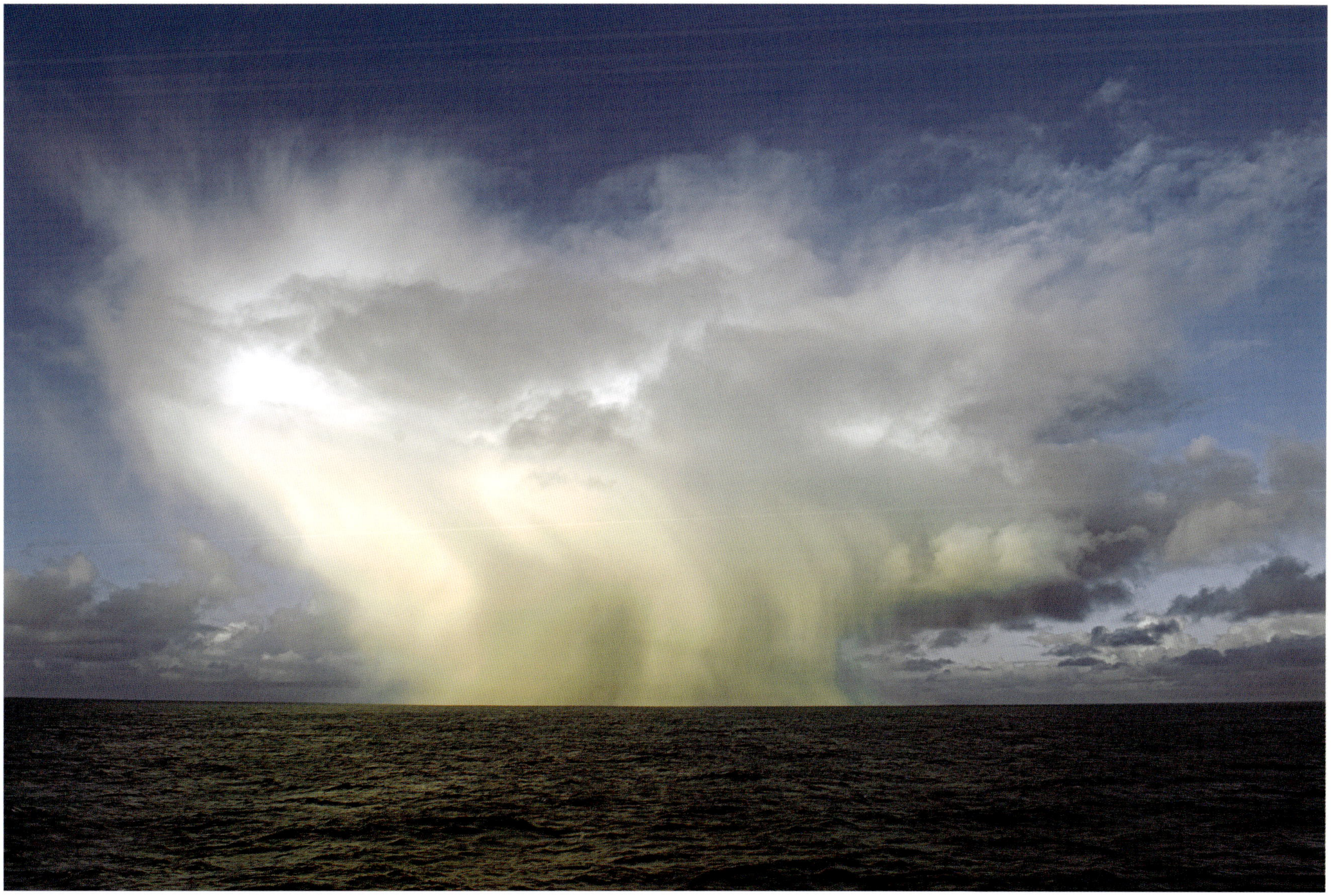

Cloudburst

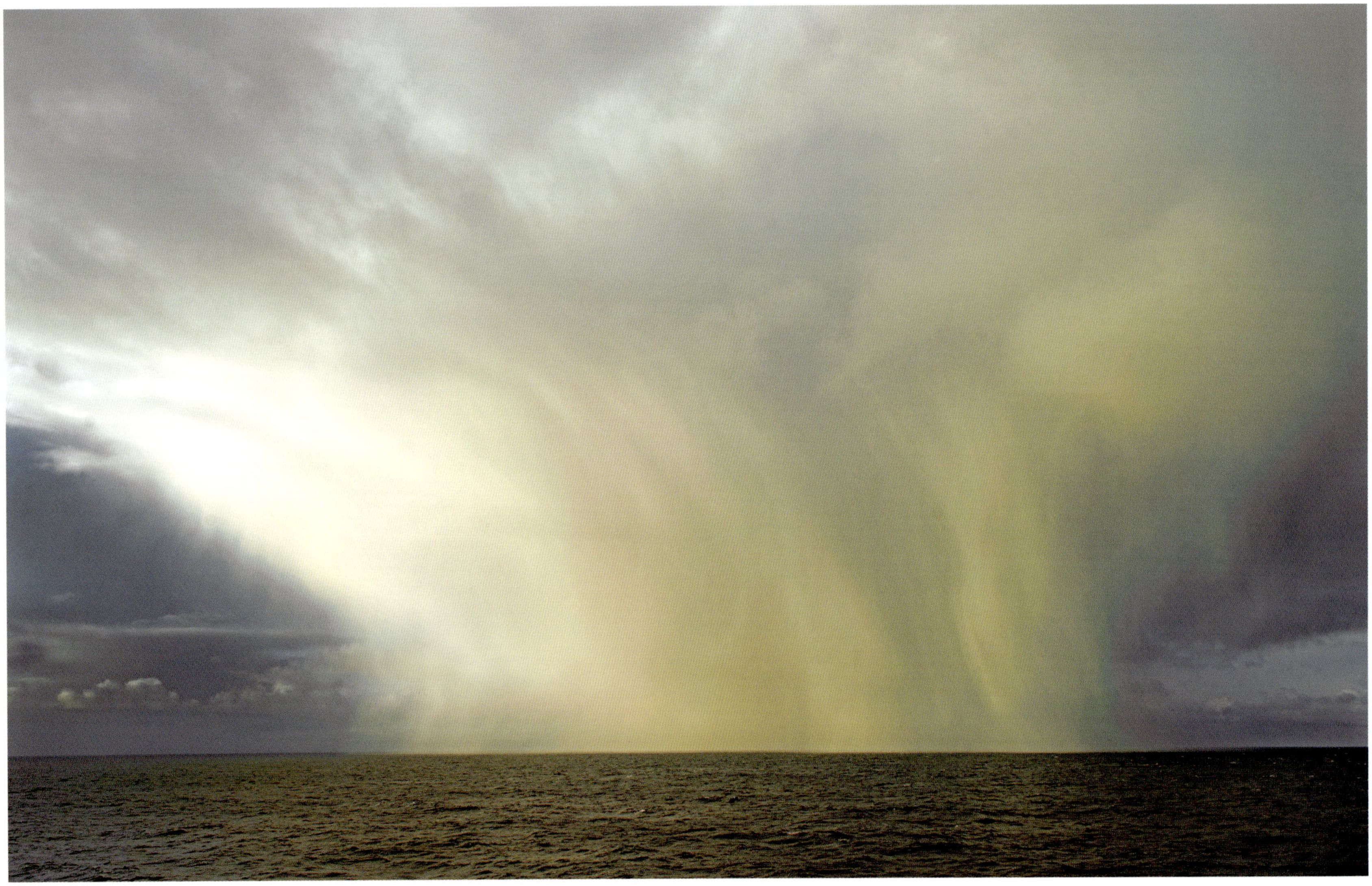

Rainstorm Drake Passage

A CLOUDBURST IS an enormous amount of precipitation released in a very short time. Over land it can cause extreme damage and destruction however over the sea it is an expression of power and beauty and utterly benign.

FOR DECADES THESE southern latitudes have had a notorious reputation for their consistently dangerous and tempestuous weather. These deep southern latitudes are known as the Screaming Sixties. Here not only do the waves roar and the gales howl but the winds shriek in a high pitch slashing and cutting through the ship's rigging. Sailors have feared it and captains have dreaded the life and death choices they would have to make in the weeks it sometimes took to bring a sailing ship around the Horn. Many ships were lost and countless lives spilled when seemingly on schedule the winds and waves of Antarctica would rise, attack, and test all those who ventured into these frigid waters. Many have at one time or another regretted entering; whether it be sailors of old in their very vulnerable sailing vessels or the seasick passengers of our modern day cruise ships.

So far I have spent about a whole month in the Drake Passage and have experienced a variety of weather conditions. However to be fair to the reputation of the Drake it should be mentioned that it also has its more serene and peaceful side. It is then that one can safely venture on deck and appreciate the variety of its character and its more subtle moods and disposition.

Why would anyone want to fly to Antarctica?

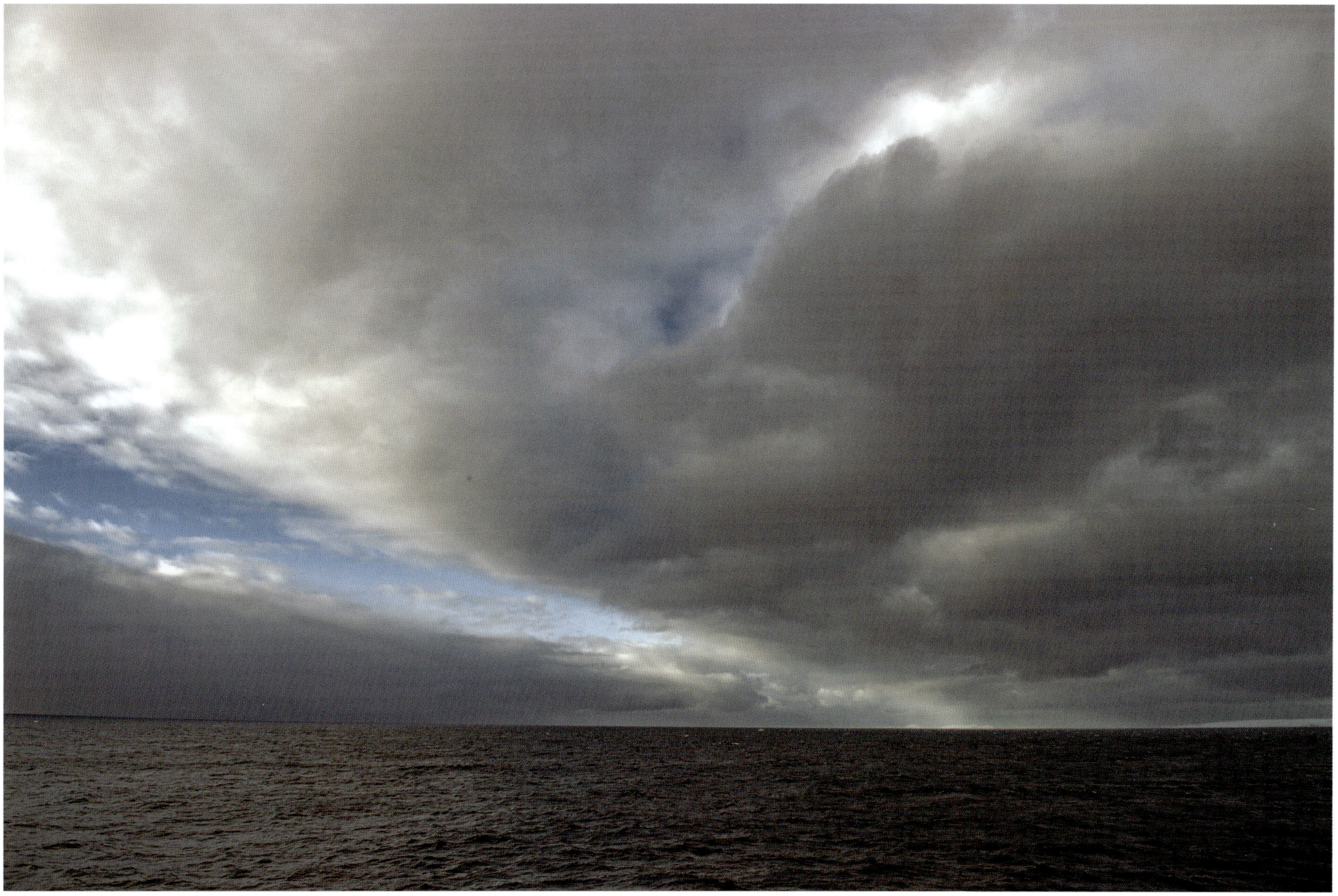

Dark sky white horizon

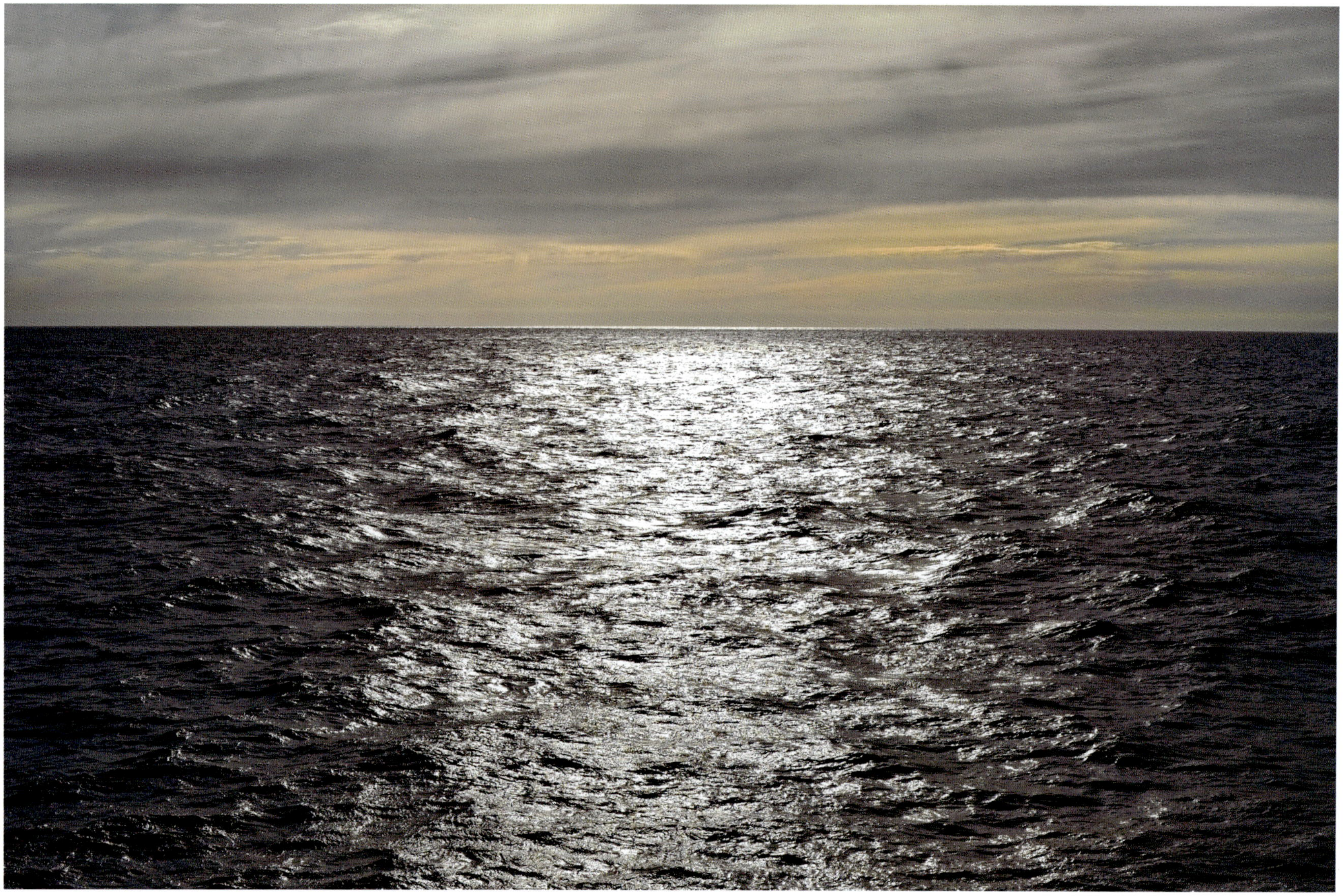

Drake Passage evening

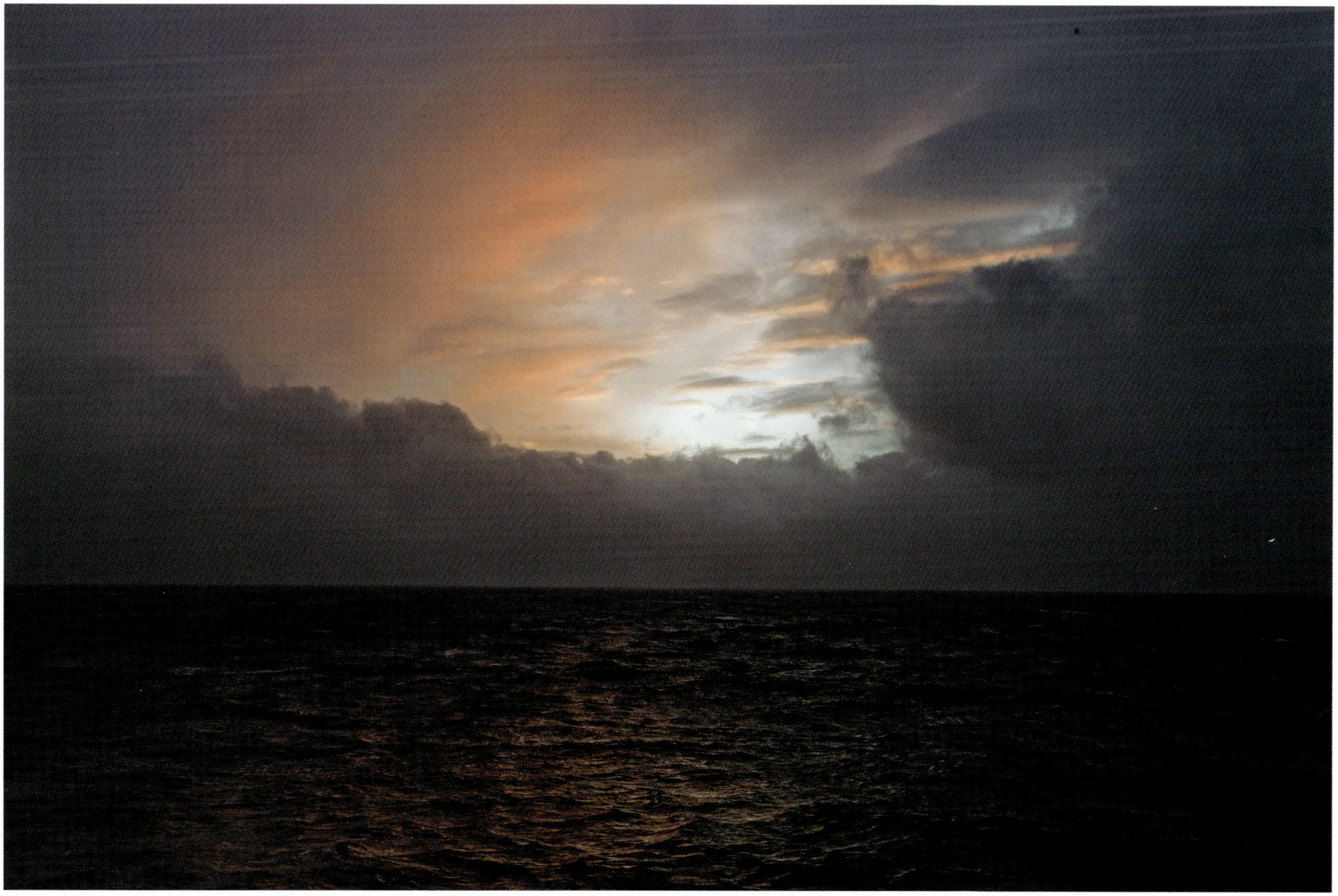

Morning

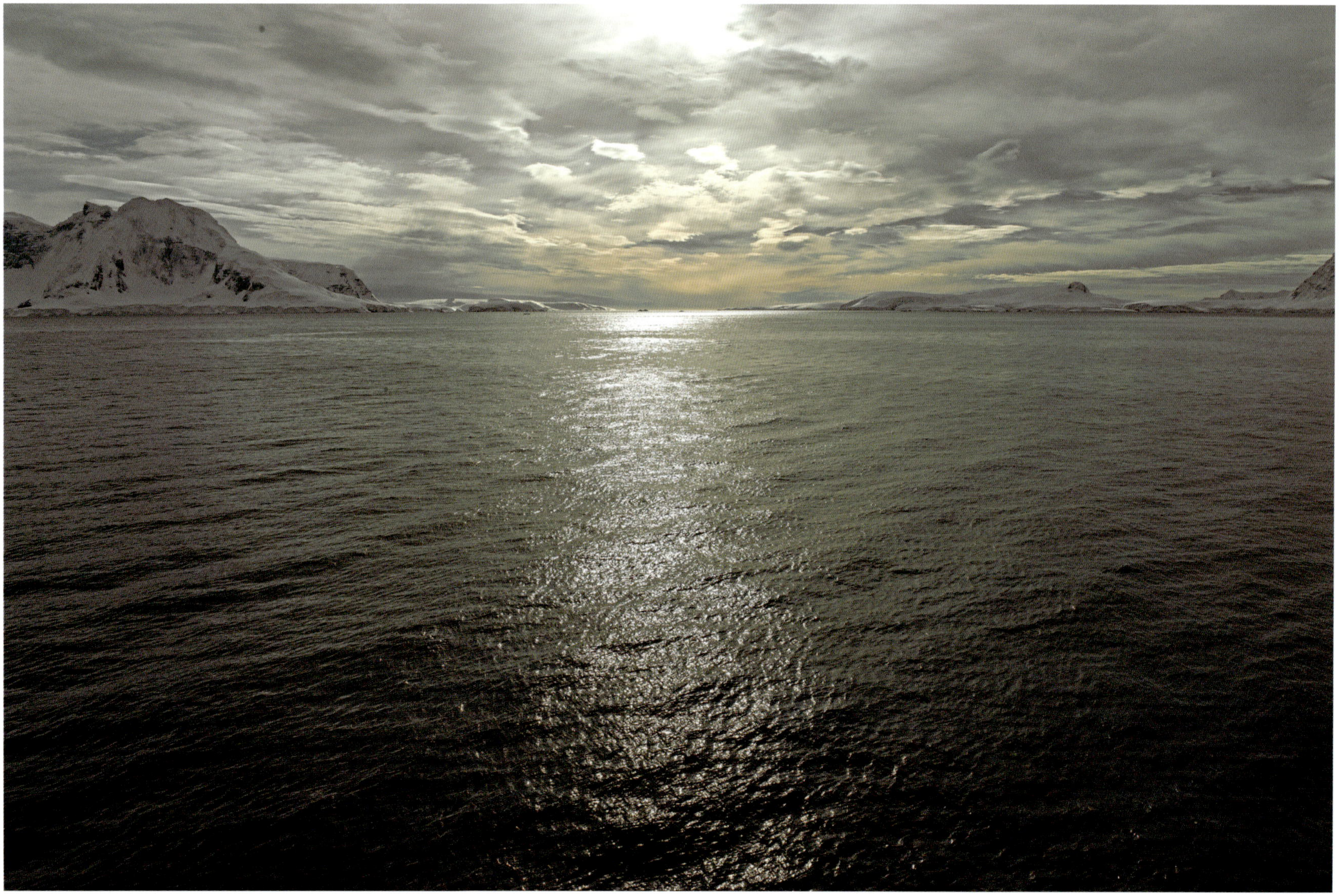

Calm water

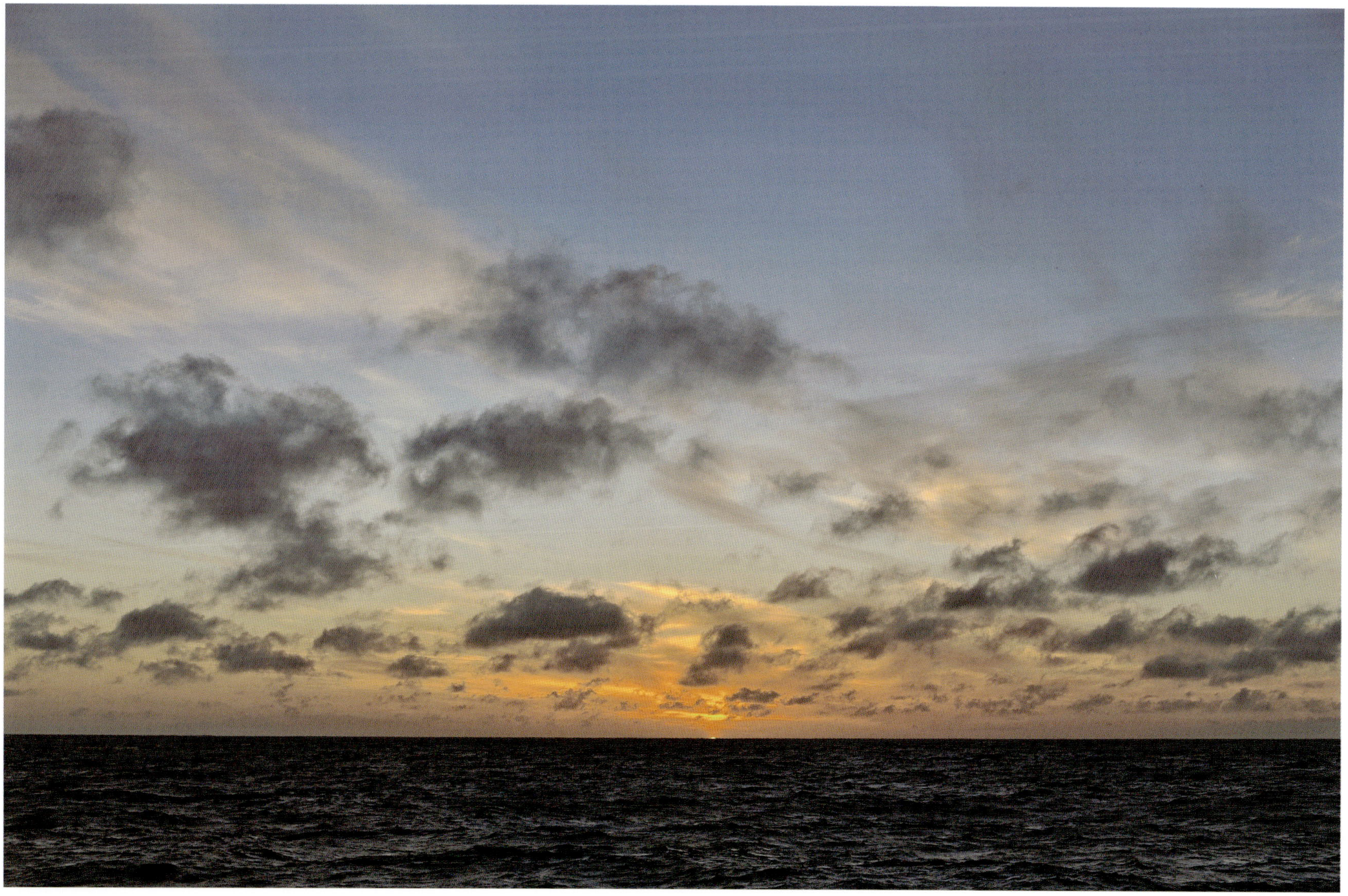

Sunrise Drake Passage

I KNEW THAT in the southern latitudes morning and evening skies could be spectacular, showing off colours that were not usually seen in the more northerly latitudes. For this reason I was always up early before the others and usually headed to the top open deck where I could easily move from one side of the ship to the other. When it was particularly windy I would choose to hide behind the great stack that would shelter me somewhat and also give off a little residual heat.

Often as I would pass the door to the bridge the captain, Sergio Osiroff, would wave me in for a welcoming cup of coffee.

He would tell me how the night went, the expected forecast for the day and what I really wanted to know; how far the ship can list before it doesn't come back. He assured me that even today's forecast would not threaten the ship. A few hours later I was happy I asked the question.

Our ship, the *Ushuaia*, is a kind of a rolling ship that gives in easily to the motion of the waves. It mimics every movement of the sea and if this goes on long enough it can bring up the best in us. Assuming one is prone to seasickness and wants to be certain to steer clear, Gravol on a strict regime should be a staple.

This condition is to many inexperienced sea travellers something they believe will not happen to them. I have seen dozens of newcomers eating heartily one meal and then be absent for the next two days. Those who have enough energy to come out of their cabins end up laying on the couches in the lounge looking tossed up like seals on a beach. It is a great equalizer and very humbling. I have been there.

The doctor on board told me there were two kinds

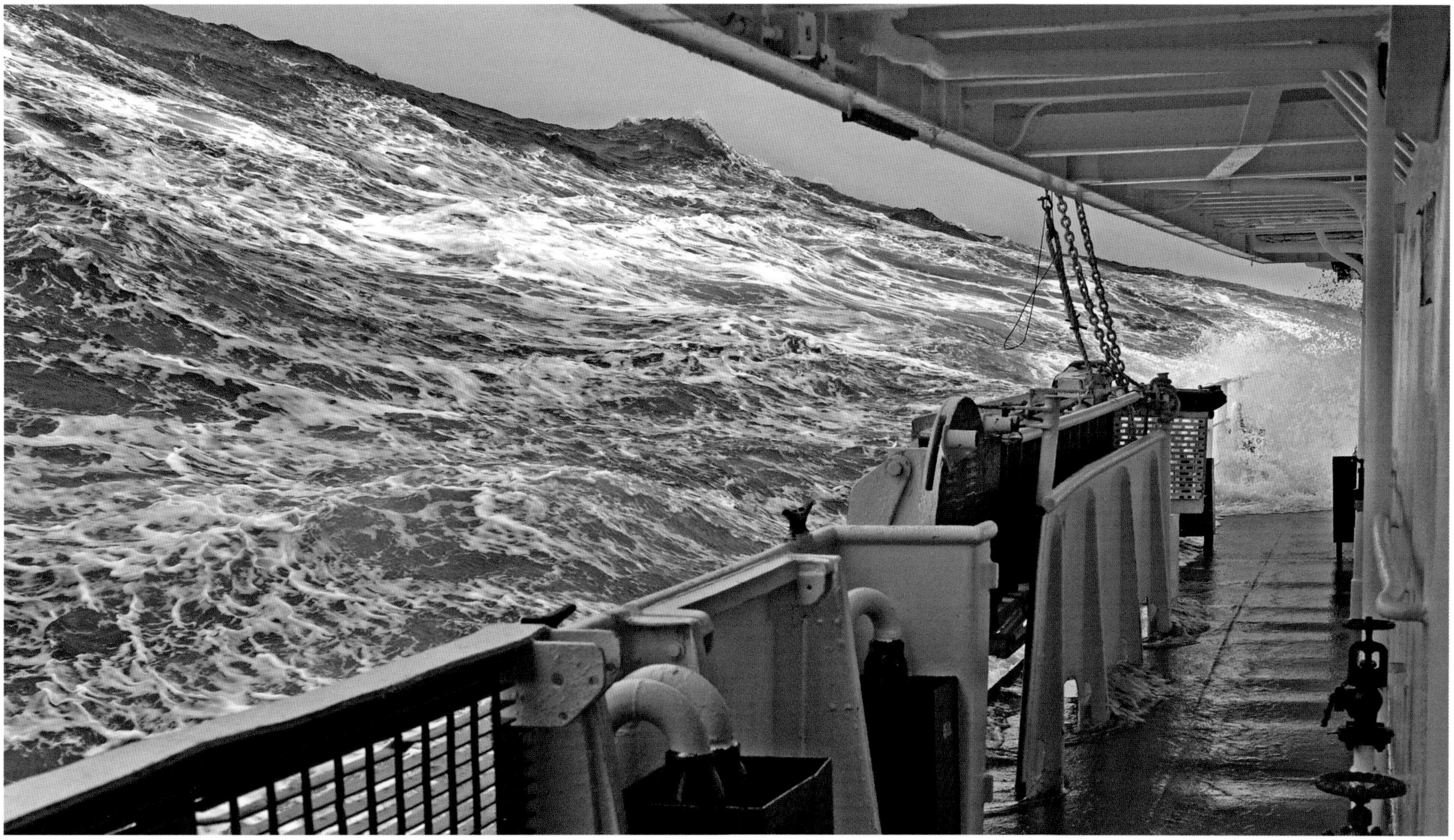

Rogue wave in a confused sea

of people: those who get seasick and those who will. No one is immune. Even the great explorer James Cook who navigated for years around the globe would succumb to the motion of the ocean and for the first few days at sea his first mate would be in charge. Cook would be holed up in his cabin. Only when his stomach settled and he got his sea legs back would he reappear and take charge. NASA still hasn't figured it out and it still is one of their major concerns. Every astronaut is vulnerable heading out into the ocean of space no matter how many times they passed the flying vomit comet test.

I look around and just by eying the quantity of barf bags distributed along every hand rail like facecloths on a clothes line convinces me to take preventative measures.

I remember coming across a passenger with whom I spoke a few hours before telling me that he was sure he wouldn't be affected. I found him leaning on the moving wall with one hand on the rail and head raised and eyes looking up in a seemingly trans like pose. Even with the light of a yellow hallway lamp his face shone ghostly white. It looked like one more move on his part and it would all be over. Reaching for a bag three feet away was too much. I asked if he needed help and with an affirmative yes from two quick blinks of his eyes I grabbed the closest bag. The swaying of the boat interfered with my finger co-ordination and I fumbled opening the bag. With trembling fingers I then handed him the opened bag and bolted.

His cabin, with a view, was on the top deck where all the ship's movements were exaggerated. Mine was at sea level with the sea rushing and breaking only inches from my sleeping head. With higher seas my port hole looked like an aquarium. Being so low it felt more steady.

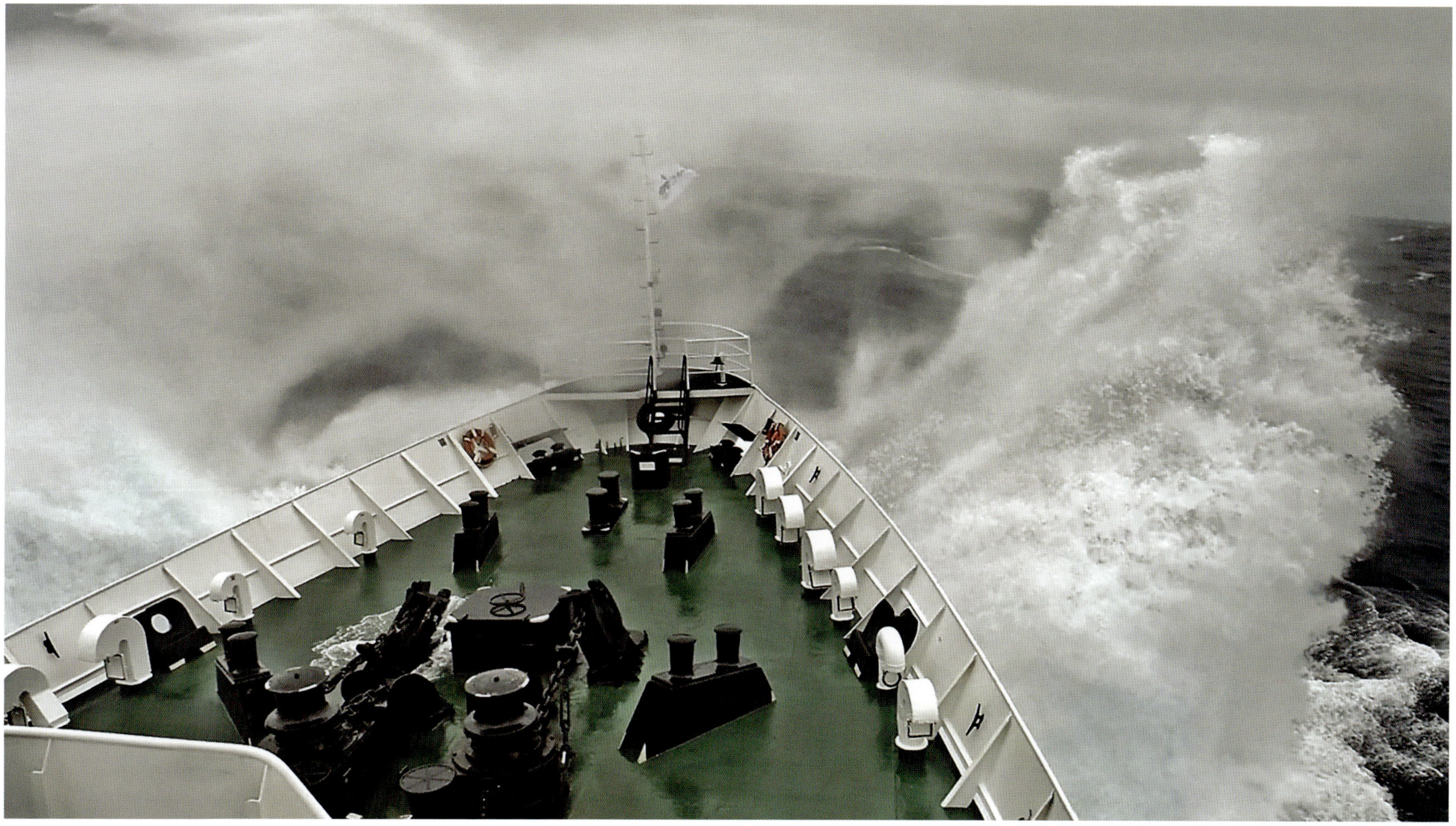

Wind around Cape Horn

WINDS ARE THE most crucial factors in making landing decisions. Frequently what seems from the deck of the ship a smooth and easy landfall is on closer inspection hampered by deep undetectable troughs in the surf. Controlling a Zodiac in such a gaping backwash without power is nearly impossible and swamping is a good possibility. One gets used to first doing a lot of necessary scouting to assess the risk before making a commitment.

With the abrupt changes in weather a vigil must always be kept to insure a safe return to the ship. I remember having made an easy and smooth landing when after ten minutes was ordered back. You sometimes question these decisions but in this case it was obvious when you looked out to sea. The water was dark and turbulent with foaming whitecaps racing to shore. It was a bouncy ride and no one regretted being back on board. This wind continued for the day and night. Our quick response prevented a miserable stay on shore.

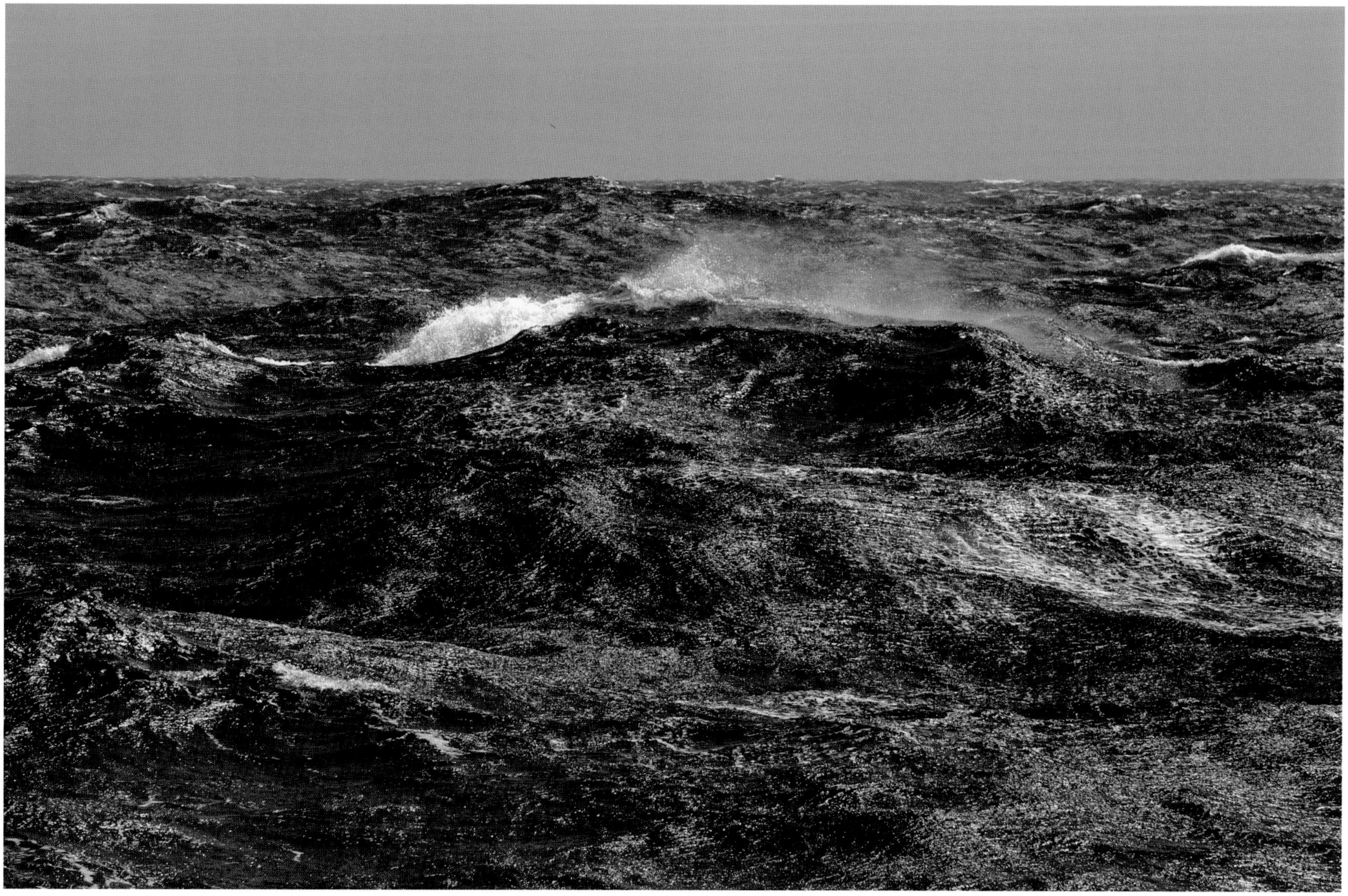

Der Wind und das Meer

ALL MORNING AS I wandered the top deck of the *Ushuaia* from one railing to another I did not expect a very memorable day. The light was harsh and the sky remained a glaring blue offering no visual relief. The surface of the ocean was shimmering with an uninspiring surface uniformly spread with splashes of light. My camera gear was still deep in my bag and I had the impression that this light was to remain constant. I sat by the rail and worked on my tripod.

As much as possible I make use of my tripod to assure a more stable and detailed image. This also gives me a better opportunity to explore what is in my viewfinder and decide on my composition. On solid land this is very easy but on a moving ship the hard working engines attached firmly to the iron hull sends a deep rumbling vibration throughout the structure. The heavy tremors can be felt under your feet even to the uppermost decks. You could not see the movement in the tripod but with a light touch you could feel it. Looking through the viewfinder with a long lens the quivering was very evident. In an attempt to solve this problem and better stabilize the camera I taped some thick sponge to the base of each leg to absorb and dampen the motion.

This succeeded but unfortunately it now turned the subtle vibrations into a sway which I figured wasn't any better so I abandoned the whole idea and tied the legs

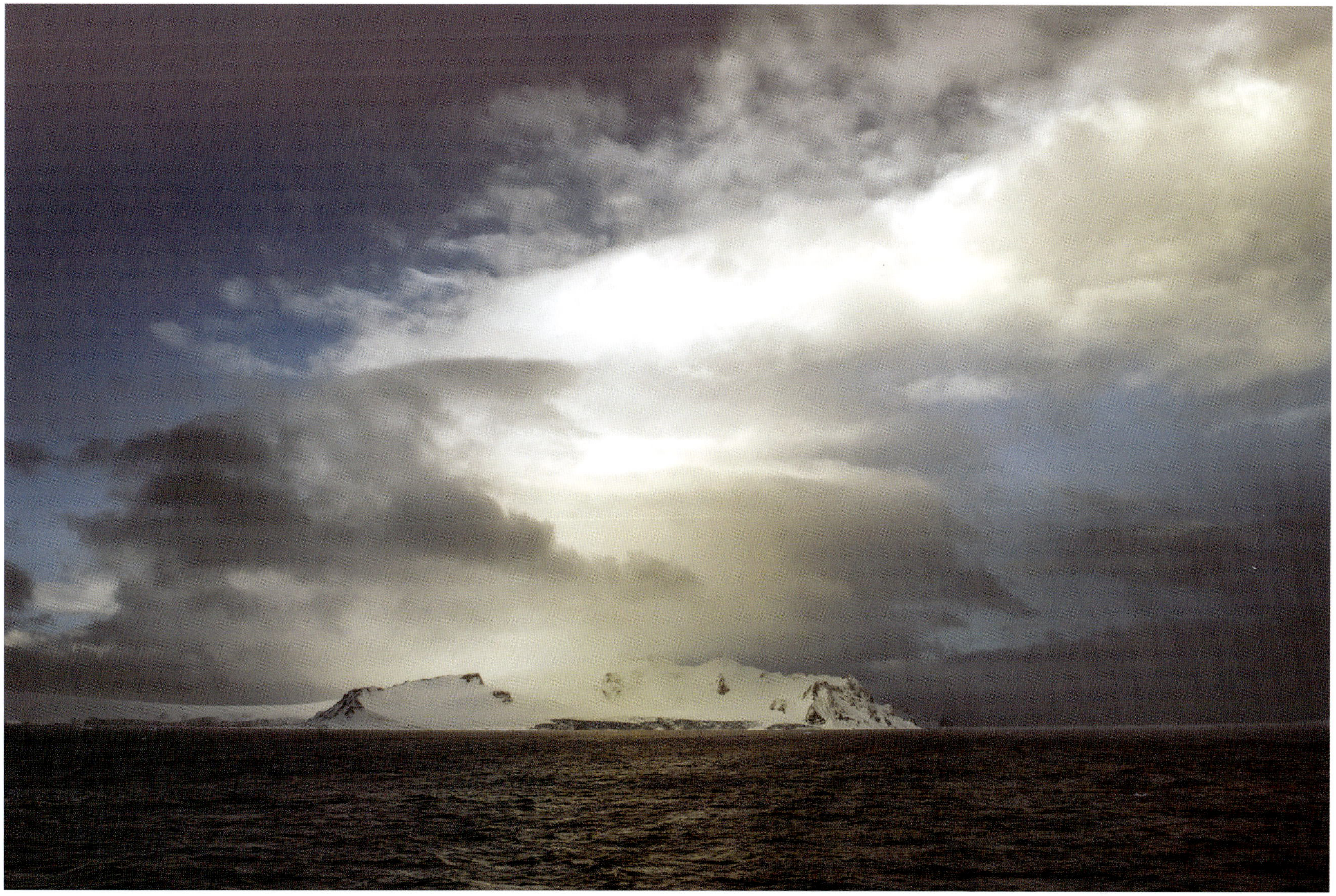

White light

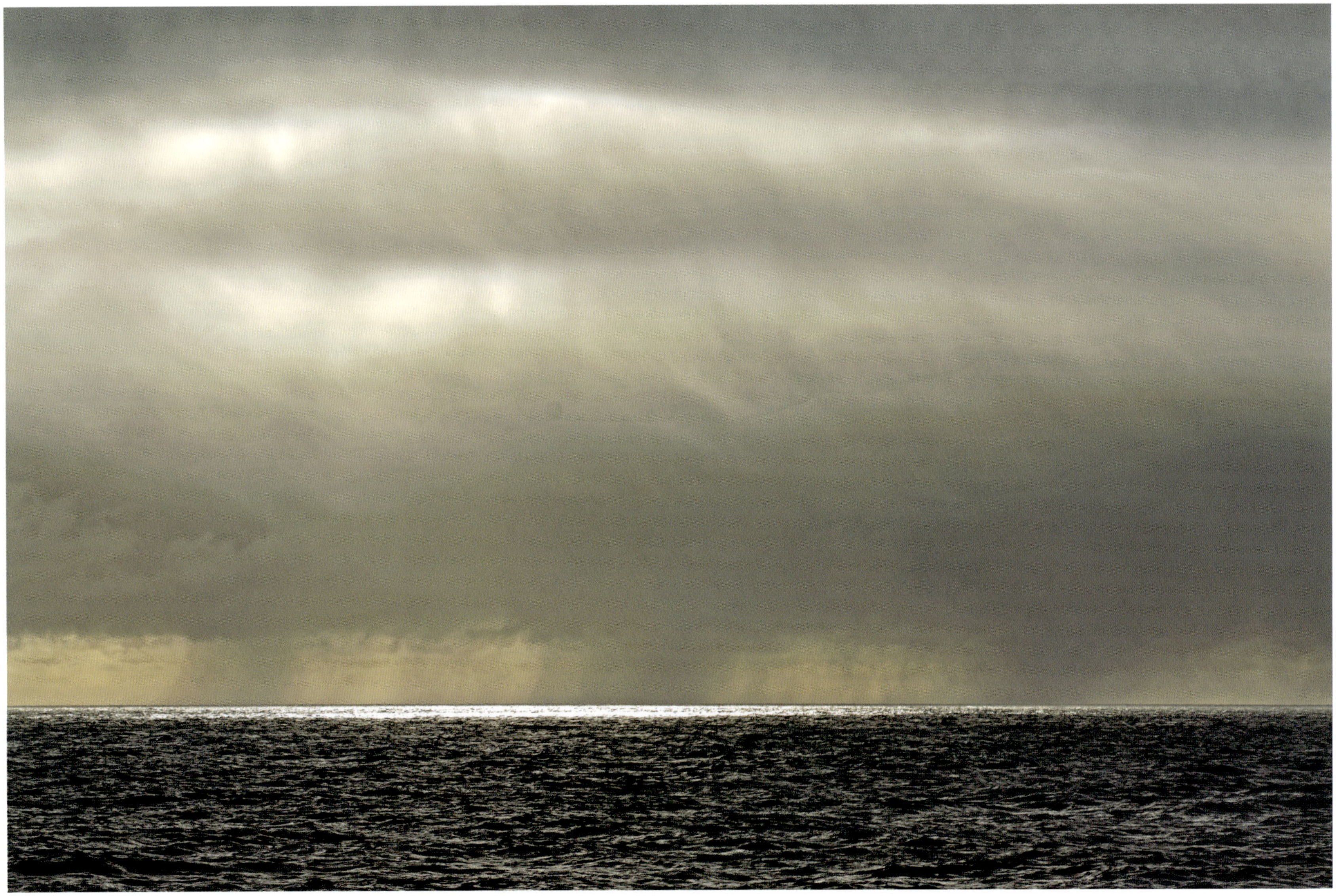

Shimmering sea

together to function as a mono pod and rested the whole contraption on my boot. It helped and with a relatively faster shutter speed and synchronizing with the sway of the ship I felt assured of improved results. On board this became my working method on all five trips. In a bouncing Zodiac however, the camera was hand held. But once again on solid land picture taking was always with a tripod.

As the day progressed the eastern horizon became fringed with clouds and the wind changed in direction and intensity. Though somewhat surprised this was exactly what I was hoping for. With this I quickly learned that in Antarctica the weather doesn't remain stable too long and taking an image should not be put off for even a few minutes. Clouds and ocean can change too quickly. With every sudden gust the variable light alters the landscape exposing it like a moving spotlight filtering through the fog and haze highlighting shapes and forms and darkening others. There is a dramatic infusion of light. What was just an ordinary scene a few moments ago now has suddenly become visually compelling and alive. Shapes are now accentuated with permeated colours and brightened form. Hues, tints and tones glide through a sweeping landscape. Changes are dramatic and exciting. In Antarctica photography is an intense, demanding and ongoing activity.

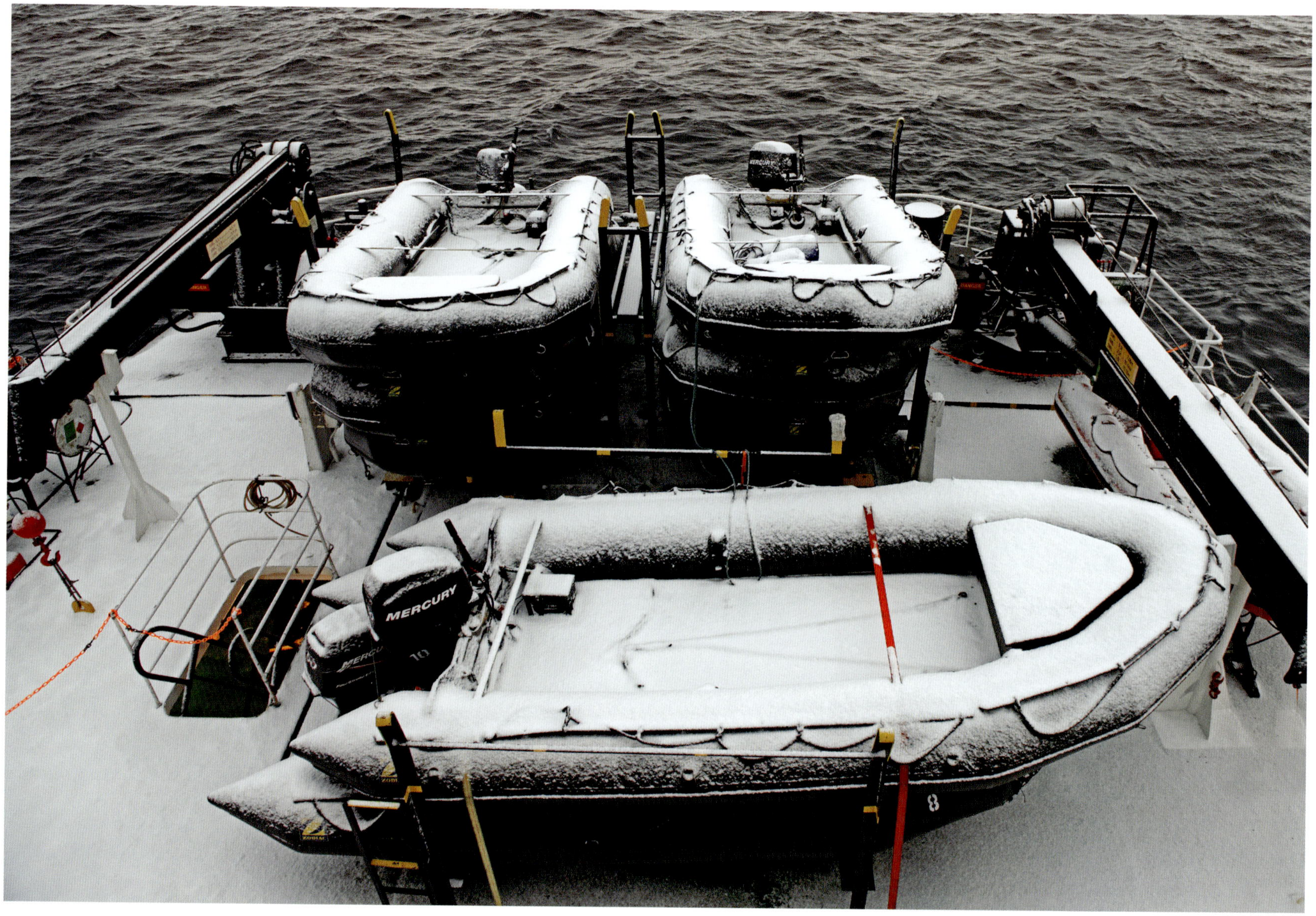

Ushuaia's *stern*

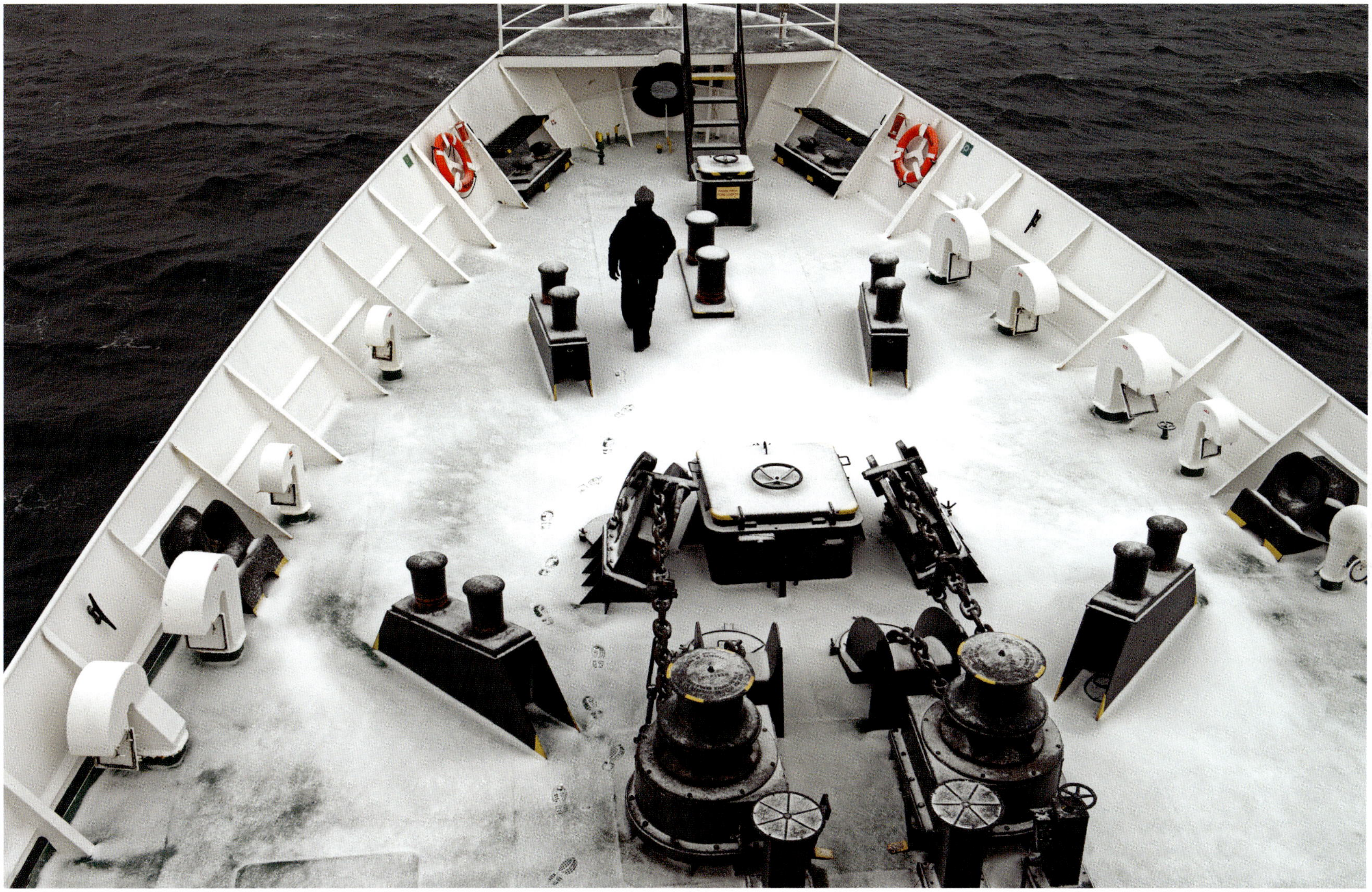

Ushuaia's bow

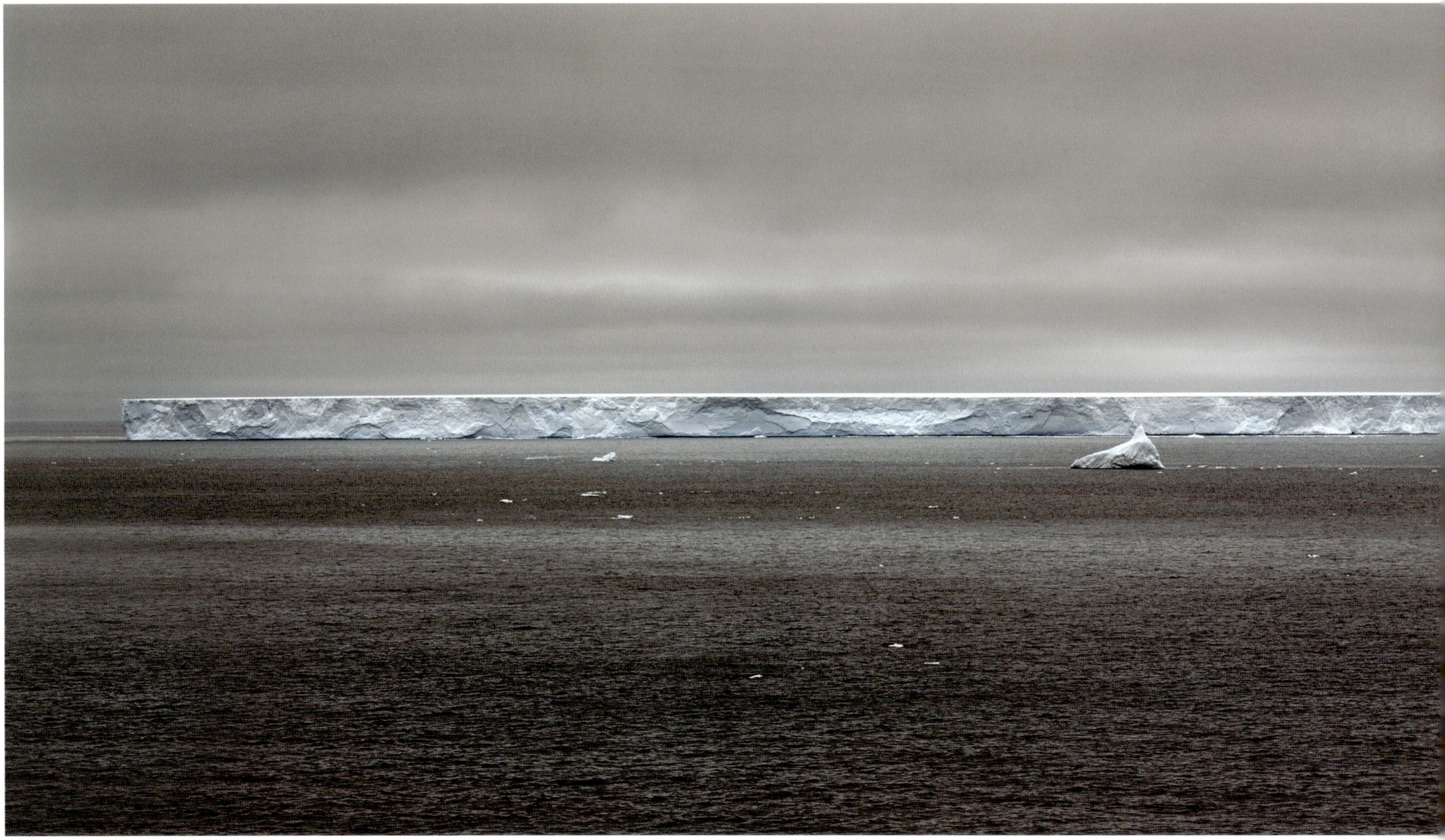

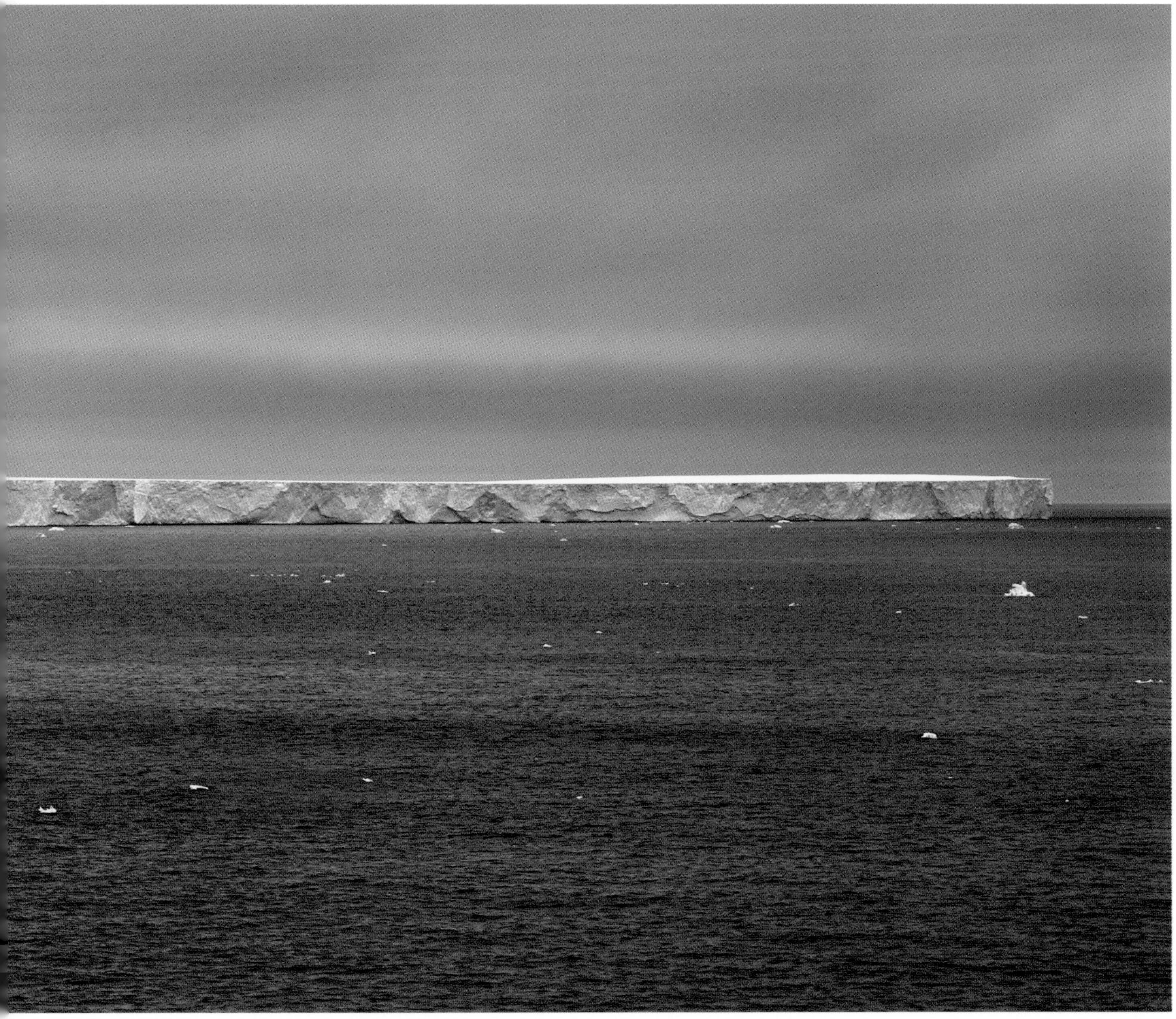

BEFORE SATELLITE IMAGERY and instant weather forecasting sailors had to totally rely upon themselves to make decisions and find their way. One of the clues they had to study was the brightness of the sky. A black or darker horizon meant open water ahead and a bright or white skyline would suggest ice and an impassible route. Often the light that penetrated the fog would be reflected by the ice and bounce back highlighting the bottom of clouds giving them an unusual brightness. This would be the direction to avoid.

Iceblink

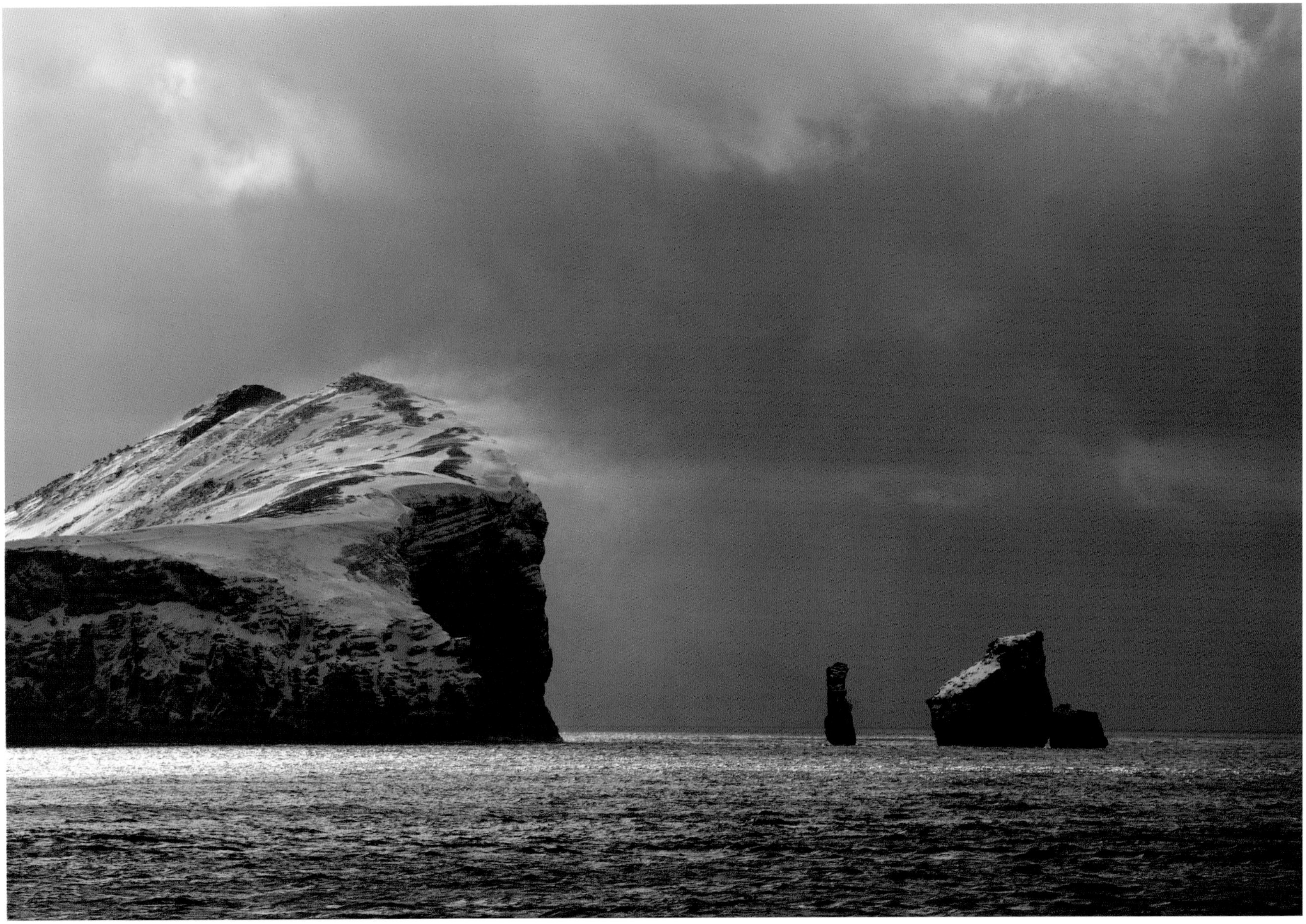

Sea Stacks South Shetlands

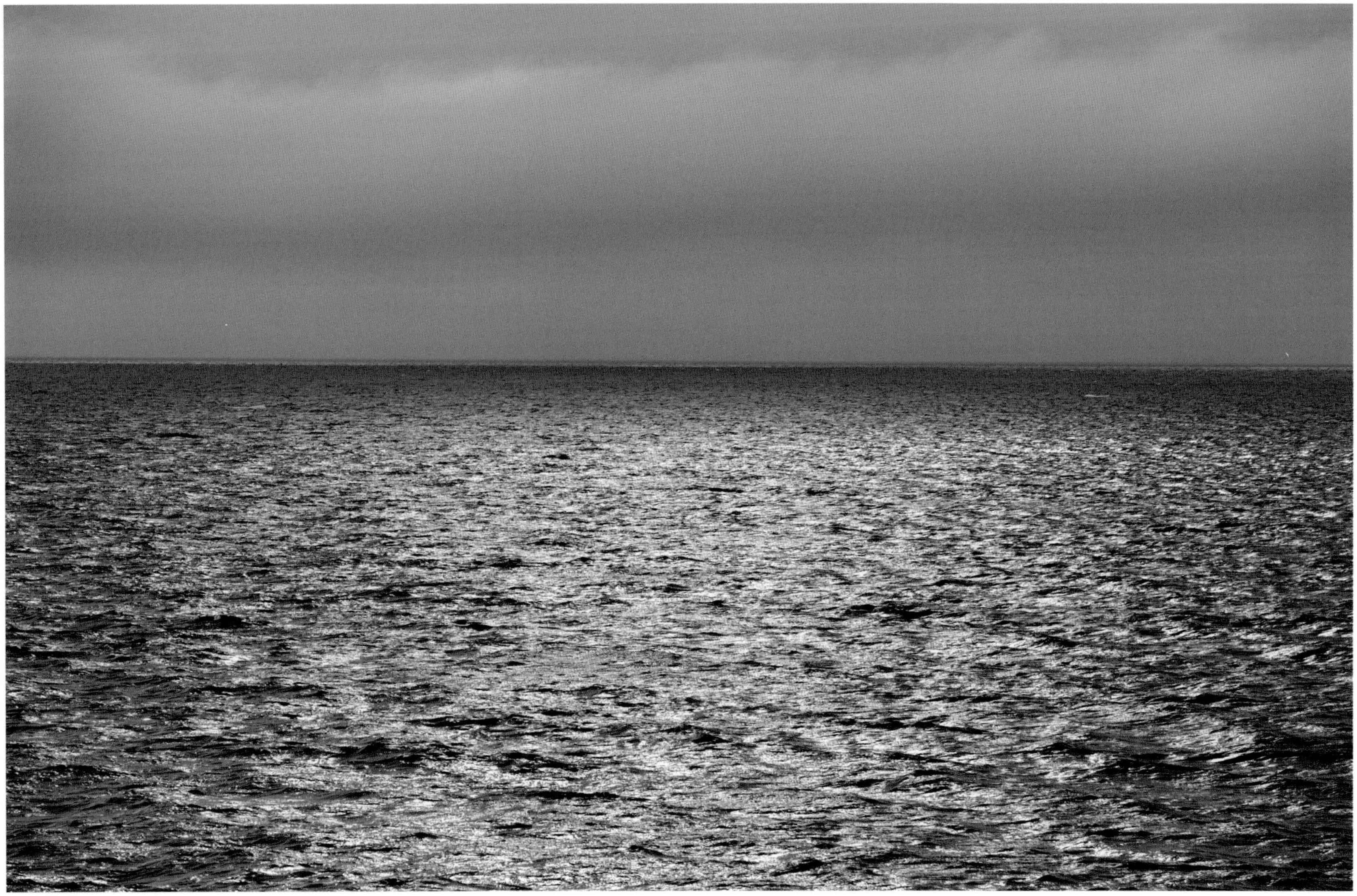

Penetrating light

IT'S STRANGE HOW sometimes things come together unexpectedly.

Before my involvement in Antarctica I was very involved with Labrador. I researched many aspects of its history and read as much as I could. One of the Labrador books that I particularly enjoyed was *Lost Lands Forgotten Stories* by Alexandra Pratt. The canoe route she had paddled was familiar to me and I e-mailed her just to thank her for telling a good story. I asked if she was writing anything else and she mentioned that indeed she was. She had just published a children's book on the volcanic explosion on Deception Island. Her father-in-law, Graham Seear, was there when it happened.

This was a fortuitous coincidence because I was now involved with yearly trips to the Antarctic gathering images and had many times visited the very place where Graham worked. I communicated with him and through his DVD learned of the event that destroyed the research facilities at Whalers Bay.

In 1967 as a young man of twenty Graham Seear began what was to be a two-year stay on Deception Island as part of the British Antarctic Survey. One night this all came into jeopardy when he felt tremors. None of the others felt them but then one day things began to sway. There was no question now that he was right and they were living over an active volcano. Overnight the temperature of the water in the bay rose 4 degrees. Graham felt something was up and started packing his bags.

It was springtime in the Antarctic and the birds

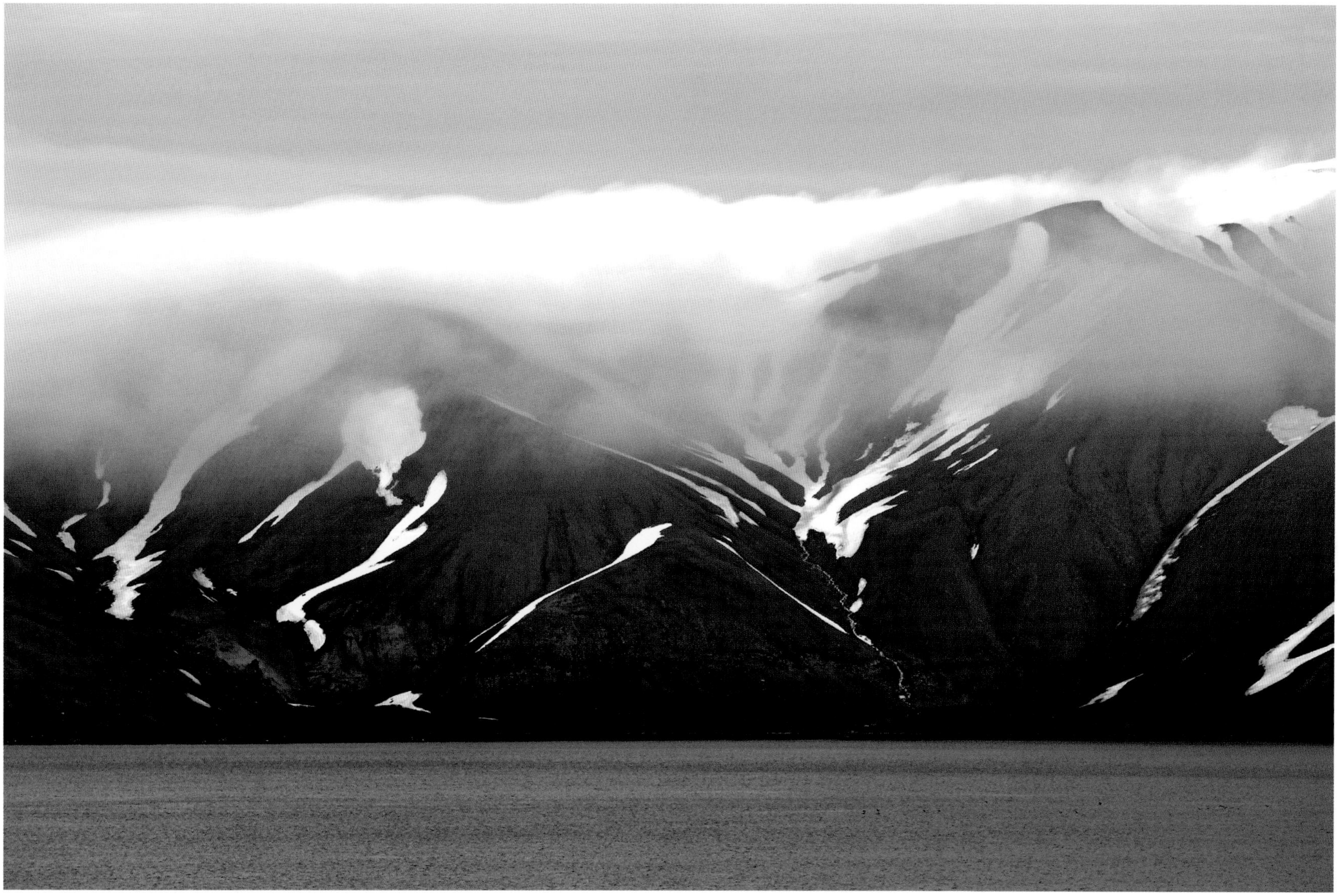

Crater rim

and seals were coming back but then suddenly they were all gone. Graham mentions how eerily still it all became. It was an ominous sign.

The tremors became stronger and louder. Ash, three kilometres away, began spewing into the air rising like an atomic cloud above the station. Noxious fumes filled the air and their lungs burned. Hot black ash, a foot deep, now covered the glaciers and visibility was down to a few metres. The men's only escape route was their forty foot dinghies but when they reached them they were high and dry. The sea was gone. With things worsening their lives were now at stake. The air, so thick with ash, caused an electric storm and the men lost all communication with the outside world. Men from the Chilean station at Pendulum Cove a few kilometres away fled their location and travelled blindly over crevassed glaciers to join the British team. Ash spread over 170 kilometres.

Eventually a break in the clouds gave them contact with a sister station on Adelaide Island and they in turn contacted the outside world. Fortunately the Chilean supply ship, the *Pilato Pardo,* was close by and had two small helicopters. After a long night and a break in the ash fallout all men were rescued. The Chilean station was completely destroyed and in 1969, because of the high risk, the British station was abandoned. Now declared an Antarctic historical site it is visited only by penguins, seals and occasional tourists.

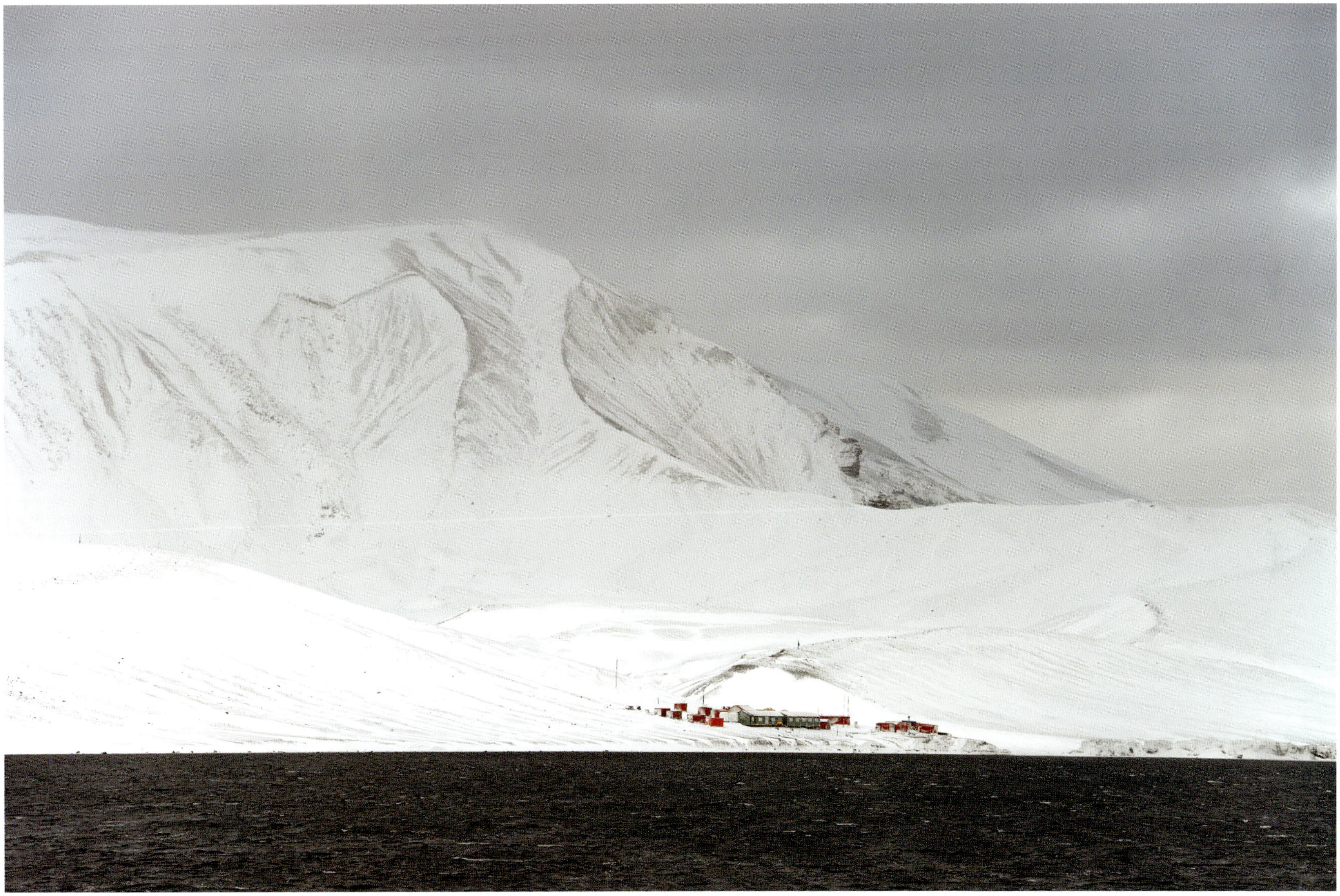

Spanish station, Gabrielle de Castille, March 2014. Deception Island.

ONE DOES NOT usually associate Antarctica with volcanoes and their presence comes as a surprise to many.

If you were asked to identify the location of this image the most likely answer would be Iceland because it is commonly known as the land of ice and fire. But here there is plenty evidence of both. Nothing that you usually associate with Antarctica is apparent. Where are the glaciers? The enormous ice caps? The penguins? The snow?

However, over this ridge and a few kilometres away on a point called Bailey Head is the world's largest chinstrap penguin colony numbering one hundred thousand and directly to the north is a dazzling chain of ice-capped islands rising thousands of feet directly from the ocean. This place is surely an anomaly. A volcano in the South Shetland Islands, Antarctica.

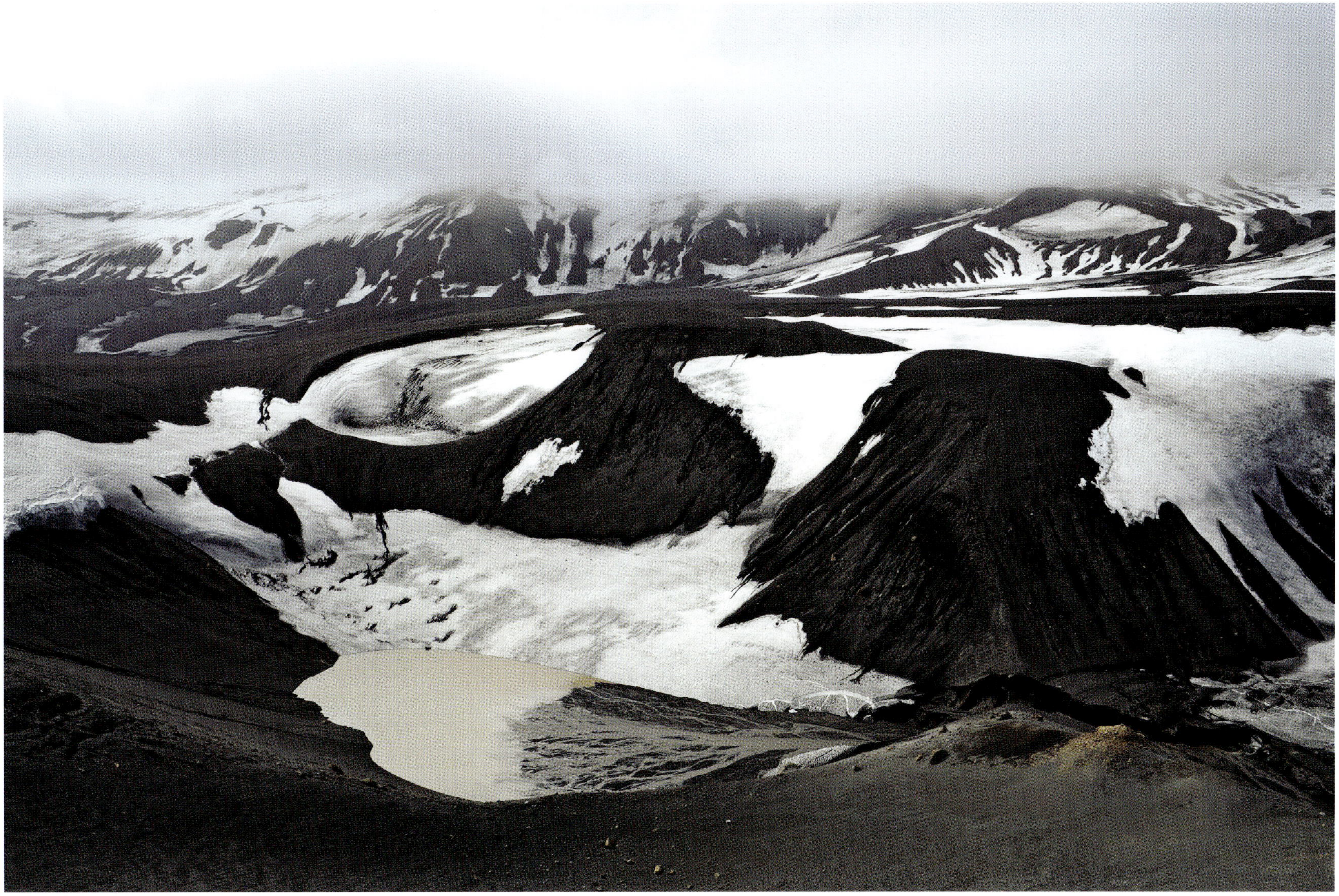

Crater in Telefon Bay

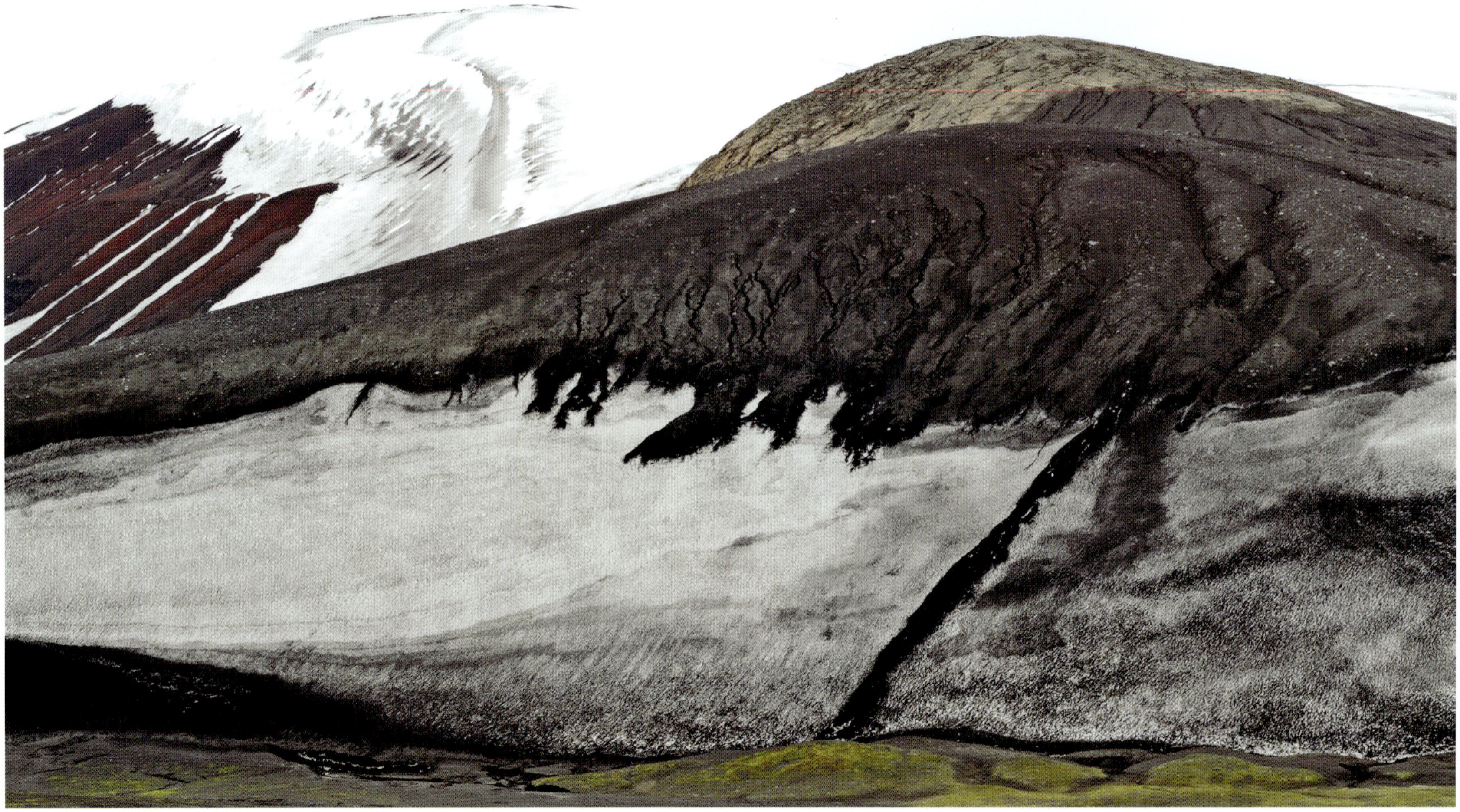

Lahar and coloured ash

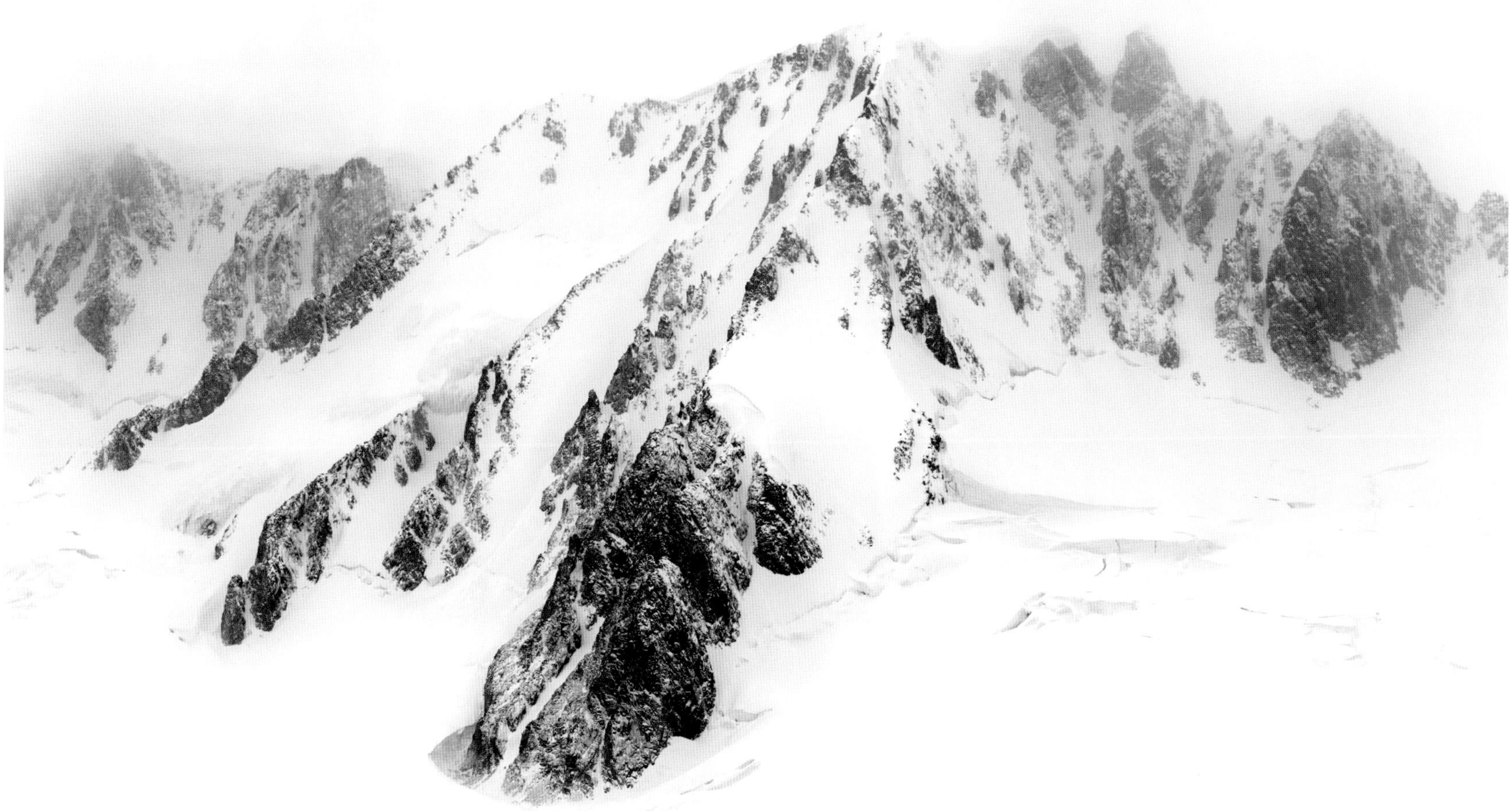

Exposed pinnacles

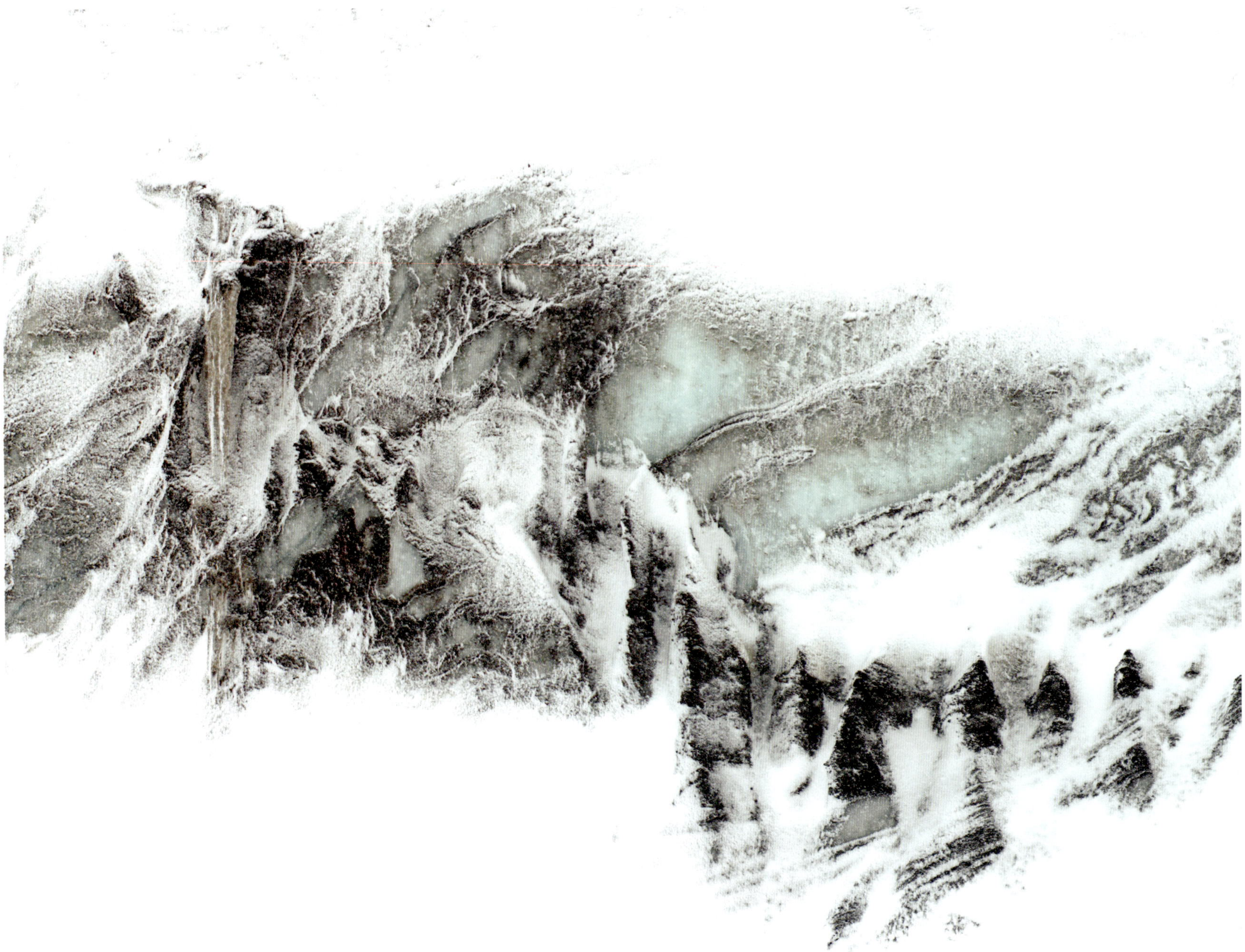

Cinders and ash

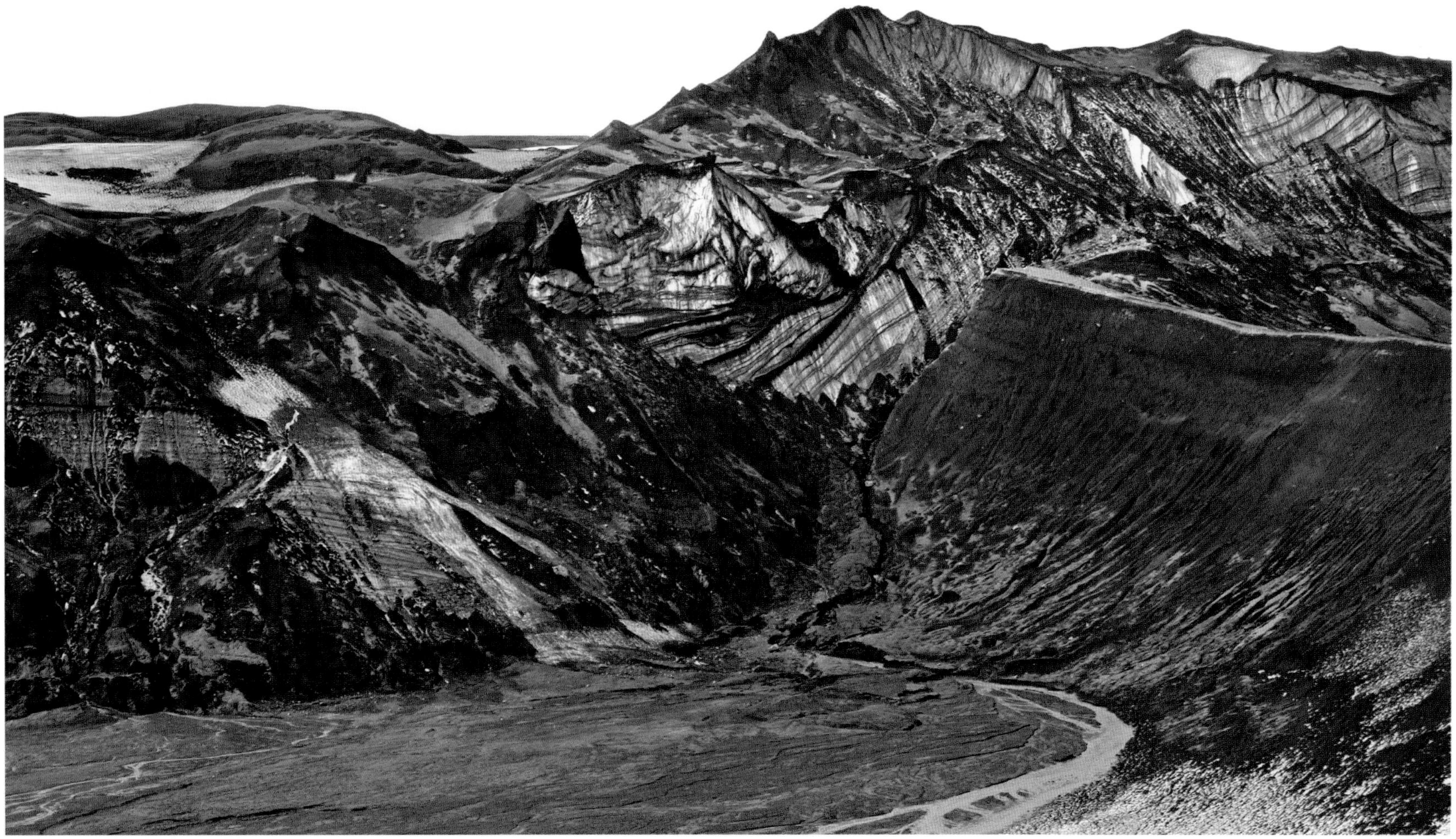

Tephra and ice

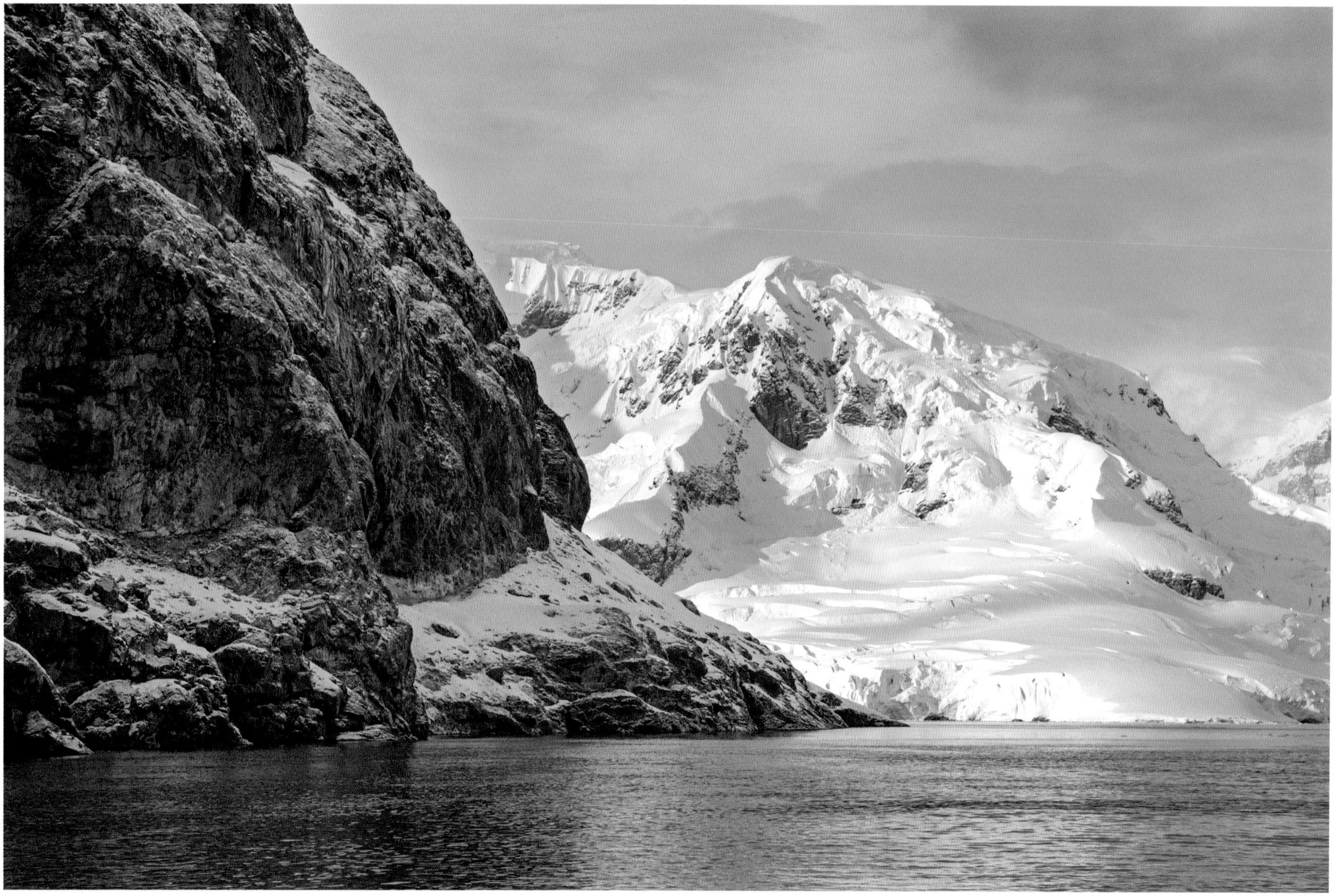

Cormorant cliffs Paradise Harbour

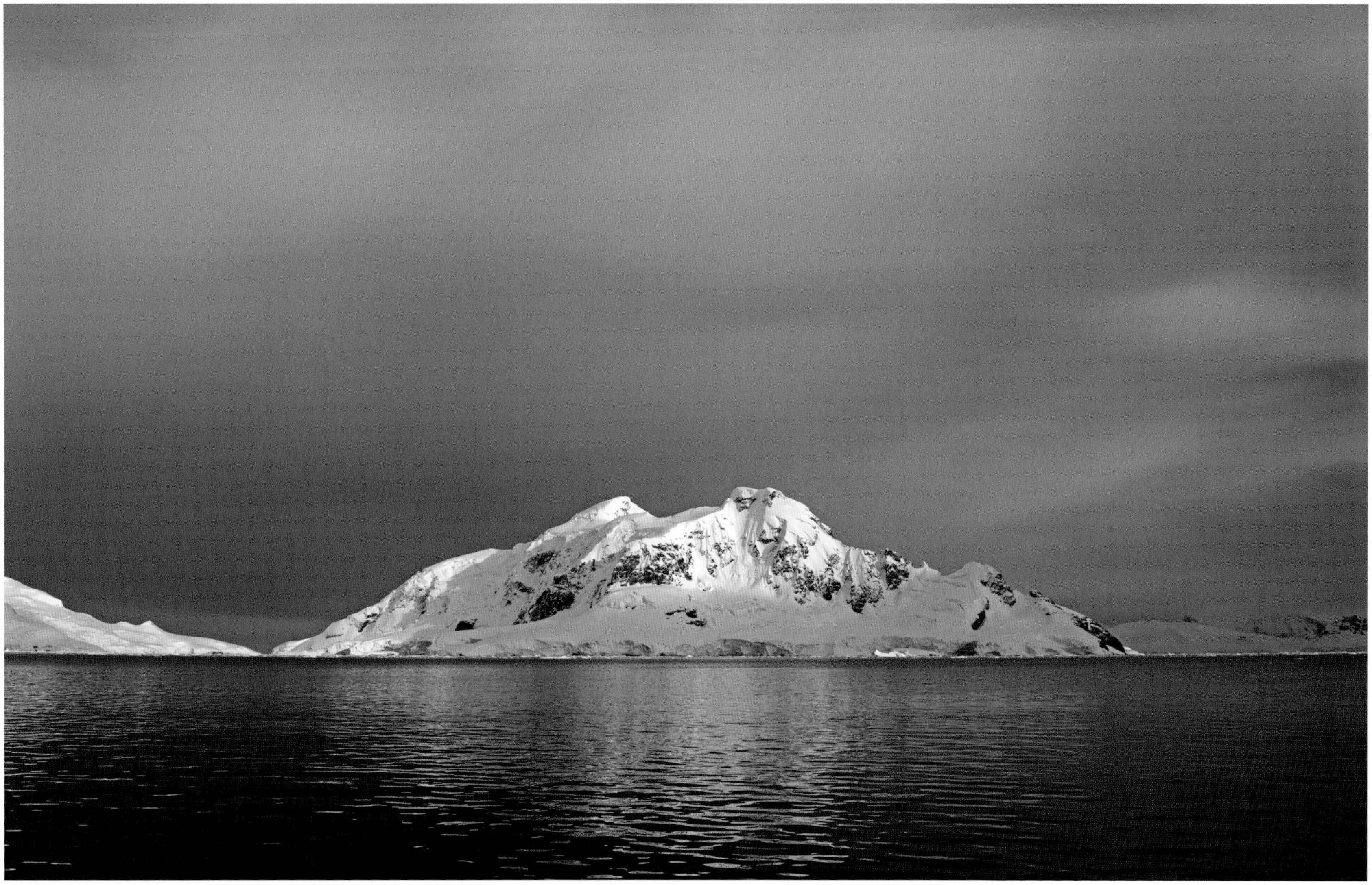

Island reflections Oscar Cove

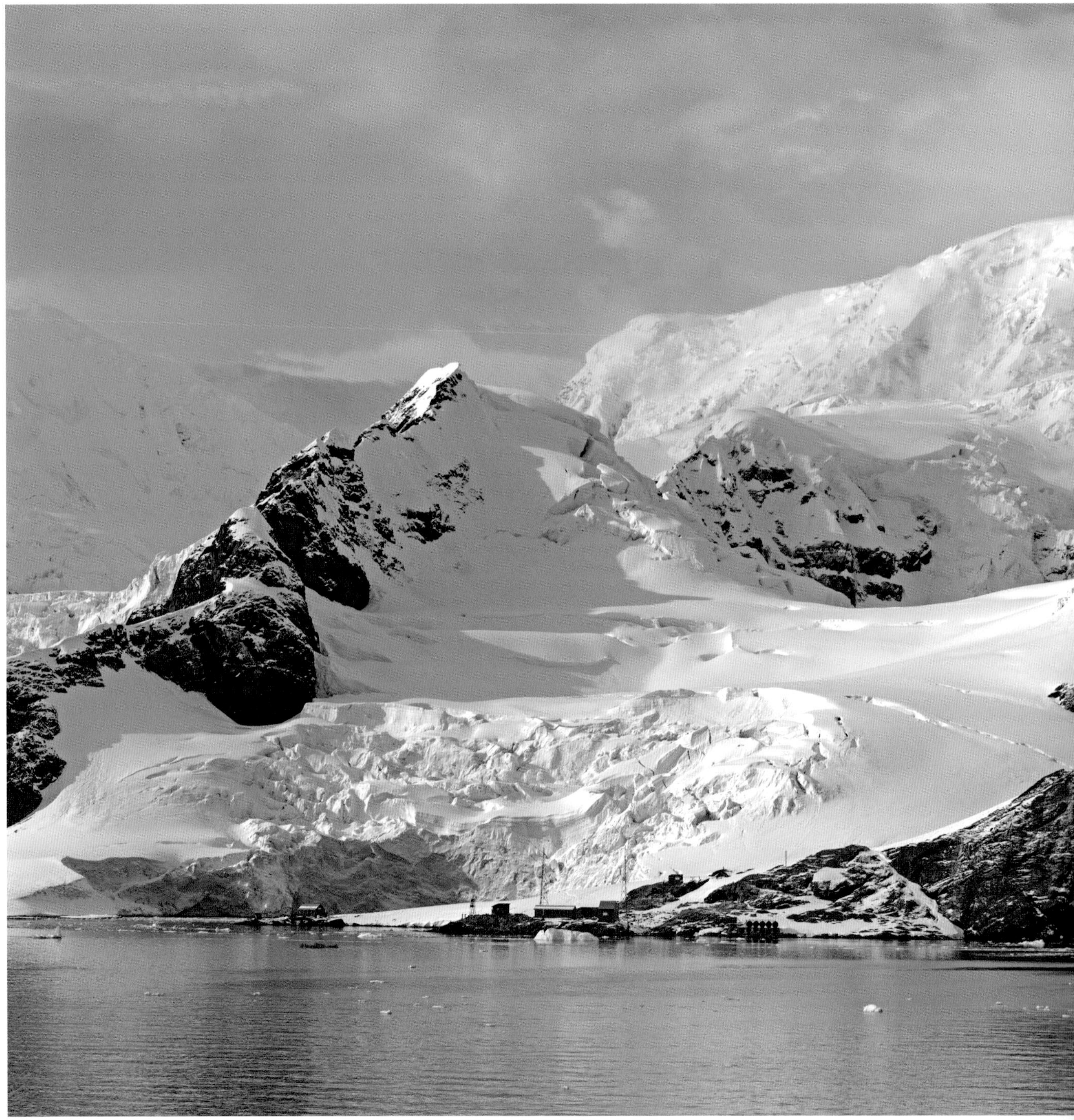

Argentine base Almirante Brown

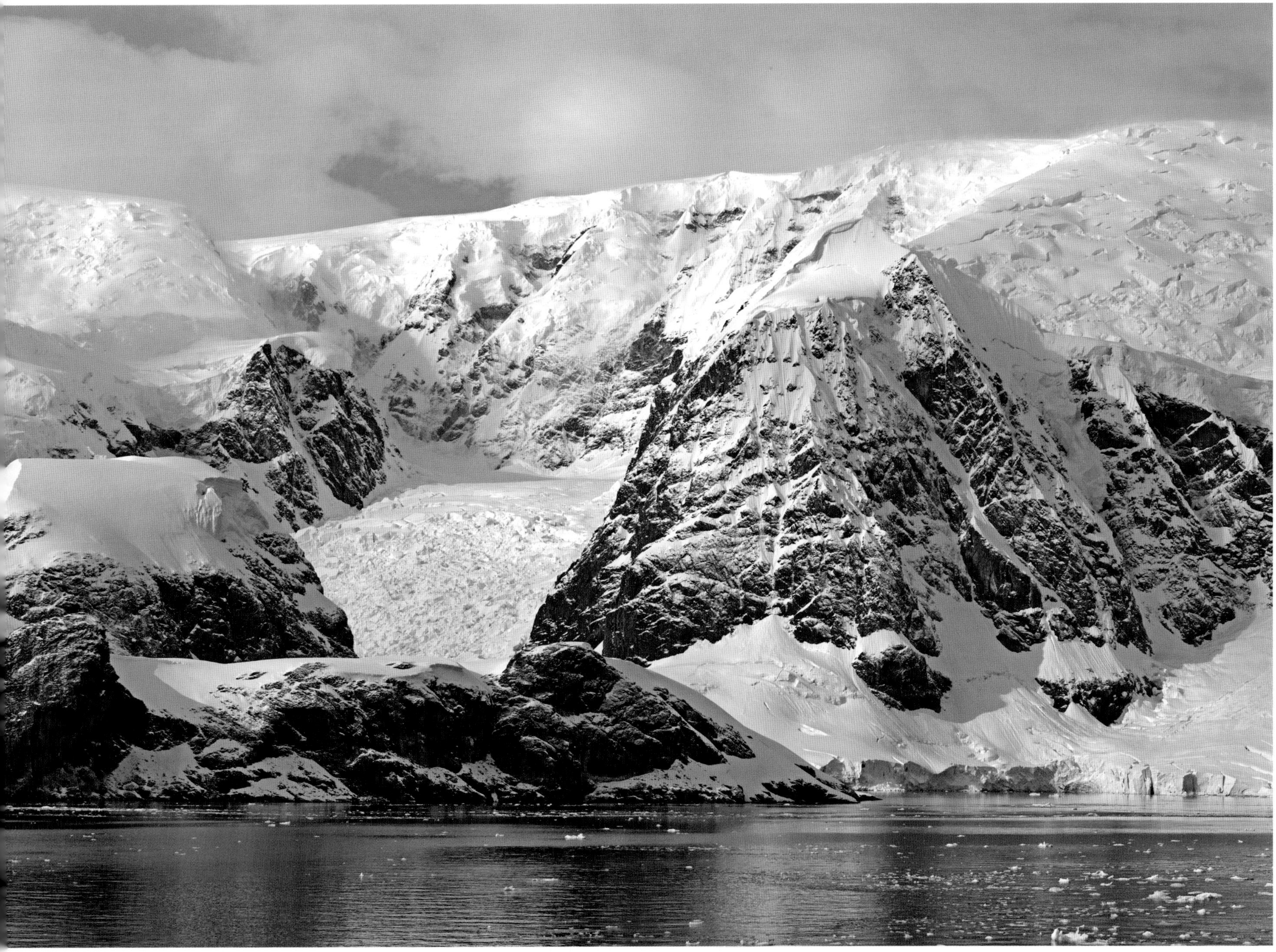

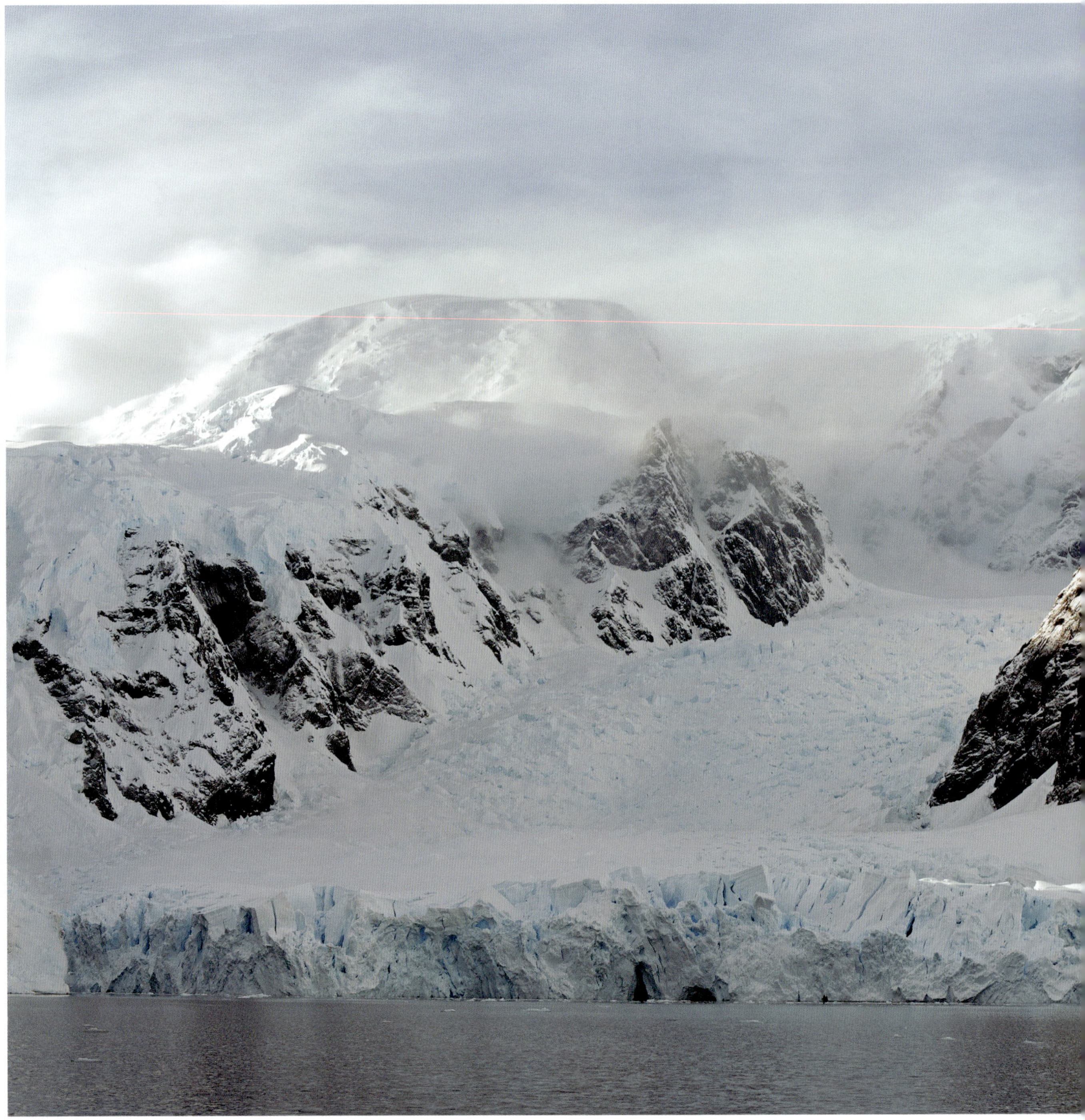

Base of Mount Inverleith Paradise Harbour

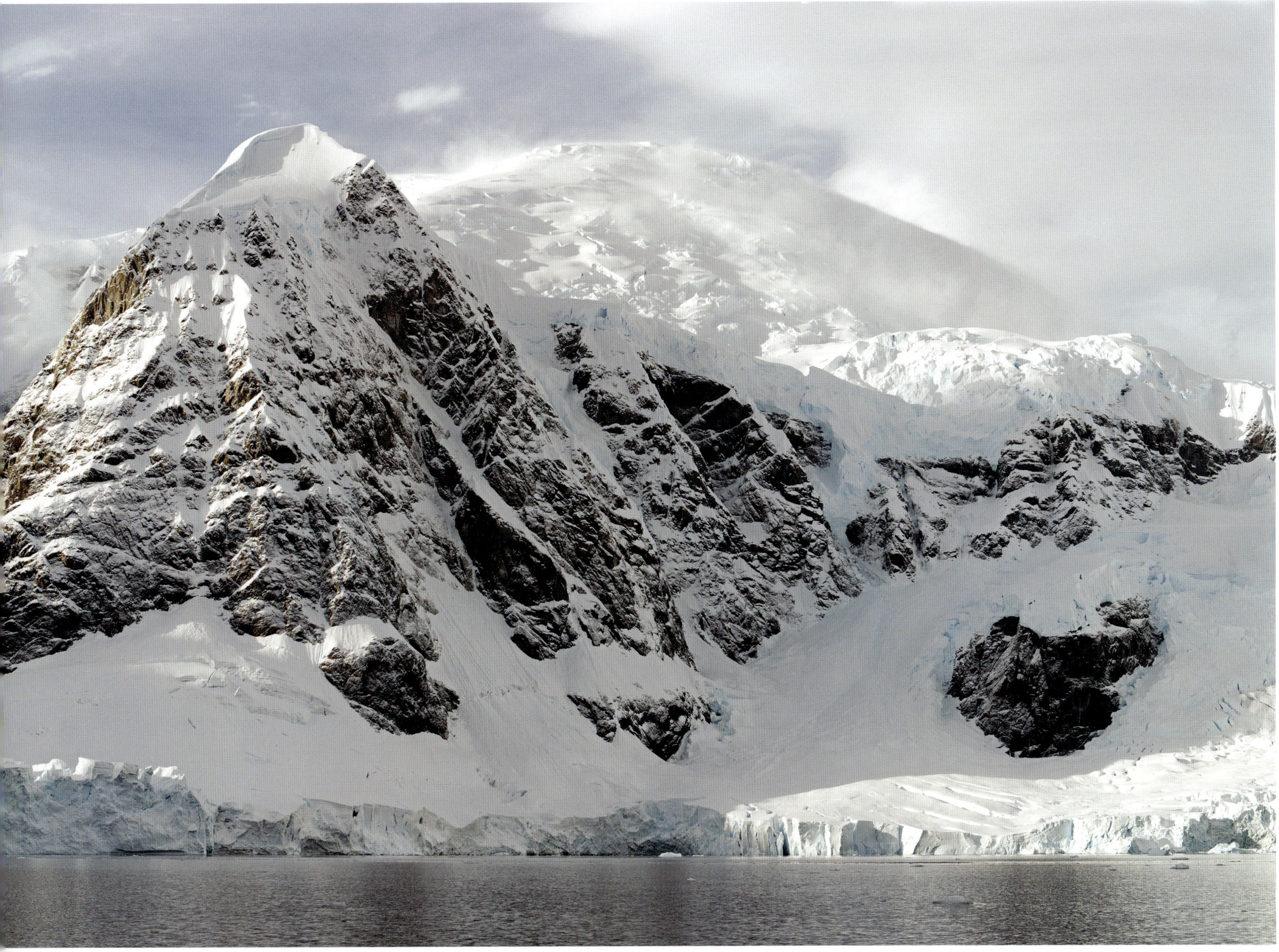

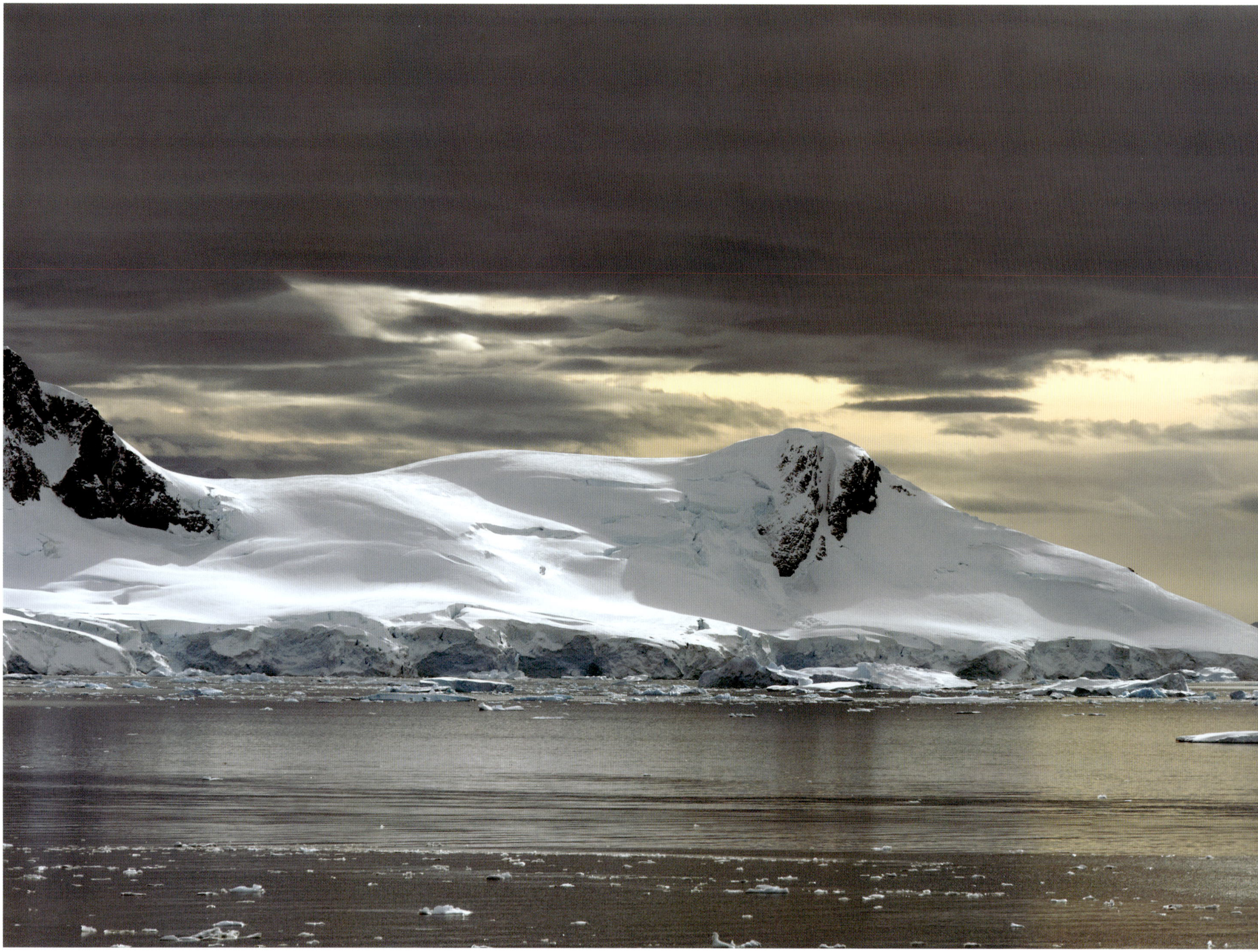

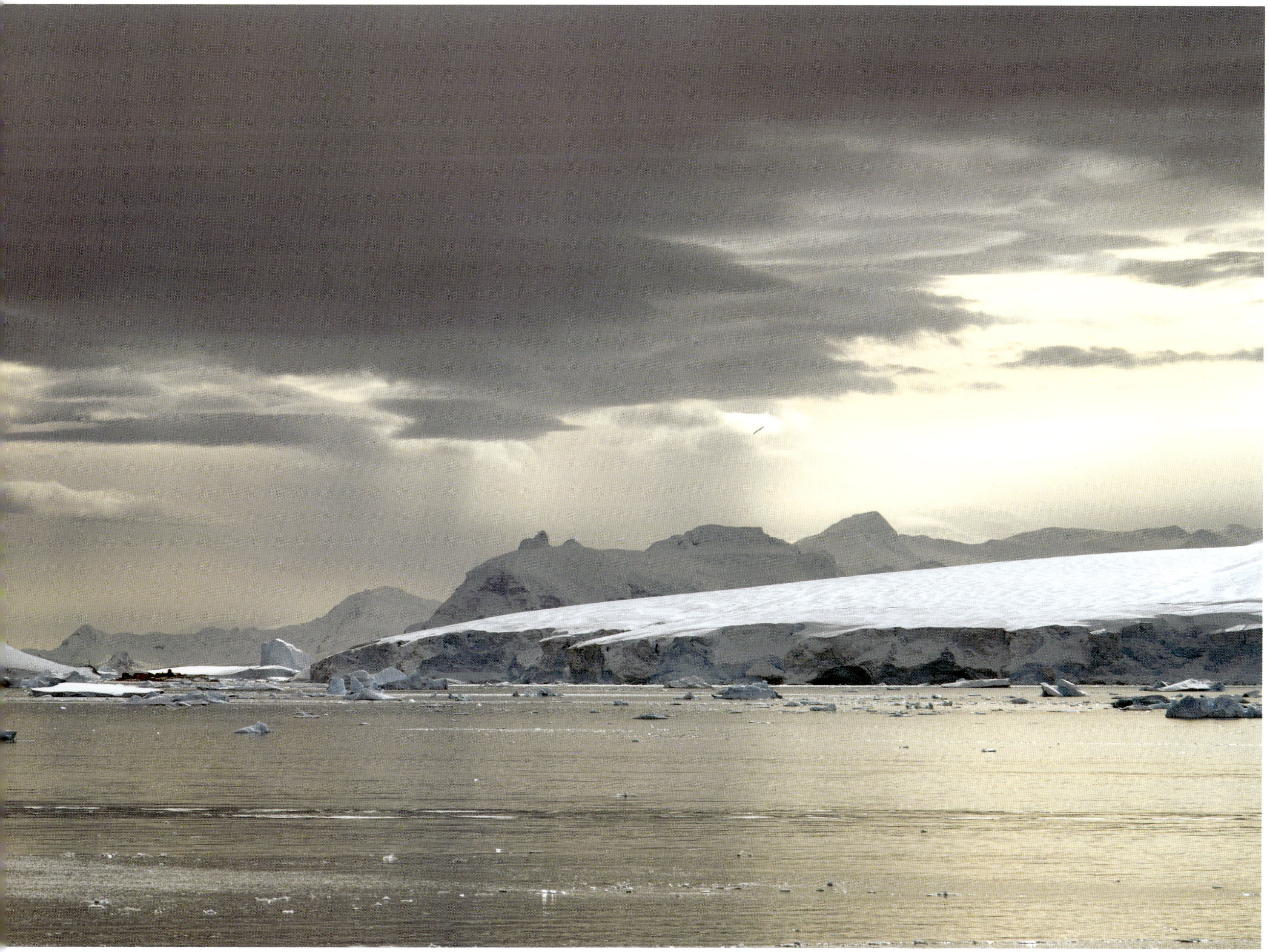

Gabriel González Videla base Aguirre Passage Chile

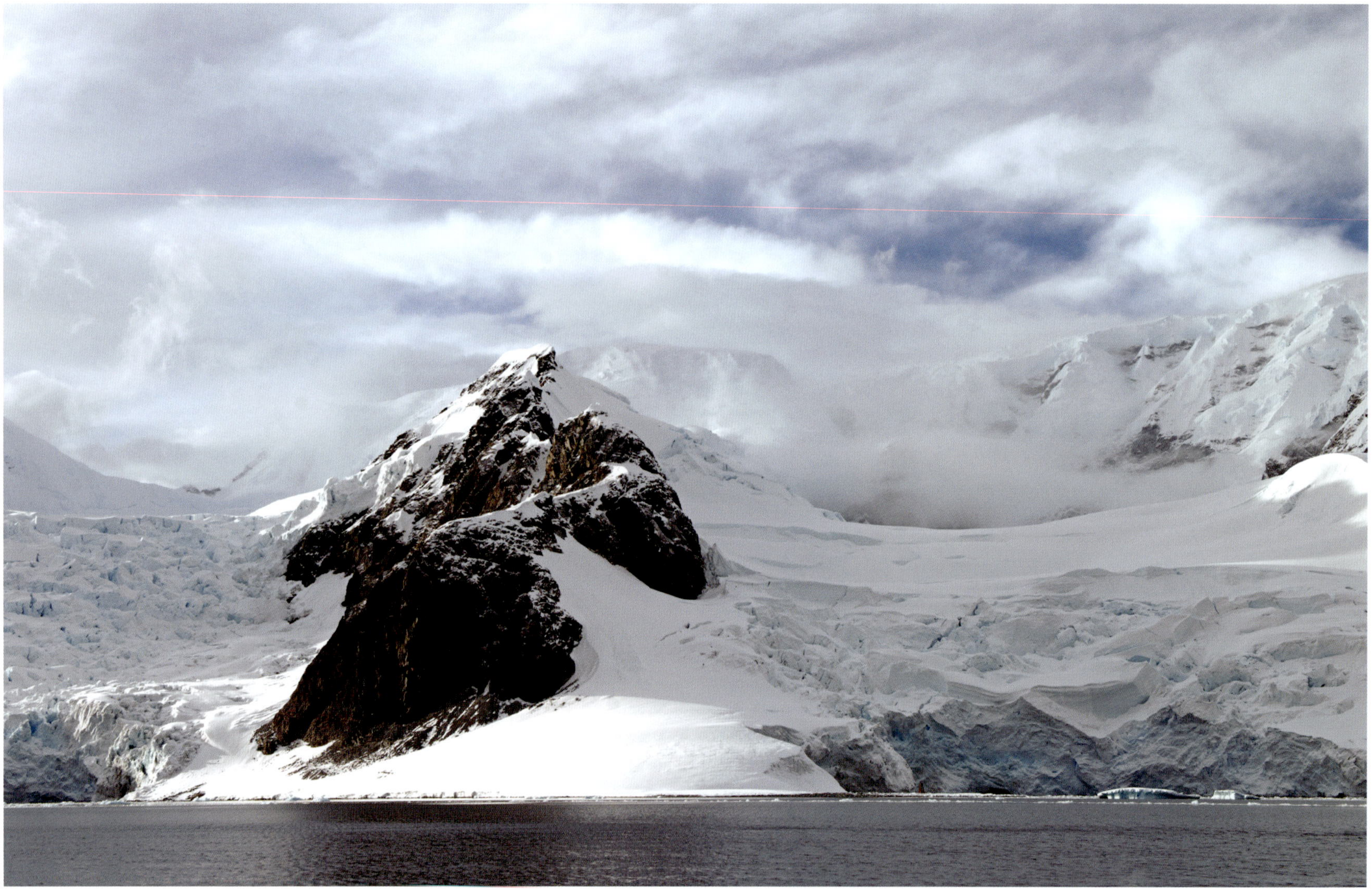

Wilderness of Antarctica

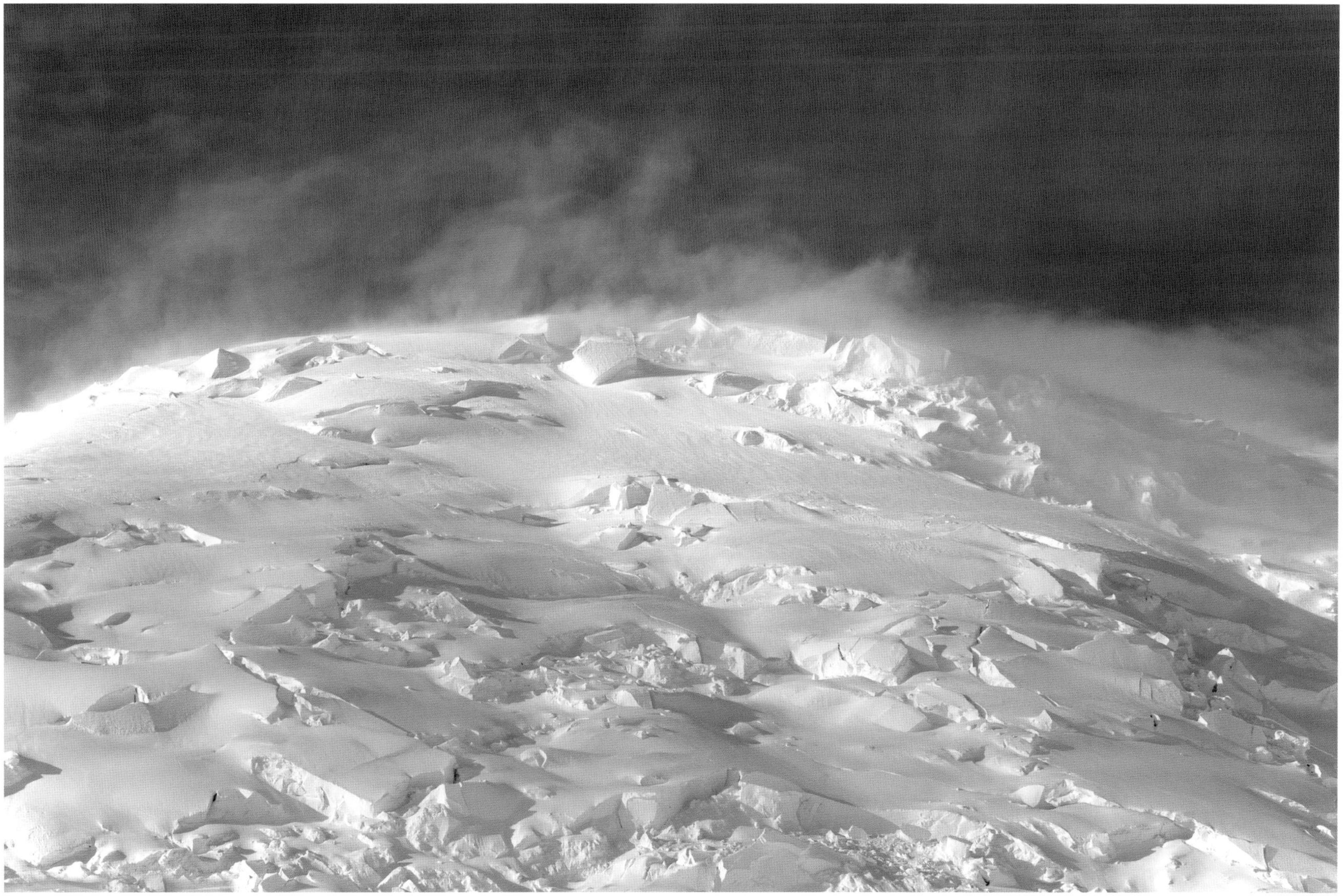

Glacier peak Mount Inverleith

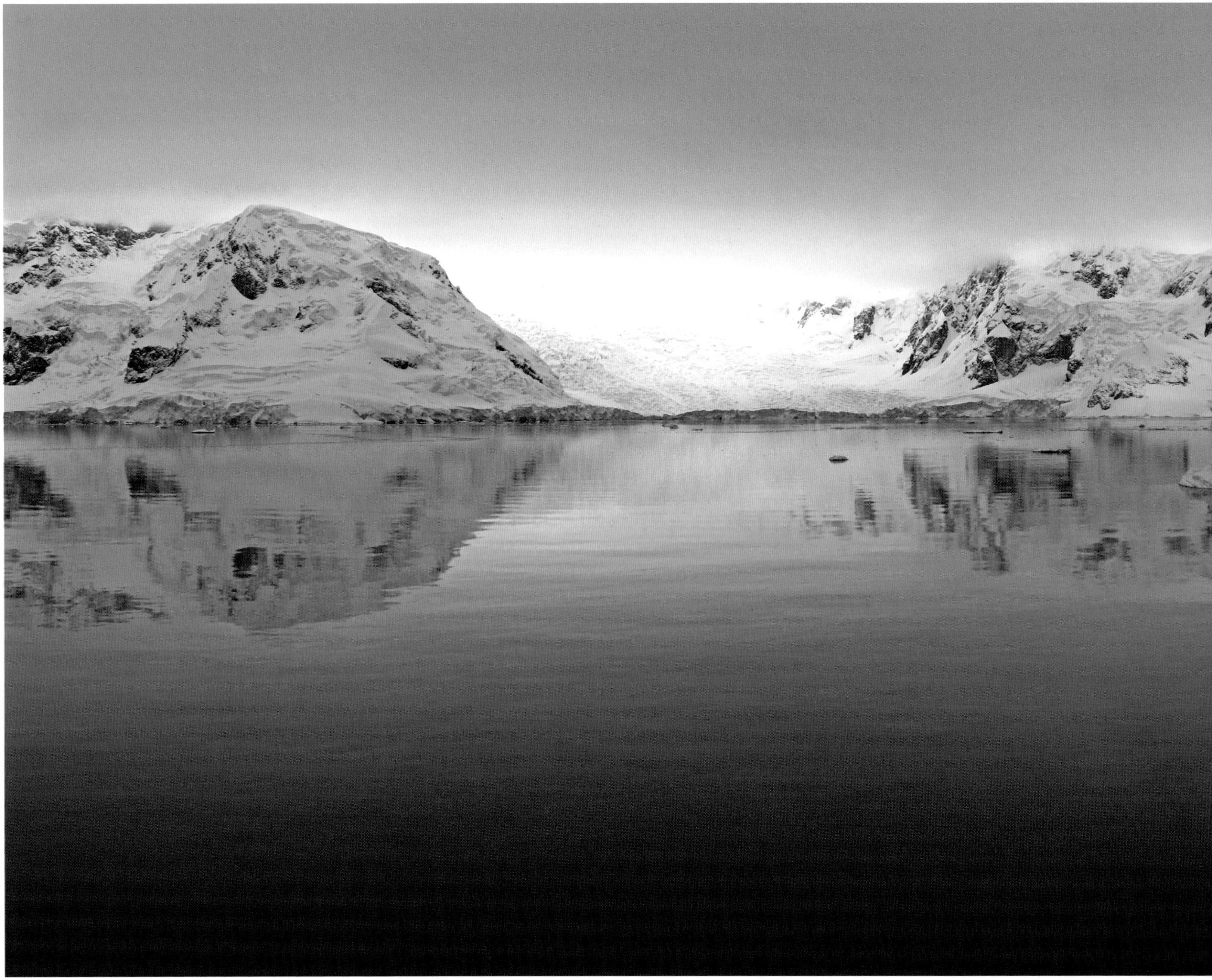

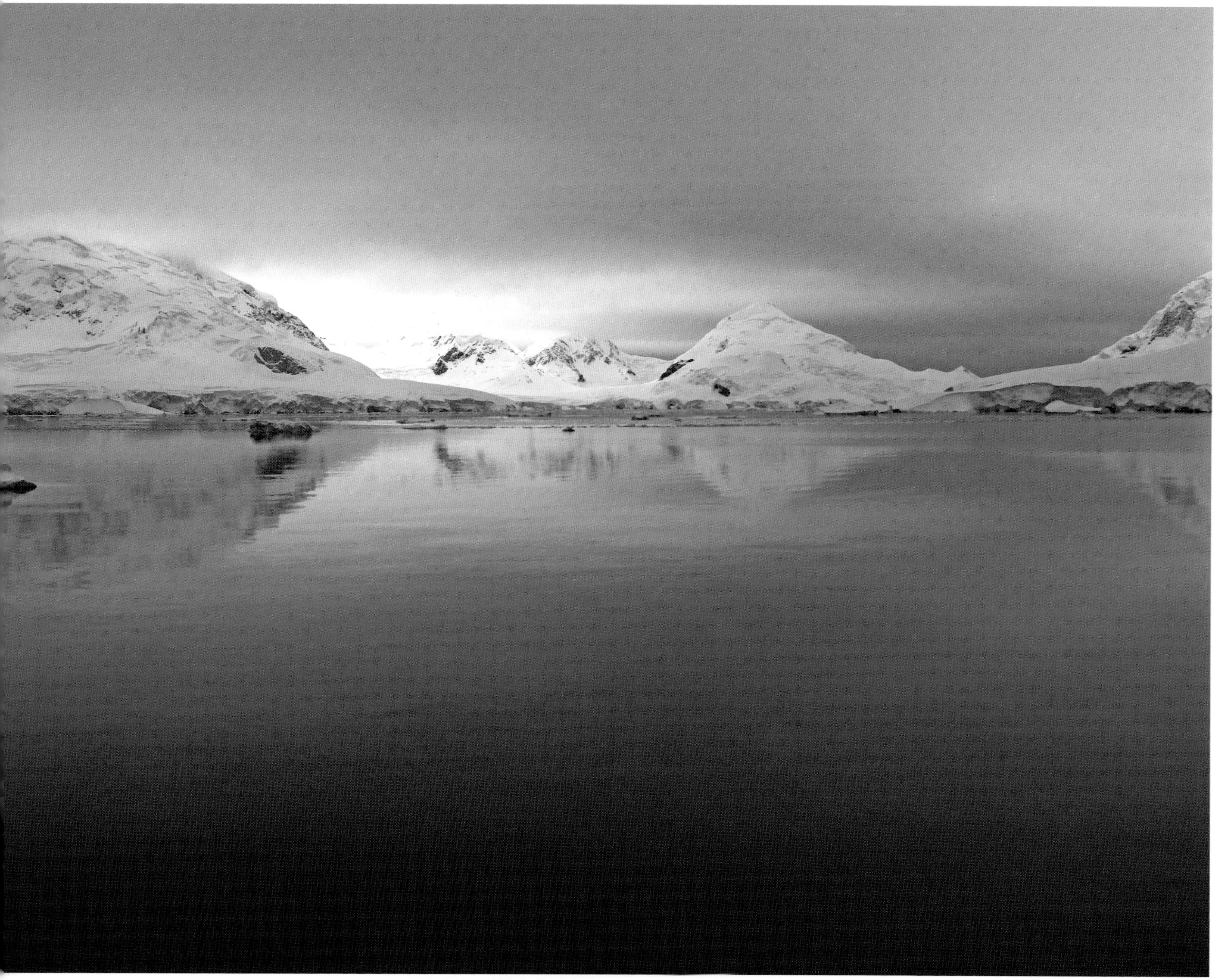

Paradise Bay

IN A LAND that is so very white and devoid of the warmer colours it is an unexpected find to come across a snowfield that appears to have been splashed with strawberry Kool Aid. Very conspicuous against the brilliant white these wide patches or long rivulets stain the snow looking more like diluted wine than a plant. Often from a distance it can be confused with the ruddy trails left by penguins. On closer inspection however it is better recognized as ice algae. Like a leaf it is filled with green chlorophyll but also infused with an overpowering red pigment that gives it the distinguishing rose colour. Four times as big as a blood cell it thrives in the frigid snow and cold salty sea. The red pigment protects the alga from the intense ultra violet rays and also being more efficient than the green chlorophyll better absorbs the sun's warming rays.

Here Rongé Island in the Gerlache Strait exhibits many Antarctic features: lee clouds that rise in a rhythmic pattern above the higher terrain, seracs spilling over cliffs, an icefall, and to the right just below Mount Tennant, patches of cherry coloured algae.

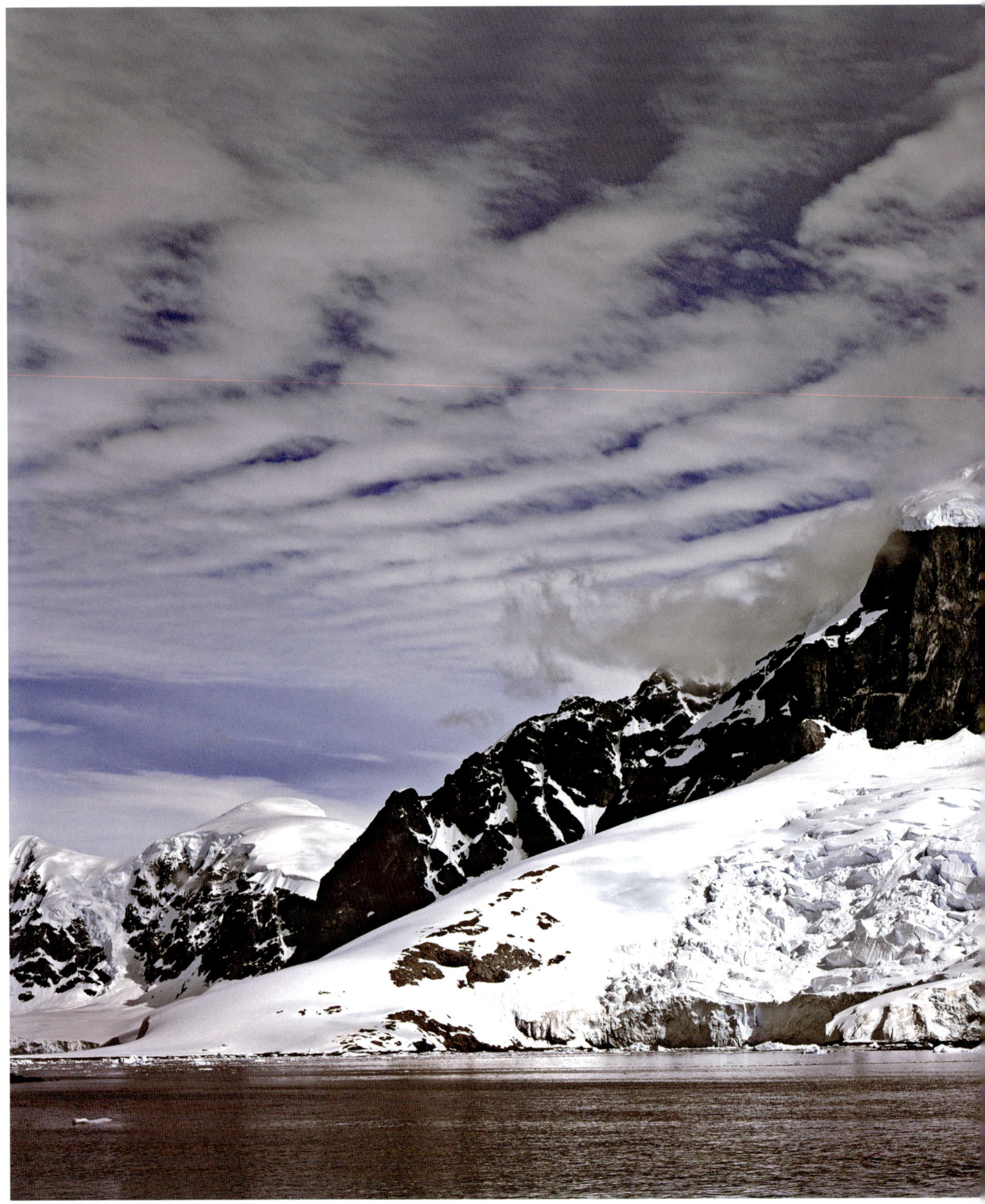

Watermelon snow

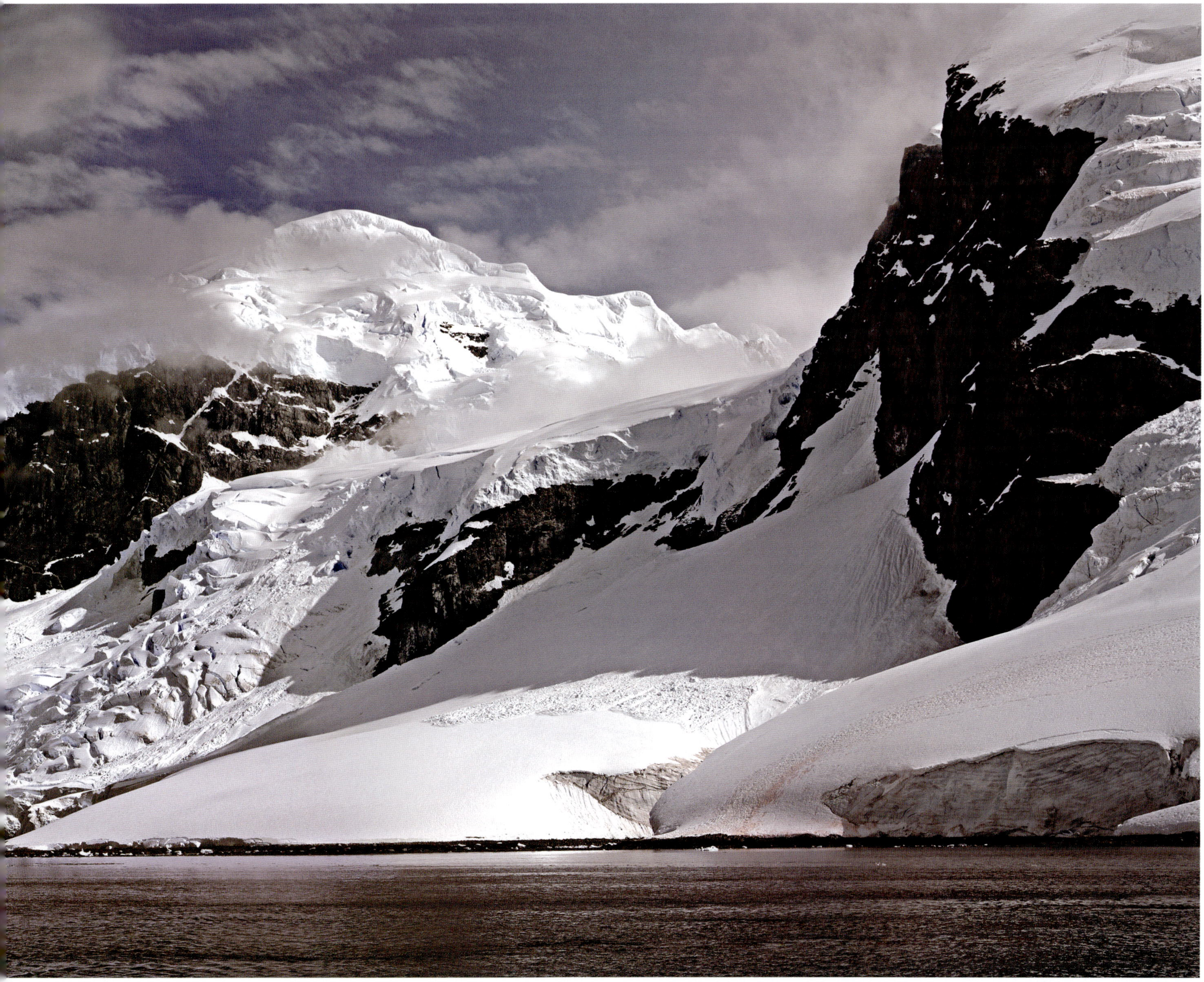

IF ONE IS familiar with the smooth lustre that coats the inside of certain mollusks then it is easy to imagine these two clouds resembling opened clamshells. Their polished mother of pearl appearance is compelling in its sculpted shape and in its smooth gradation of tones ranging from the very bright to a dark grey. Their shape and size alone make them the prime elements in this image.

On further study however it should be noted that what really makes the image come alive and bring it past simple documentation is the slant of the larger cloud tilting downward towards the horizon to another rising cloud of similar shape and quality. This fortuitous alignment makes this an image that now not only depicts the character of these captivating clouds but also imparts a sense of where cloud formation might take place. This animates the scene and gives it purpose.

The perspective in this image is particularly engaging because it very much looks like an aerial photograph. Yet, it was taken with a regular 50mm lens only about ten feet above sea level.

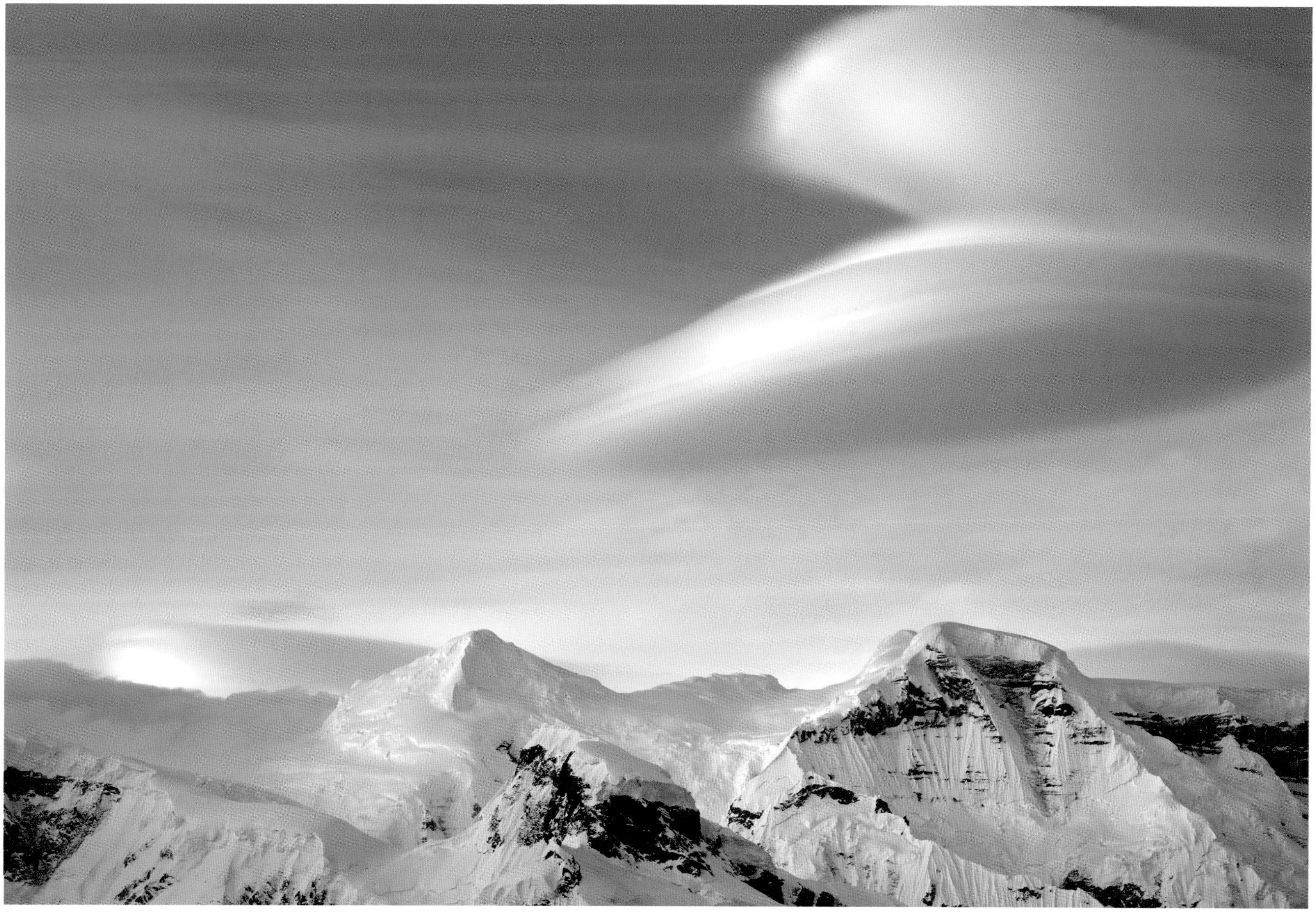

Lenticular clouds, spinning winds

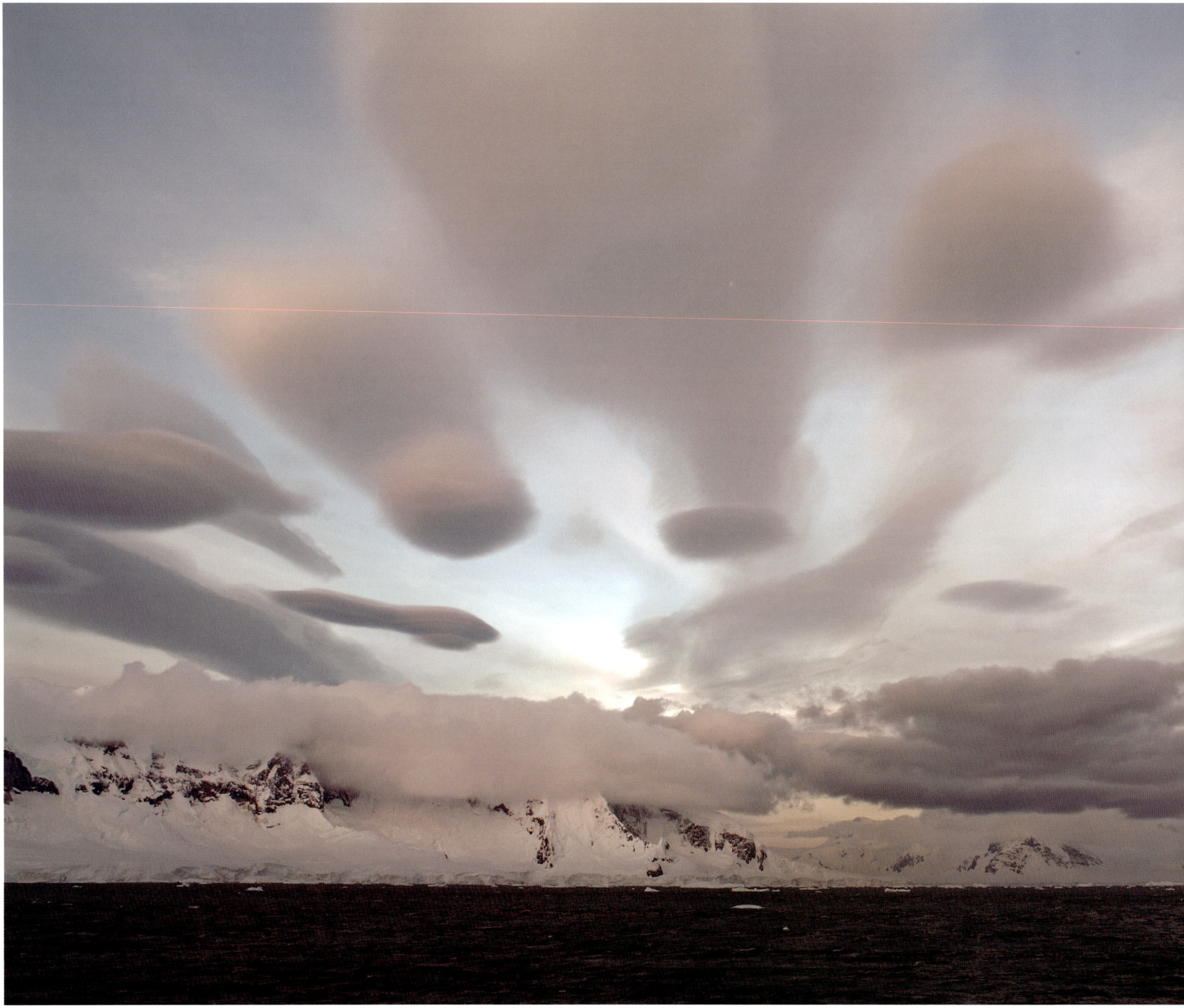

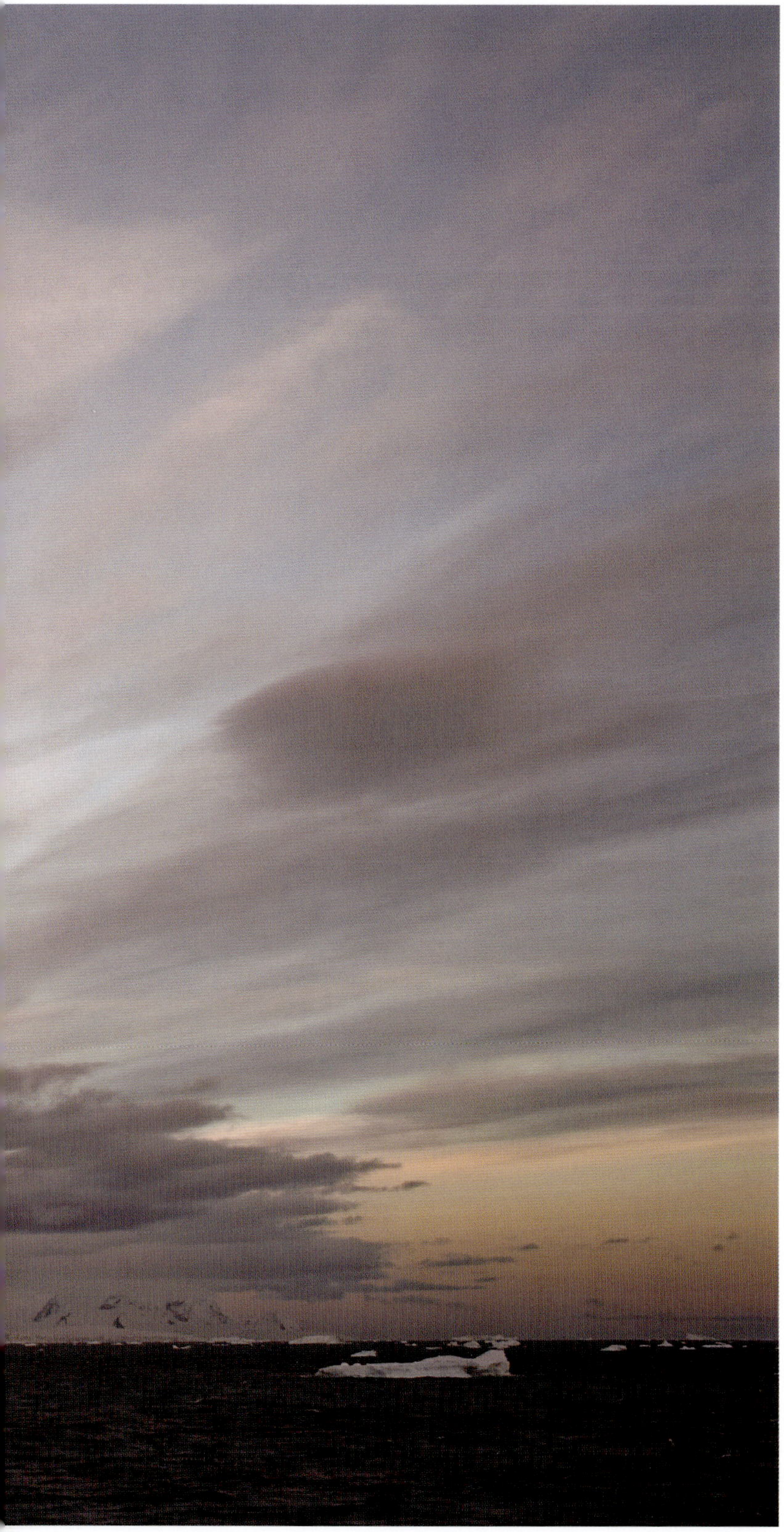

THESE VARIED CLOUDS continually heaved and stretched at different heights but held their positions for a long time. It was only the movement of the ship that seemed to change my perspective. I had taken a few images already but wasn't satisfied because even though there was a good mix of shapes there was not enough variety in colour to separate the forms. The image seemed flat. A bit frustrated I hoped and waited for the sun to emerge from behind an obstructing cloud. Fortunately the ship's advance was slow and the clouds in my viewfinder held steady. An agonizing five minutes later and the higher underlit clouds took on a light shade of pink and then slowly the white horizon turned warm. Different portions of the sky higher up retained their darkness while others lower down filled with a delicate new light. Moments later another sun-blocking cloud moved into position and the snow turned grey and the glow was gone.

Twilight colours

SEEMINGLY DEVOID OF any terrestrial life, binoculars make visible colonies of penguins standing at attention in rows farther up inhospitable slopes. They are not everywhere but have chosen their bare nesting sites with careful selection backed up by eons of experience. A beach protected by immense bluffs in the centre of this image makes this, in their eyes, a suitable landfall.

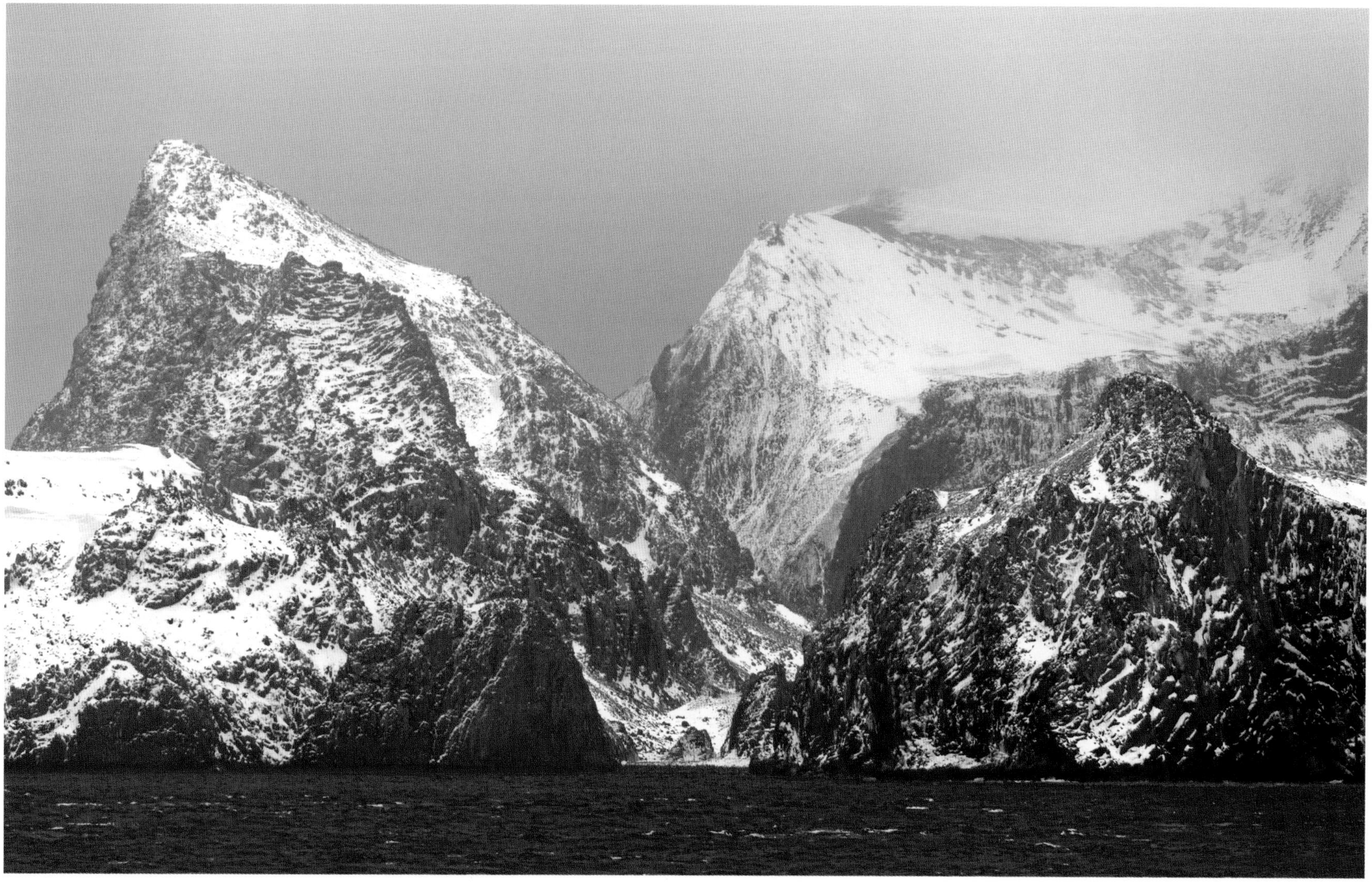

Rugged shore South Shetland Islands

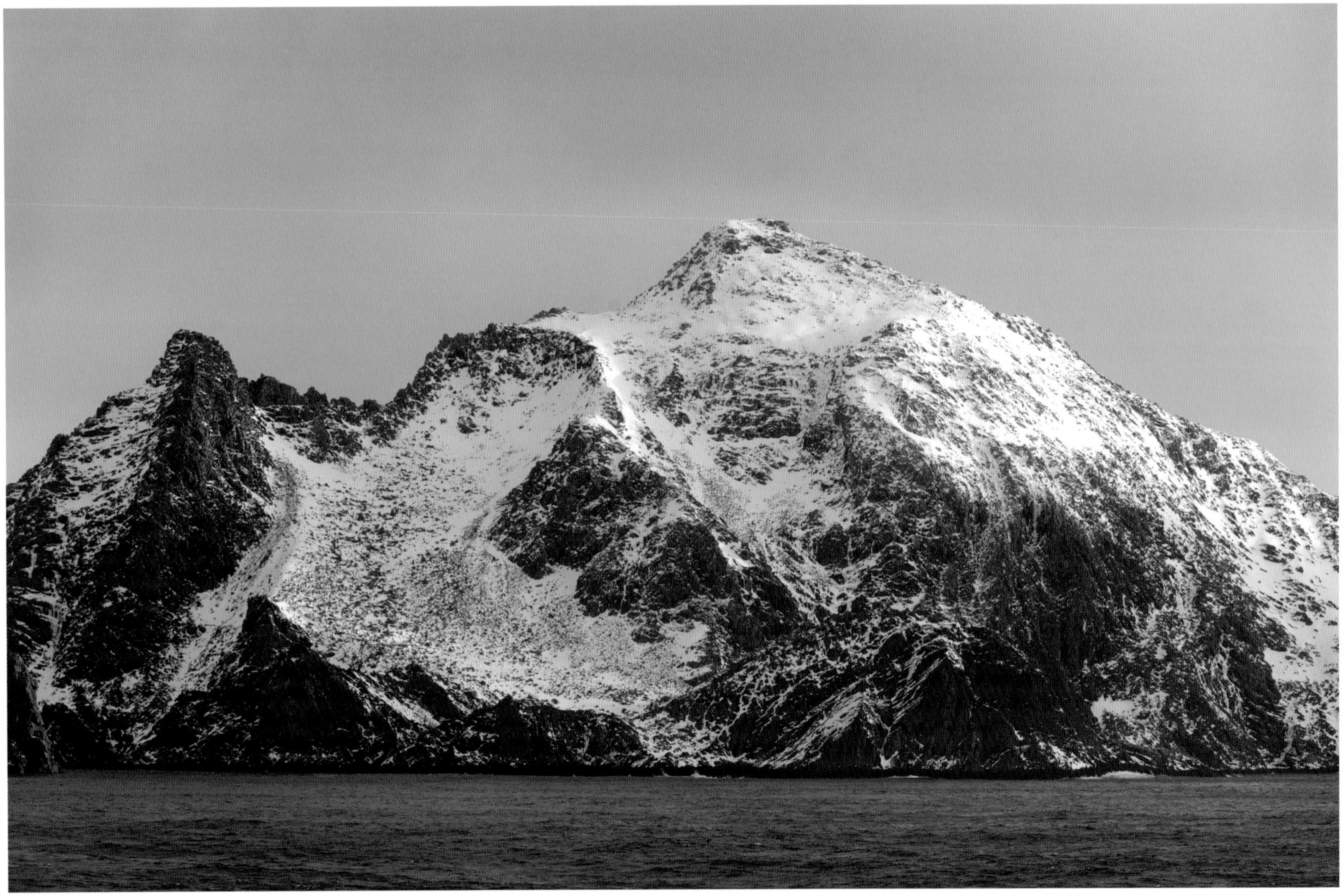

South Shetland Islands detail 2014

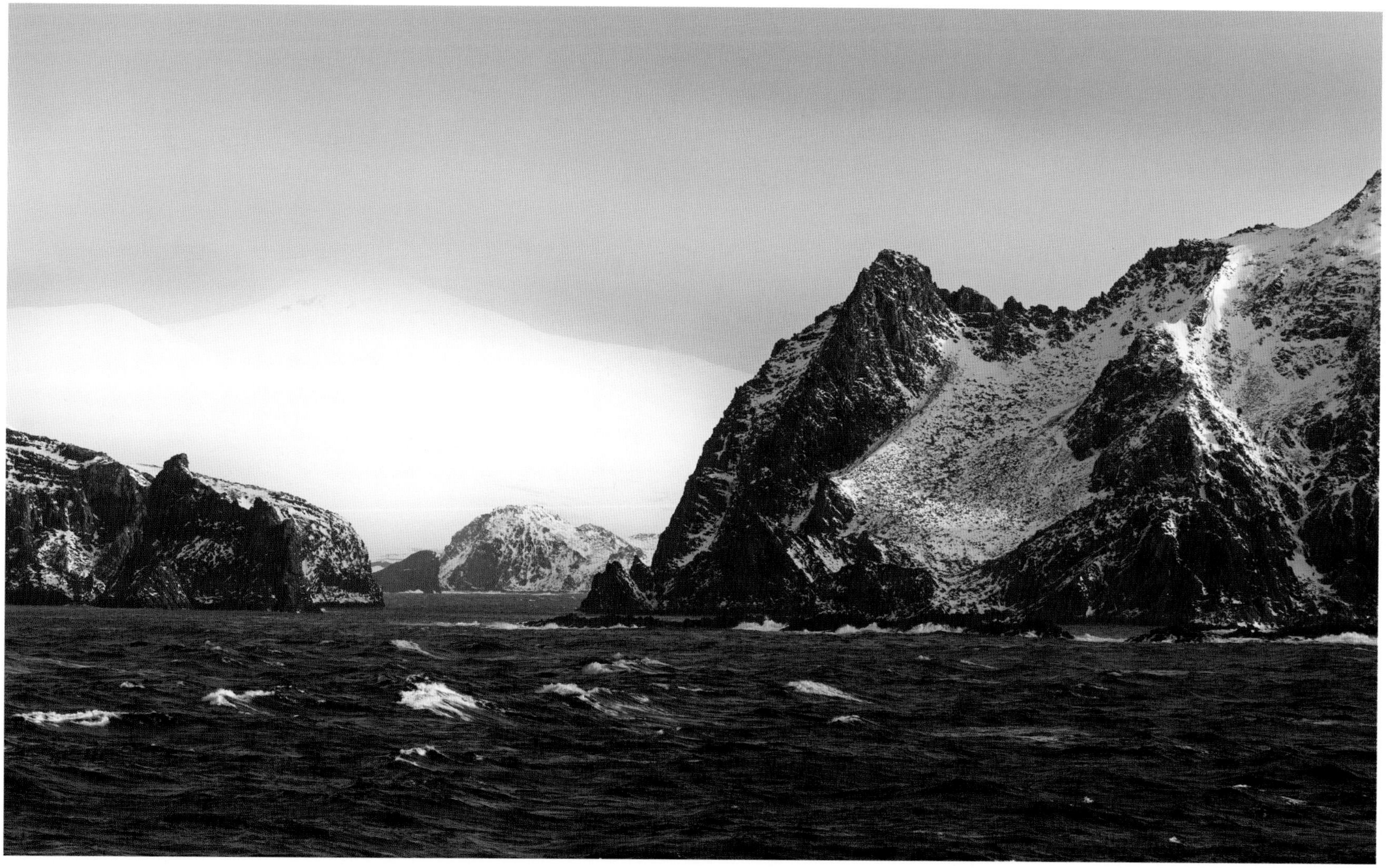

Outer South Shetlands

THIS WAS A time I will never forget. We had just pulled the Zodiac high onto the pebble beach and as usual I removed my life preserver, strapped on my camera bag and began to consider where to go. Everyone headed in different directions but generally in the direction where most of the penguins were gathered. I work best alone and wanted to concentrate as much as possible on my photography so I moved away towards a distant mound.

The mood of the day was dark and the light subdued. These were my preferred conditions where shadows were softened with muted highlights and the sky was filled with textures and substance. The atmosphere was dramatic and I hoped to benefit. Sinister mountains surrounded me and in between the rugged peaks menacing clouds descended abruptly. Conditions looked harsh and all felt ominous and cold. It felt at the moment that I should not be here and nothing else living belonged or could even survive this land.

I mounted the ridge and a scene emerged that absolutely stunned me. It was as if the curtains of an opera house parted and revealed a grandiose setting. Suddenly all my apprehensions changed. For a minute I stood in disbelief trying to understand the surroundings: mountains, sky, snow, ice, rocks, penguins. All were so harmoniously linked. I was wrong. I had so easily imposed my fears and needs. Something does live here. And happily.

Penguins brayed loudly as I set up my tripod. I worked nervously. Clouds gathered and stretched. Light penetrated the glaciers emitting a soft cyan and snowfields brightened and darkened. It was hectic. Busy penguins hustled and jockeyed for position fussing over empty nests. Other individuals seemed to move about aimlessly. I waited patiently for the best arrangement and with heart pounding I released the shutter. I lingered on that mound for a long time.

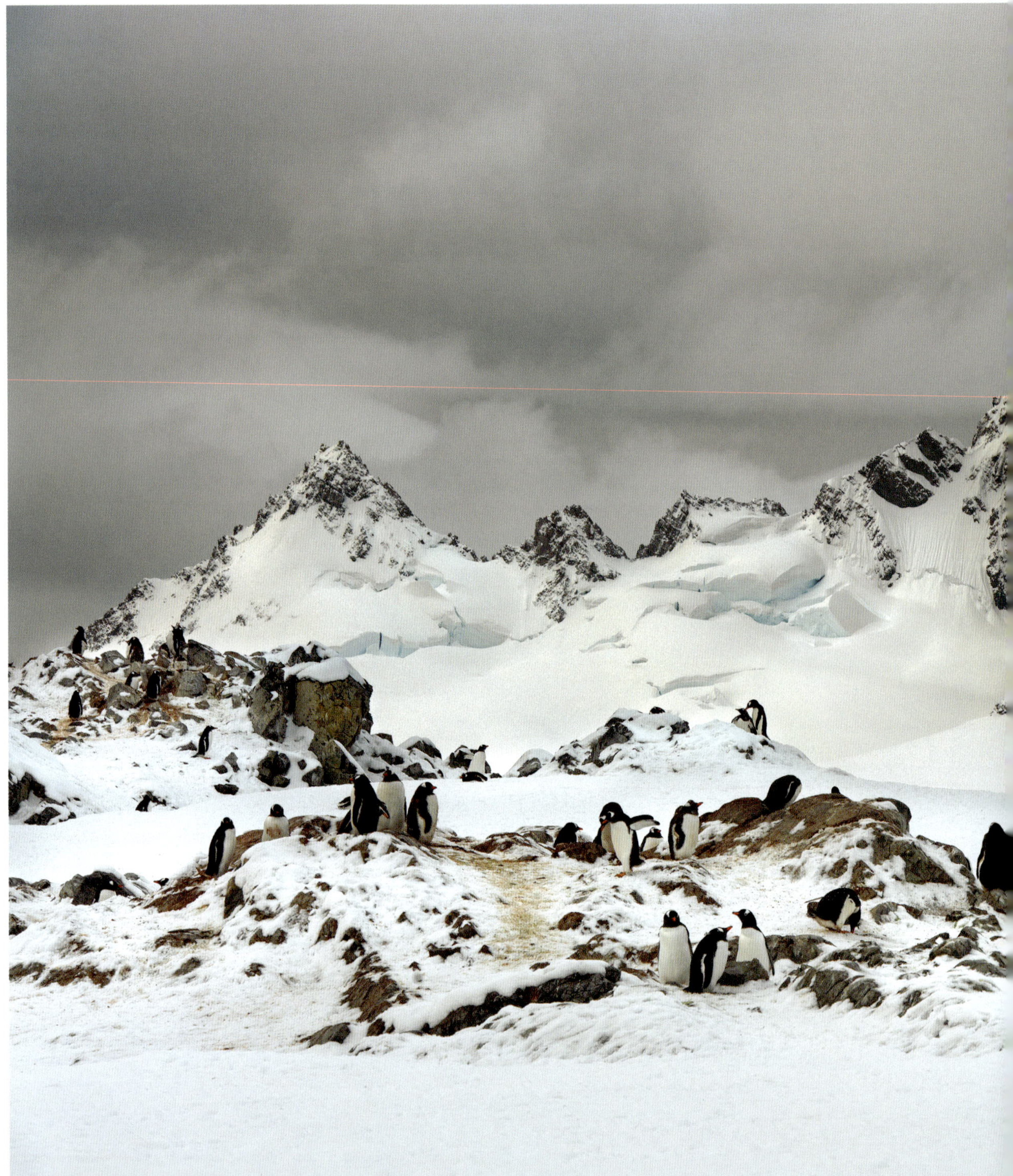

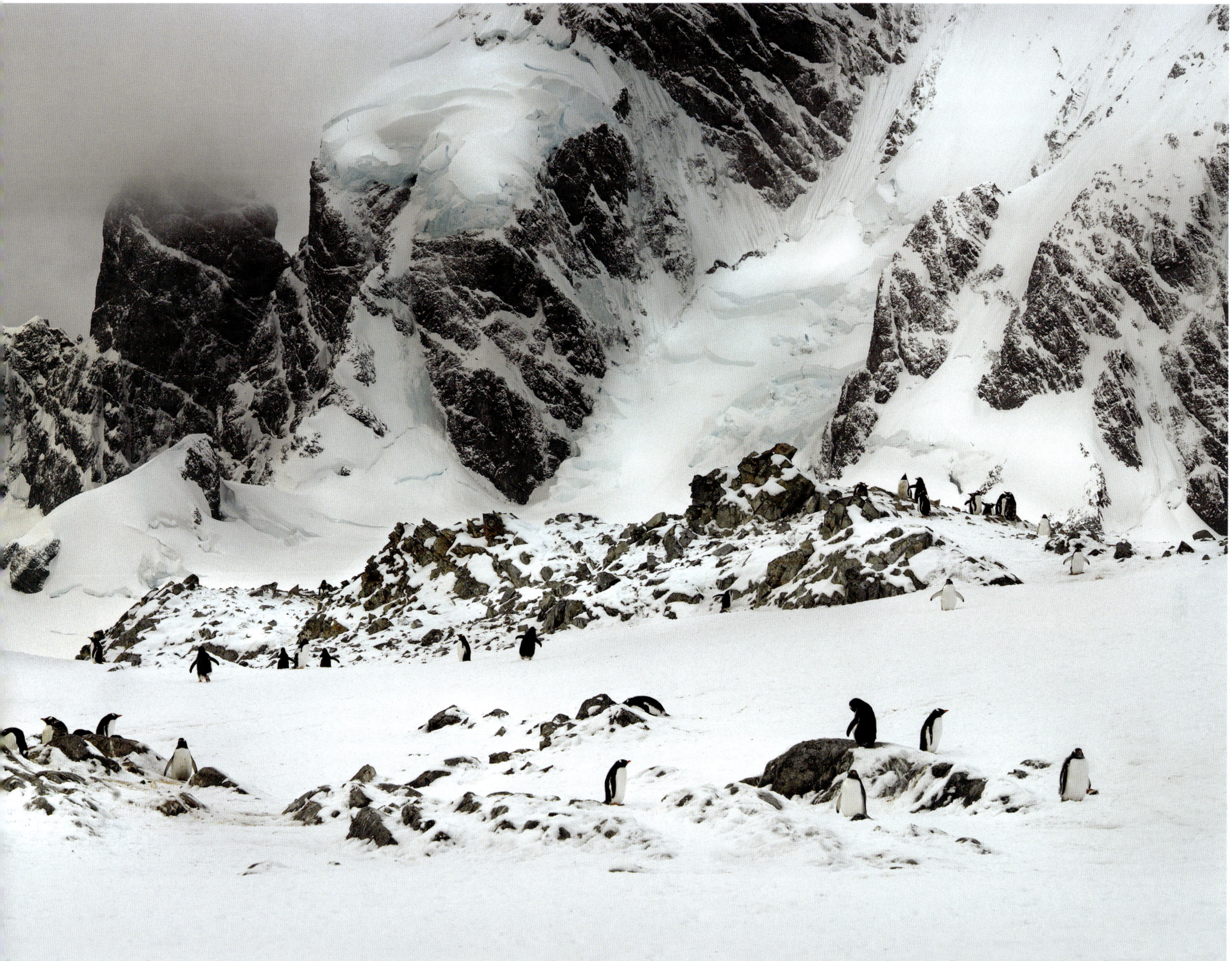

Penguins at home

HERE ON ELEPHANT Island is the site where one of the most compelling sea stories began.

Bursting nearly vertically out of the sea this island monolith reaches 2000 feet above a deep ocean. Its shore is but a series of cliffs interrupted by glacial tongues forcing their way into the sea. There is hardly a beach between the rocky barriers and landing anywhere even on a calmer day is a challenge. Ocean swells from persistent storms at sea approach this barrier and wash up the steep shore denying any foothold. This was just one of the problems Shackleton and his twenty-seven men faced when they attempted to land here after they escaped their sunken ship. Their initial attempt was to cross the Weddell Sea and reach the Antarctic continent but in an unusual cold season they became trapped by sea ice. The unyielding mass ensnared their ship the *Endurance*, and crushed their only home. The journey would now be one not of exploration but survival.

Abandoning the condemned ship they hauled three lifeboats over the decaying ice and sailed whenever they found open water. They tried to reach Paulet, a small volcanic island at the tip of the Antarctic peninsula, but again the ice prevented any progress. Their only route of escape now was to the east towards Elephant Island. After five months of uncertainty drifting on a disintegrating platform of ice and only living in tents, they covered about 3220 kilometres. Finally reaching open water they were now able to launch their tiny crafts. After days of difficult sailing they reached Elephant Island on a point called Cape Valentine. They scrambled like drunken men onto a very exposed pebbly beach but were enclosed, surrounded by inaccessible towering rock.

The thankful men were so overjoyed and relieved they played with the pebbles letting them run through their fingers and like children on a beach played and scrambled on the solid ground. They had not been on terra firma for nearly seventeen months. Shackleton however soon realized this was not a place to wait for rescue. The beach was beneath falling rock and the shore nearly disappeared with every rising tide. Storm swells would wash them away. The men would quickly substitute

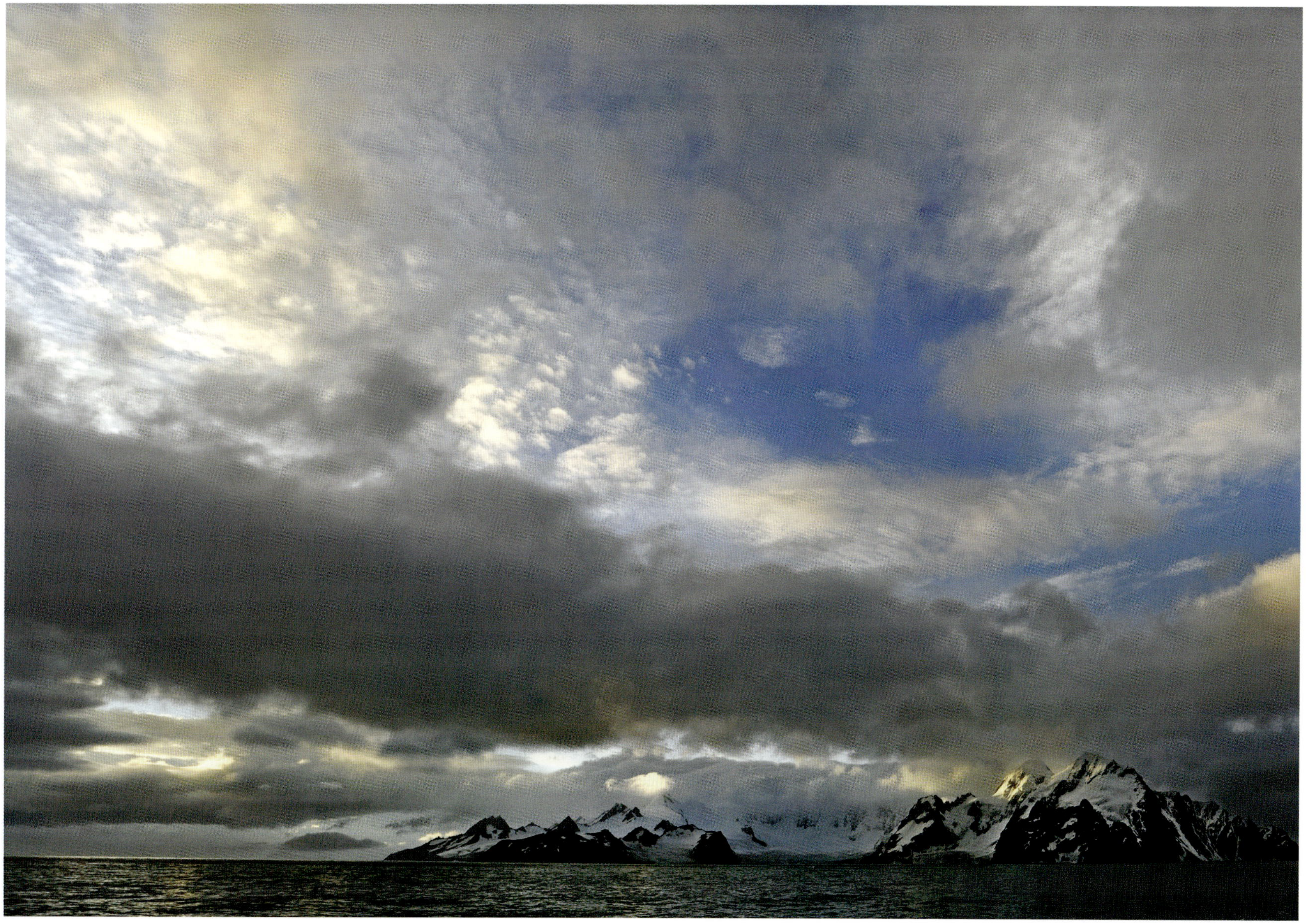

Coloured sky Elephant Island 2011

the name Elephant Island with a euphemistic "Hellofanisland"

They decided they had to move and after a couple of days left and sailed a dozen kilometres to the west finding a spit of rock above the tidal swell and the unpredictable waves. Still there was nowhere to go. They were again trapped. Surrounded by calving glaciers that dropped into the sea, Point Wild was for now their only hope of rescue and this is where they would have to wait.

Nobody knew however where they were and rescue was highly improbable so Shackleton with five of his crew left the twenty-two men behind and set sail in a twenty-two foot open boat. With only rudimentary equipment to guide them they aimed for the only place they could expect help. It was the island of South Georgia one thousand three hundred kilometres away and that was over some of the world's stormiest oceans.

After fifteen precarious days at sea they sighted the island but were on the wrong side. Two more days of nearly disastrous sailing and not able to get to the other side, Shackleton and two of the stronger men decided to land and negotiate the island's never before traversed mountains. Without a proper map and mostly travelling non-stop for two sleepless days they reached the whaling port of Stromness. Shackleton recognized the first man and greeted him but the man only responded "But who are you?"

The only thing now on Shackleton's mind was the rescue of his stranded men. Using different ships three attempts ended in failure but on his fourth on the Chilean tug the *Yelcho*, Shackleton was finally able to rescue his men. All had survived the one hundred and five day ordeal. Today, surrounded by penguins, and raised on a pedestal above the regular surf on a narrow spit of rock, the bust of Captain Luis Pardo, who negotiated the rescue, faces stoically out to sea. It was here on the same site where the twenty-two stranded men desperately holding on to hope waited out their uncertain fate.

In the annals of naval history this is recognized as one of the most extraordinary ocean voyages ever made.

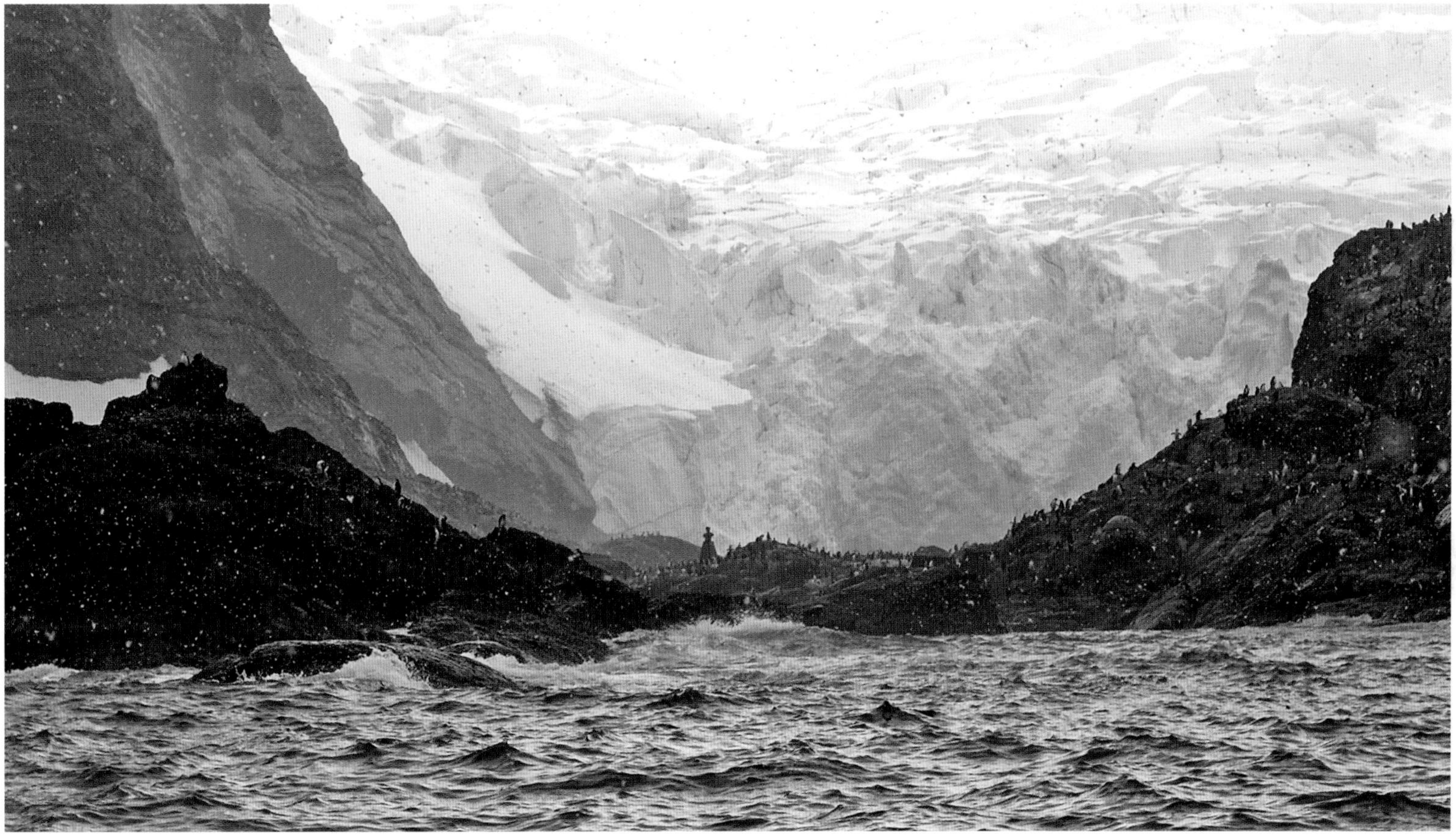

Bust of Captain Luis Pardo

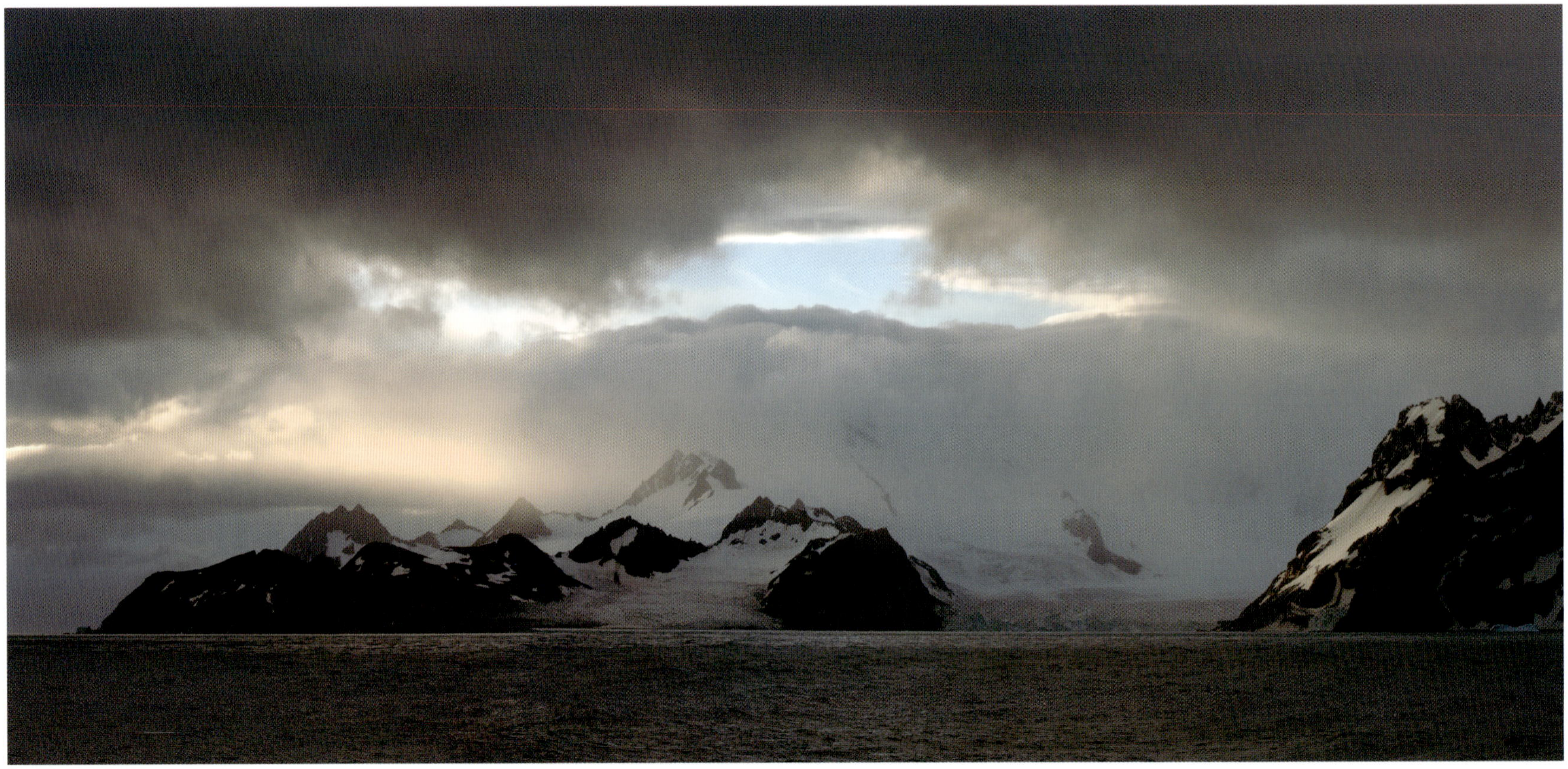

Opening sky Elephant Island 2012

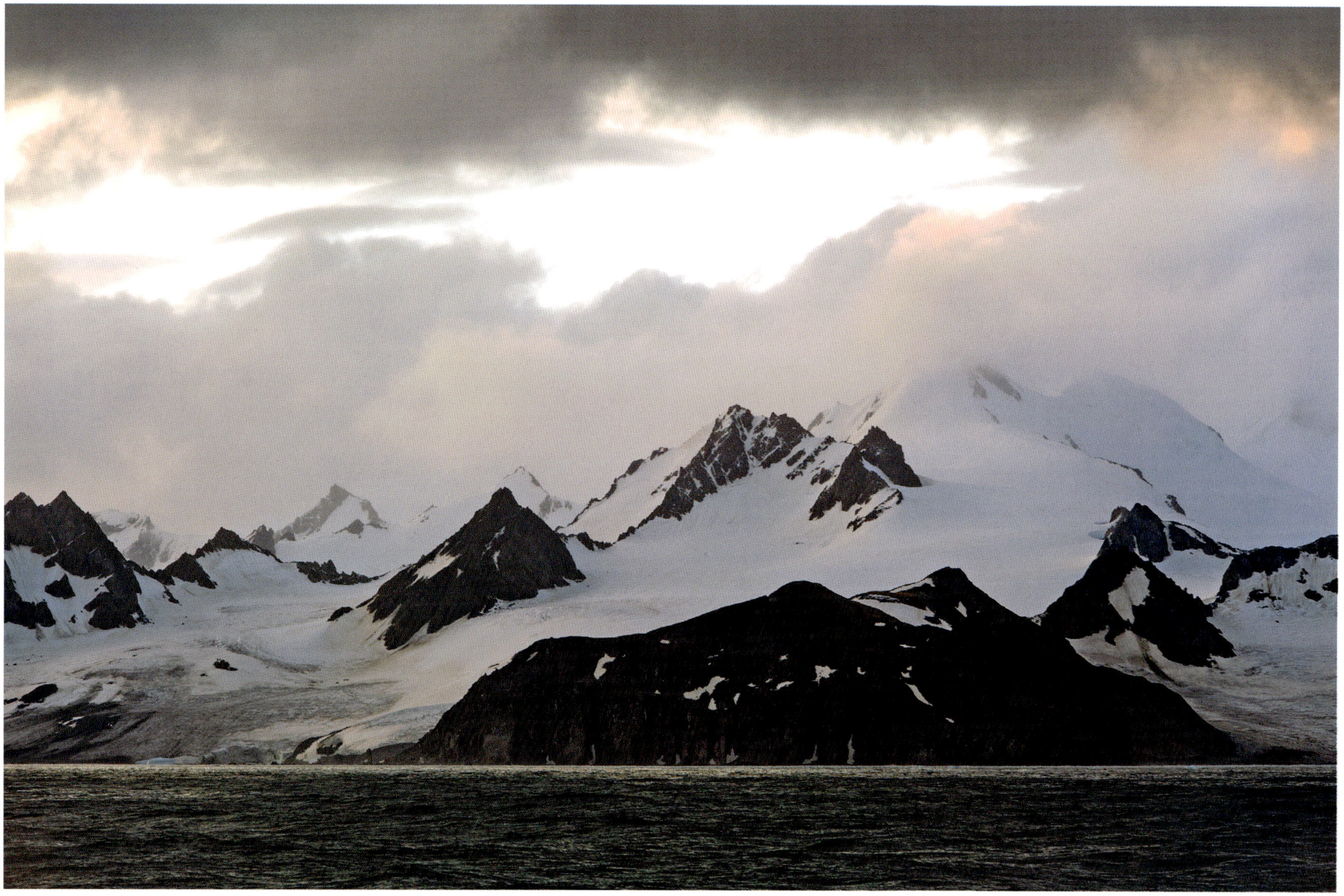

Soft light over a tinted sea. Elephant Island.

WELL-TRODDEN AND MARKED with russet these deep paths weave their way from the ocean into a single path to their nests high on top of this snow covered hill.

I saw these three gentoo penguins descending from the steeper part of the trail and decided to set up my tripod and wait to see what would happen. They continued their way down towards me seemingly unaware of my presence. I expected them to veer and change paths but strongly committed to the sea they proceeded. Only the last one gave out a croak as he stopped to nip my pants. Younger penguins are more curious and investigate anything new. Often when I was absorbed under my dark cloth examining my viewfinder I would feel a pecking on my pants or the tripod. Older ones would move on but the juveniles would linger and gather sometimes jockeying for position to test the new product. As much as you know that they are just birds they are wholly compelling and one cannot help but be captivated by their appearance and demeanour. Before I left for this trip I hardened and half-heartedly pledged not to take more penguin pictures and concentrate more on other Antarctic trademarks. However it took only a few appealing encounters like this for my resolve to weaken. Once again like everyone else I remain enamoured.

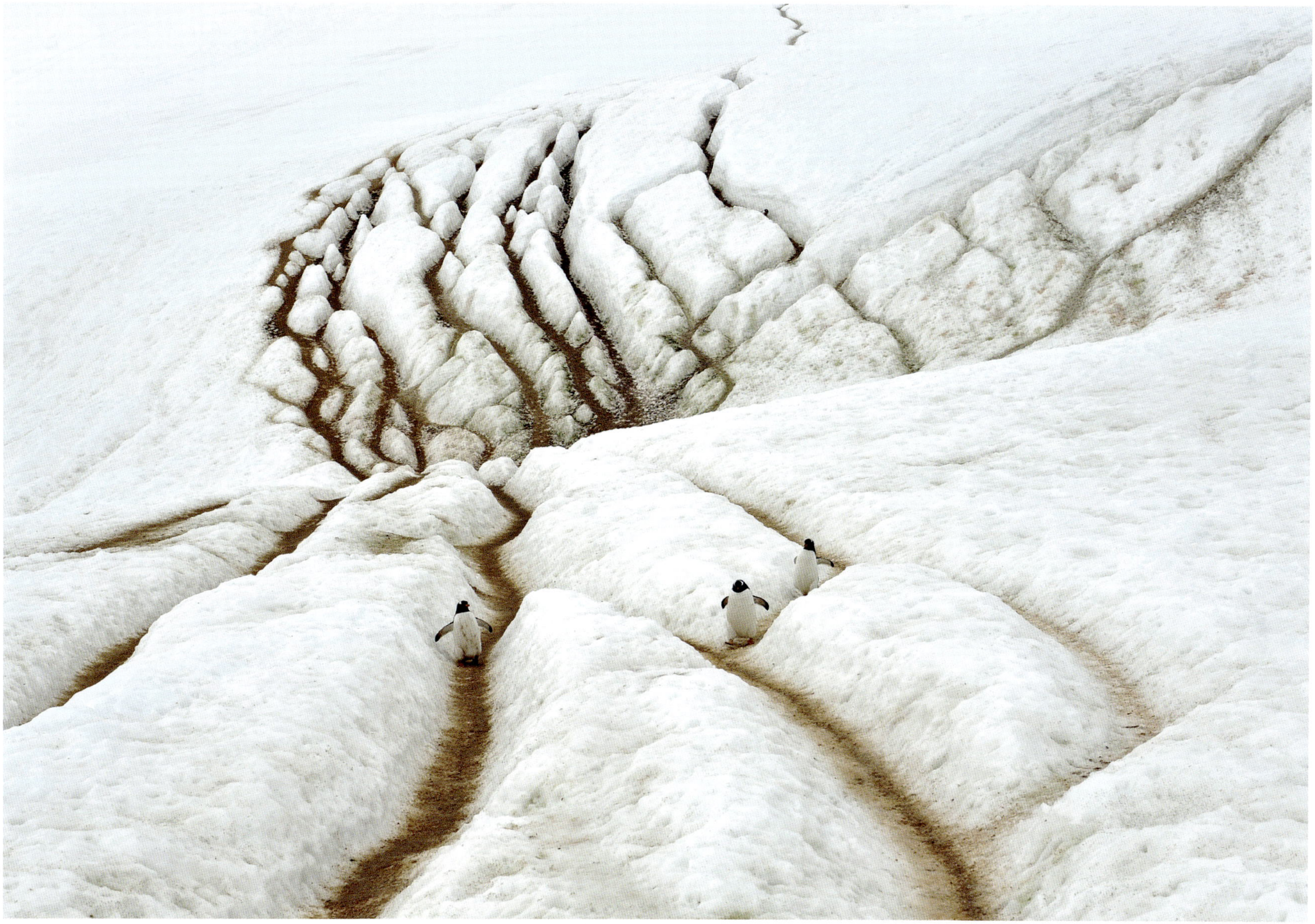

Penguin paths leading home

WITH OVER NINETY-SIX percent of the continent covered in snow, bare rock is a precious commodity. Here on Danco Island these tireless climbers rest between steep ascents to their nests. Like footballs with a beak they can spend most of the day recharging after a long forage at sea feeding on fish or shrimp. Some of the pathways over the outcrops have been used for centuries to the point where their claw marks are visibly etched into the rock.

There was a stillness to this place that I found nowhere else. Even the colony seemed subdued and their usual braying sound was intermittent. The morning sun pierced the clouds revealing a flat and steely sea surrounded by the purest white. In this quiet moment encircled by a dreamlike scene, even among the guano and decay, I could think of no other place I would rather be.

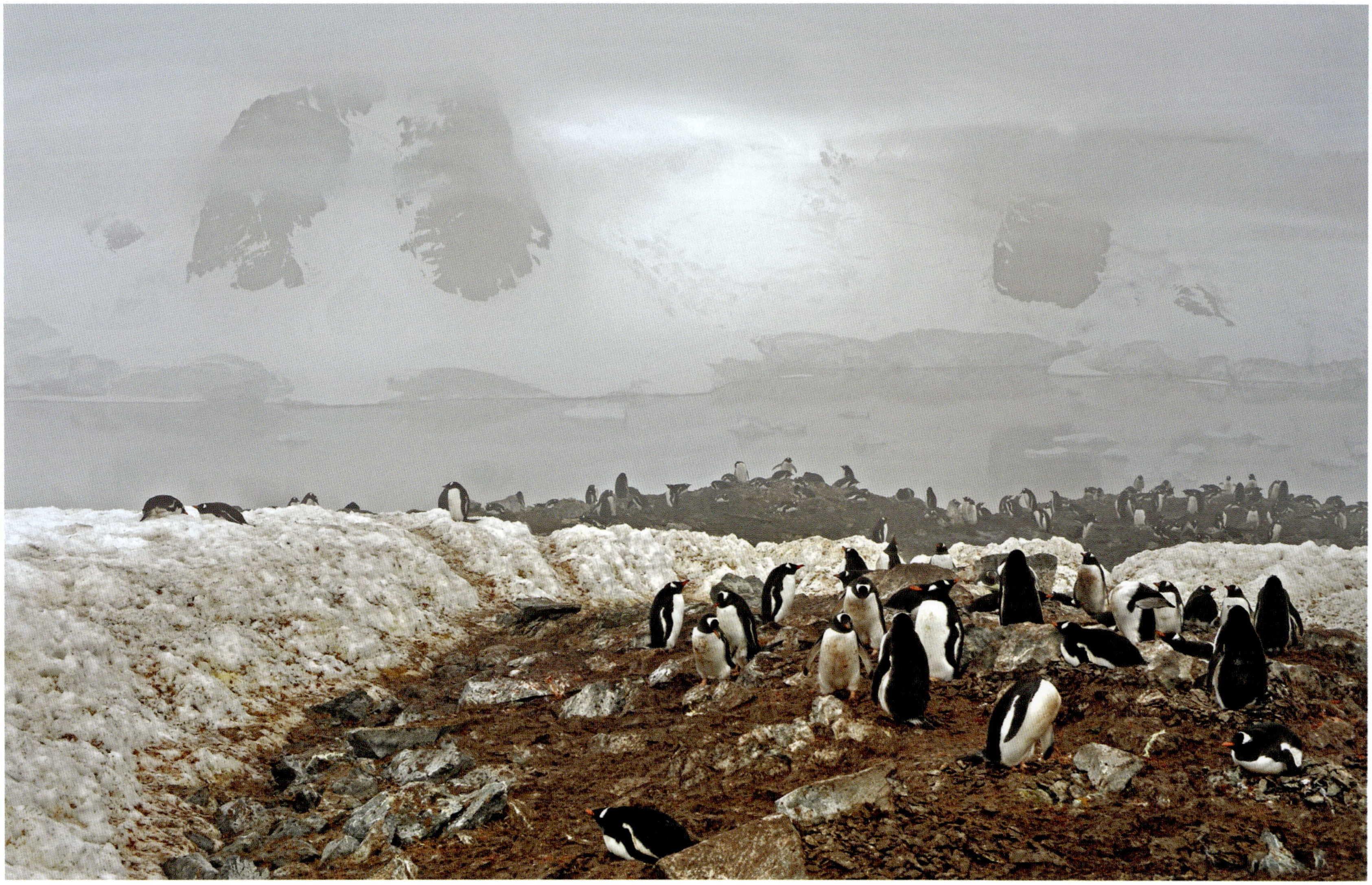

Gentoo penguins Danco Island

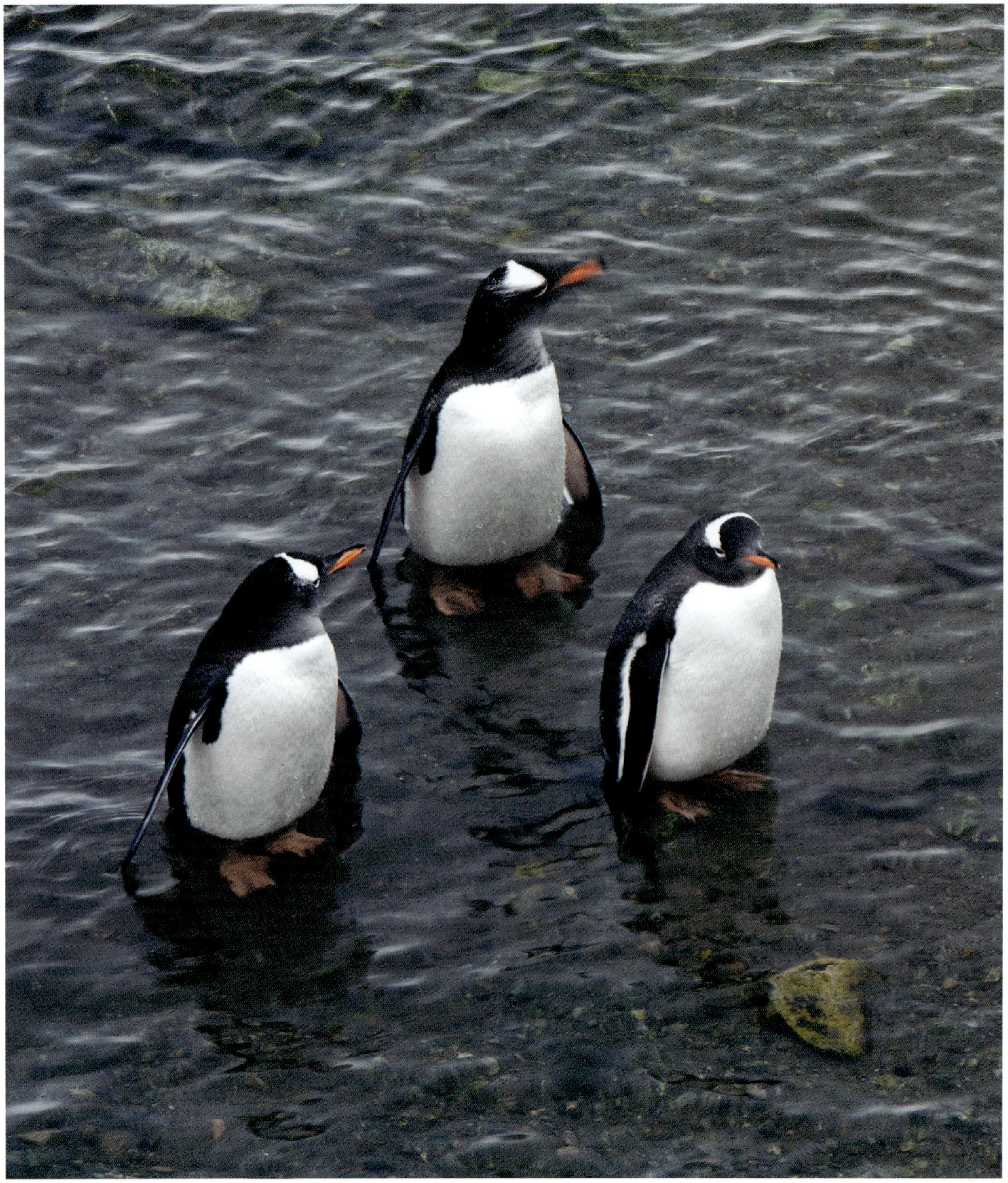

Hesitant gentoos

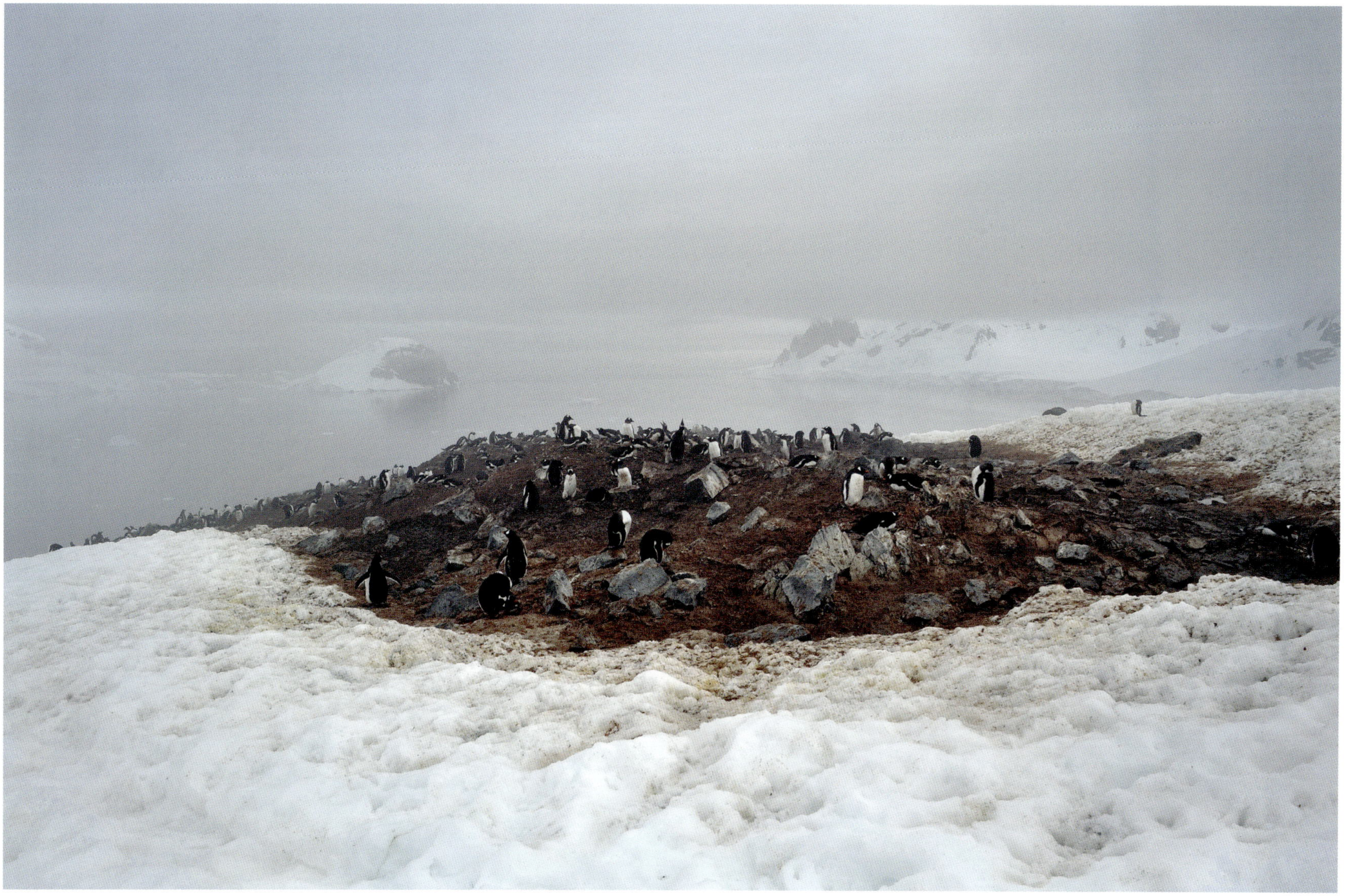

Hilltop nest gentoo penguins Danco Island 2010

PENDULUM COVE HAS a two hundred year history of whaling, science and volcanic eruptions. One of the more notable visits was in 1829 by Henry Foster. He was the captain of the *Chanticleer* and used this base to measure the earth's gravitation. A pendulum was the crucial instrument to accomplish the task. Other measurements were also made in other parts of the world and by comparing results map makers could figure out to some degree the shape of the earth. It would improve navigation.

As we walked the bending bay I was told by a geologist that Foster's 2000 hours of work was useless because the iron content in the soil affected the pendulum's swing. The unfortunate captain later lost his life somewhere in Panama drowning in a river.

The water filled caldera is named in his honour.

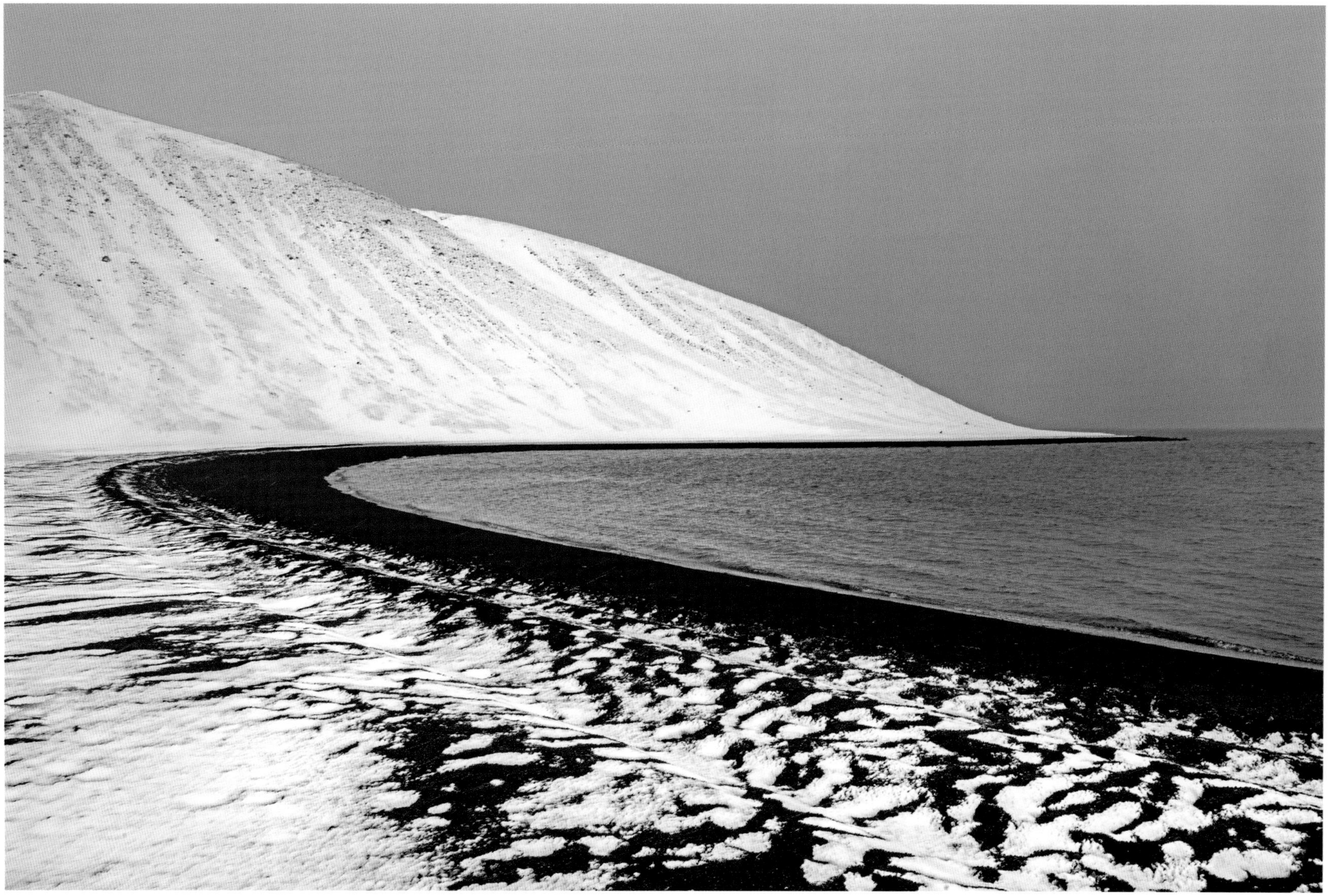

Pendulum Cove Deception Island April 2014

I WAS NEVER more surprised when I was hit on the shoulder by what felt like a ten pound bag of wet potatoes.

We were slowly moving along this frantic shoreline totally absorbed by the loud racket and boisterous activity. I was sitting on the rubber edge of the Zodiac with my camera focused on the shore when a heavy object hit me from behind. It pushed me forward and amongst the screams of the three others on board a most bewildered penguin was trying to regain its footing. Suddenly there at my feet was this confused bird flailing its dripping wings trying to get back out. Its first attempt to get a grip on the smooth rubber failed but with a small spring it hopped onto my boot and vanished over the side. It disappeared as quickly as it came.

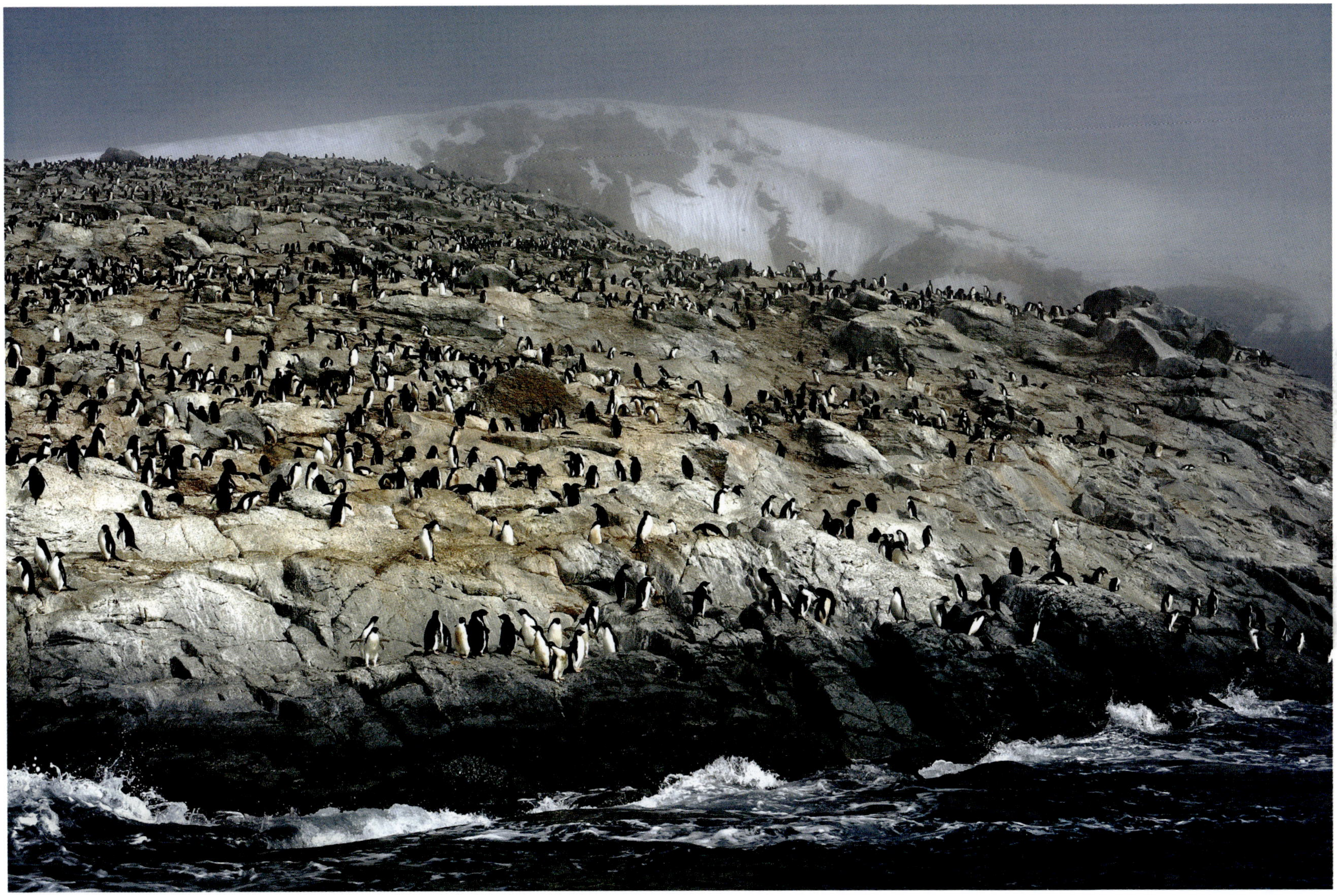

Adelies on Heroina Island

SEEMINGLY SAFE THESE few penguins on the ice know better. Besides the occasional orca, leopard seals are their biggest threat. They can easily lift themselves six feet out of the water to snatch any stragglers lingering at the edge. Only the leopard seal's more restricted movement on the ice gives the penguins a chance of escape.

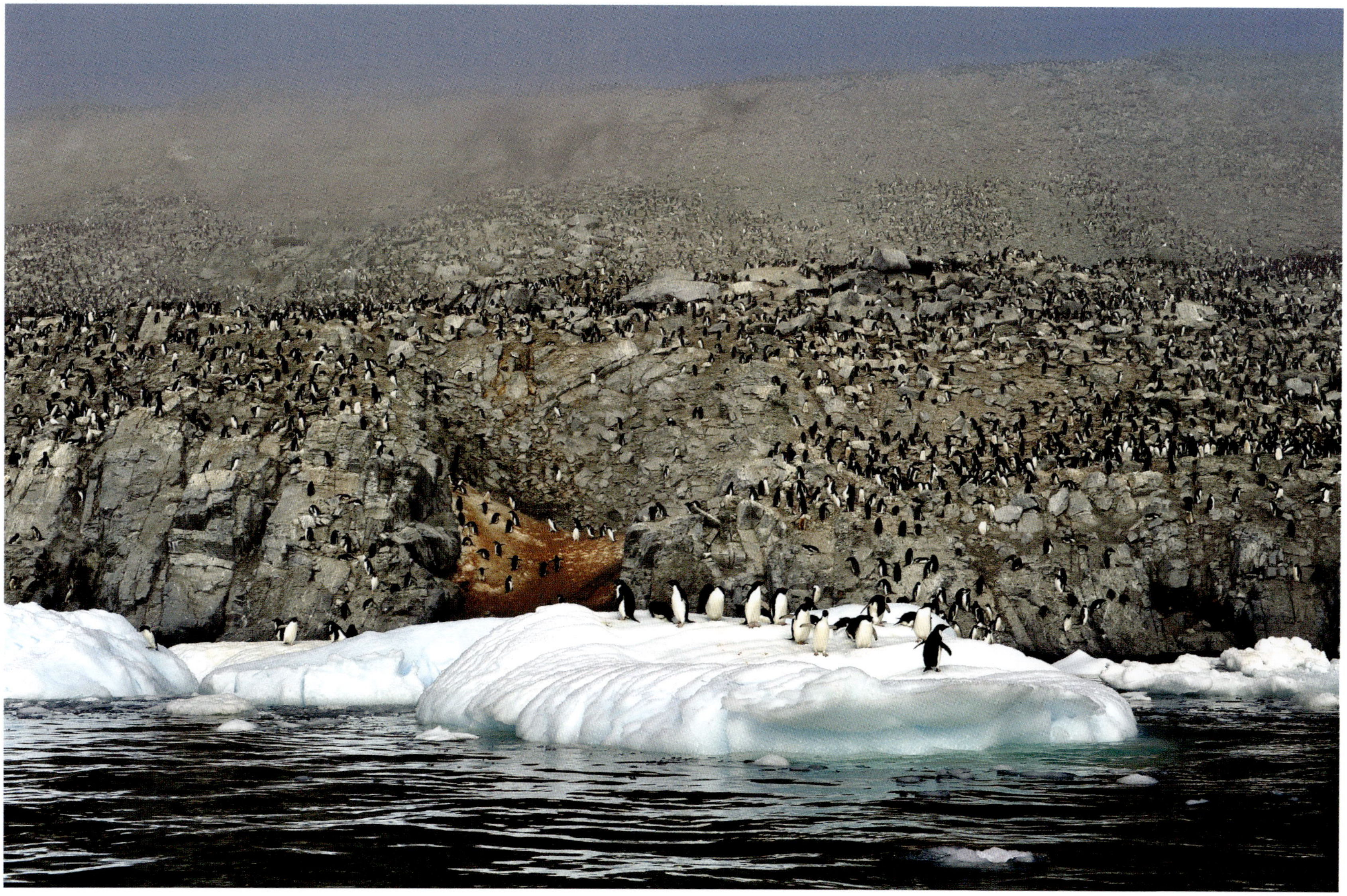

Adelie cliff in fog

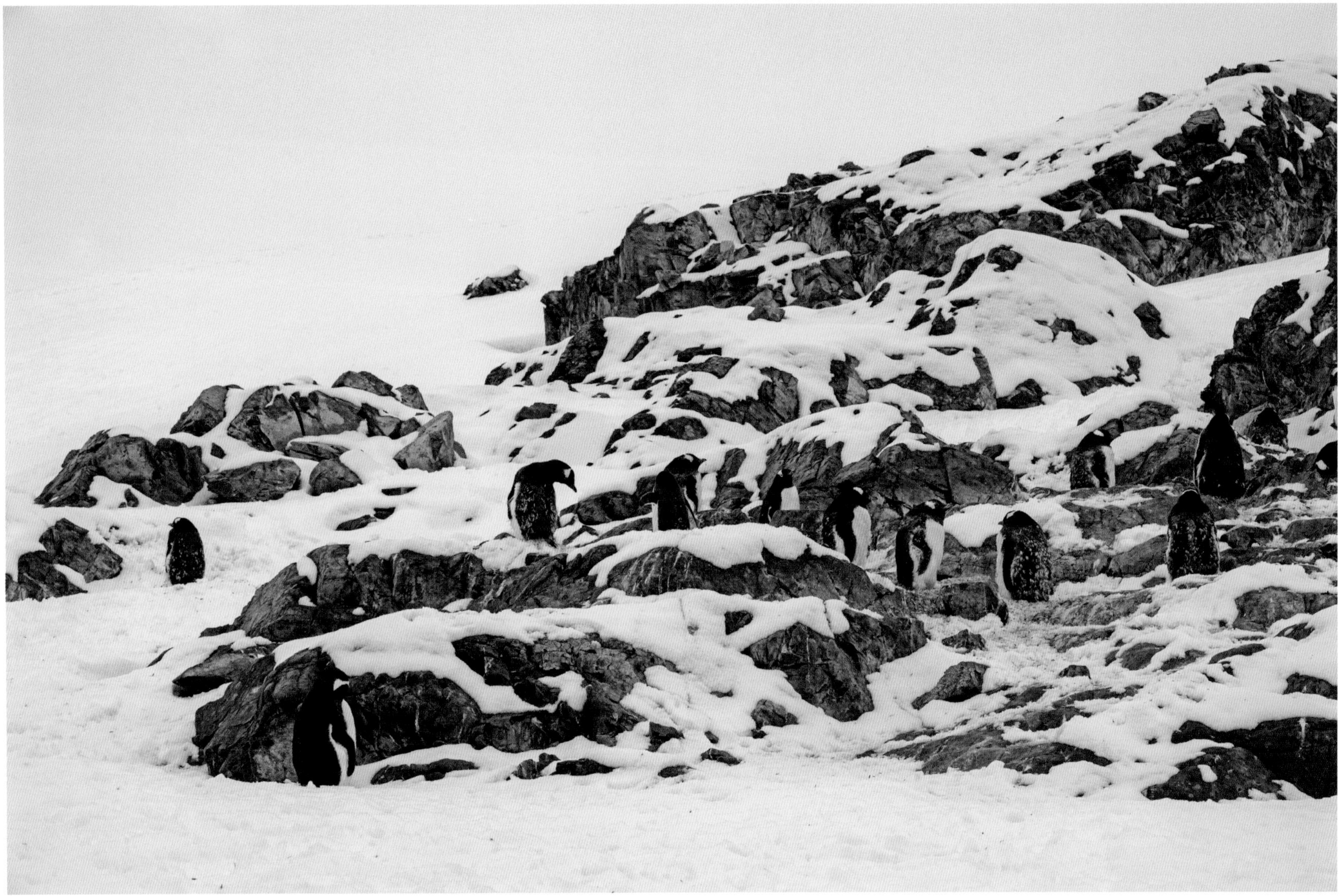

Molting penguins

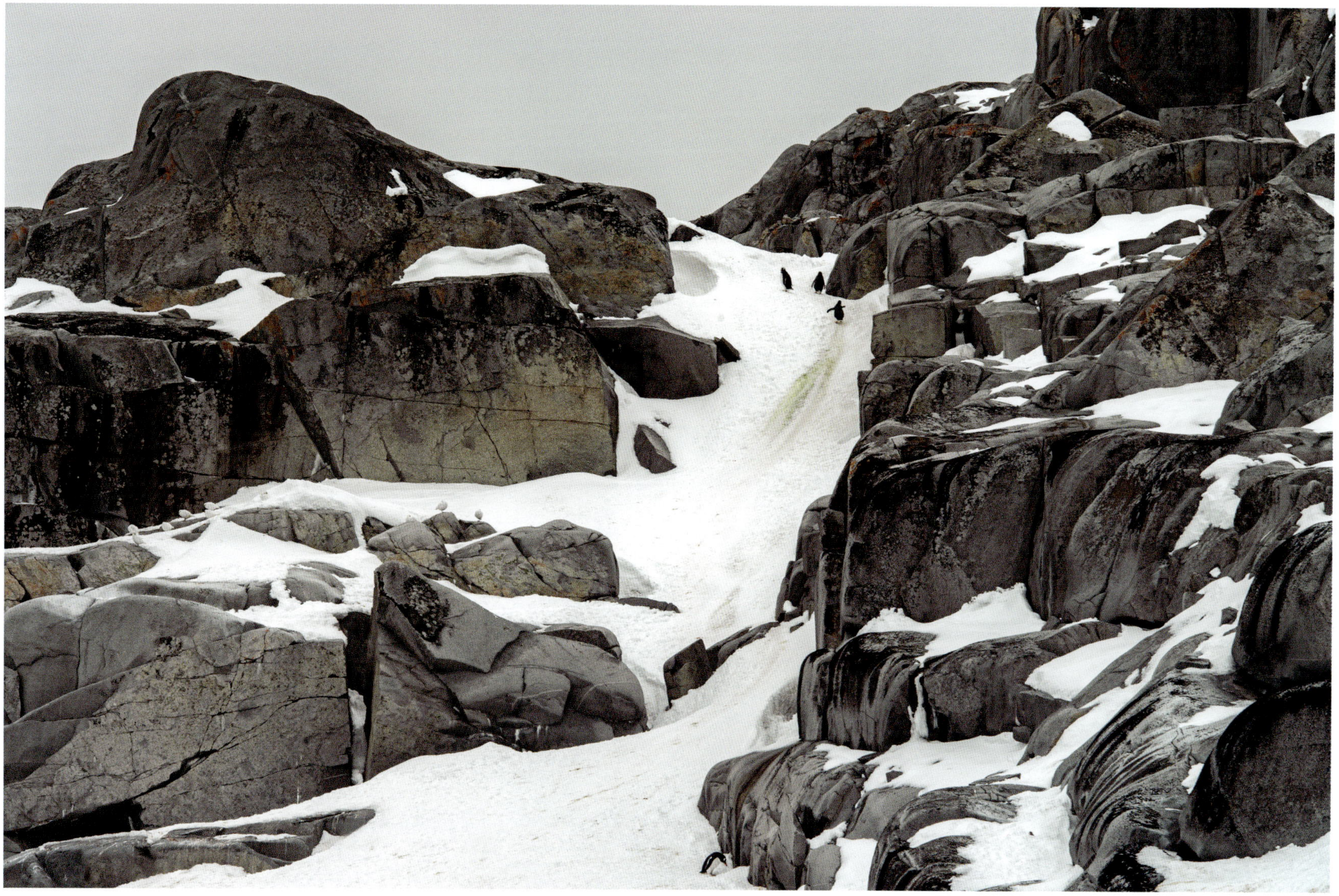

Three penguins nine sheathbills. Useful Islands.

THESE ISLANDS WERE discovered by Adrien de Gerlache on his 1897 expedition. They first appeared as the Useful Islands on a 1927 chart.

I remember landing on this cluster of rocks speculating about the weather. The ride from ship to shore was already choppy and my glasses were covered with moisture. I didn't get to wander far before we got word from the ship to hustle back. The return trip was as expected: wet and rough. In conditions like these the captain always turned the ship so the gangplank was on the leeward side protecting us from the wind and making the first step out of the Zodiac less risky.

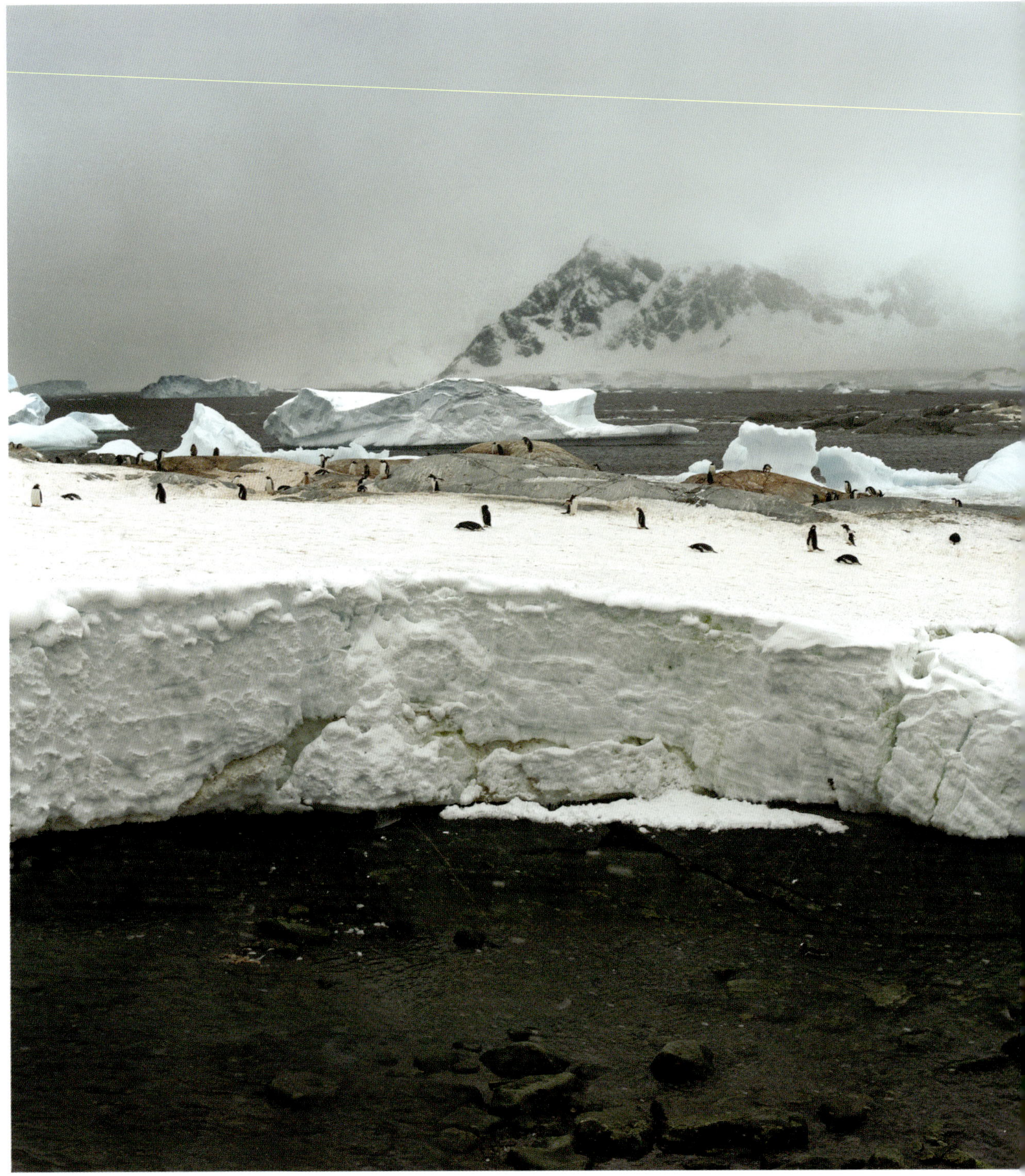

Storm over the Useful Islands

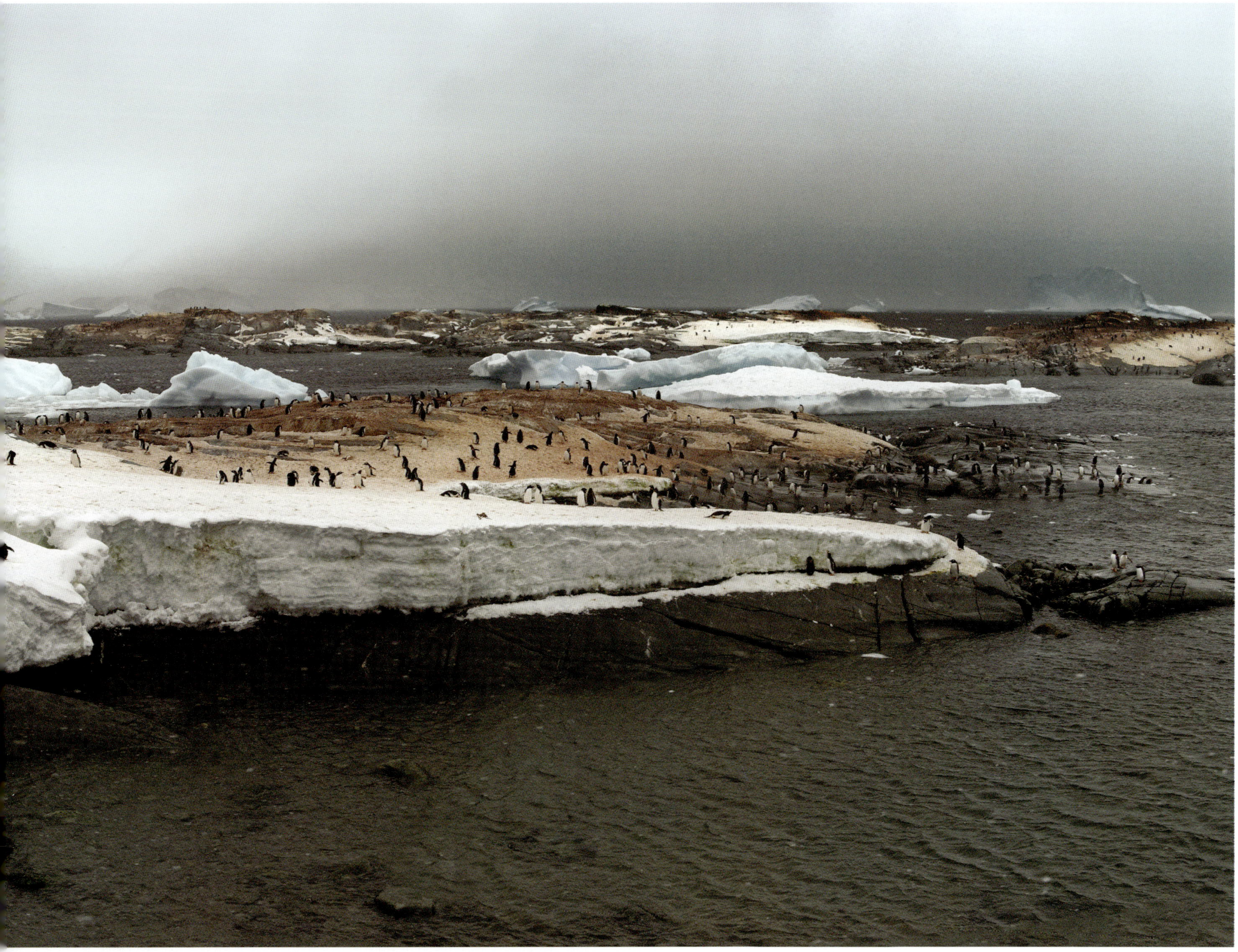

THE FOG WAS much heavier the morning we approached Heroina Island. There were only a few bergy bits floating by and the land we hoped to see was not yet visible. A familiar stench however penetrated the air and the sound of the estimated million residents was muffled by the heavy air. Every few moments I would glance in the direction of the increasing light hoping to spot some of the scenery as the ship moved cautiously through the thinning haze. Finally the ship dropped anchor and in the mist we began to explore by Zodiac. Weaving amongst the rafts of ice heading in the direction of the ceaseless commotion penguins began to porpoise along side shooting out of the water onto any sizeable ice pan. Those that failed their first attempt nervously doubled their efforts and after an uneasy floppy landing began an agitated shuffle away from the edge. Nothing about the situation was relaxed. Every penguin looked concerned and swayed back and forth with heads bobbing in every direction. Something dark was in the water. A leopard seal.

I first saw it coming out from under the ice moving towards us and then disappear beneath the Zodiac. It turned abruptly underwater and emerged beside another ice pan. Its horse sized head with studying eyes peered in

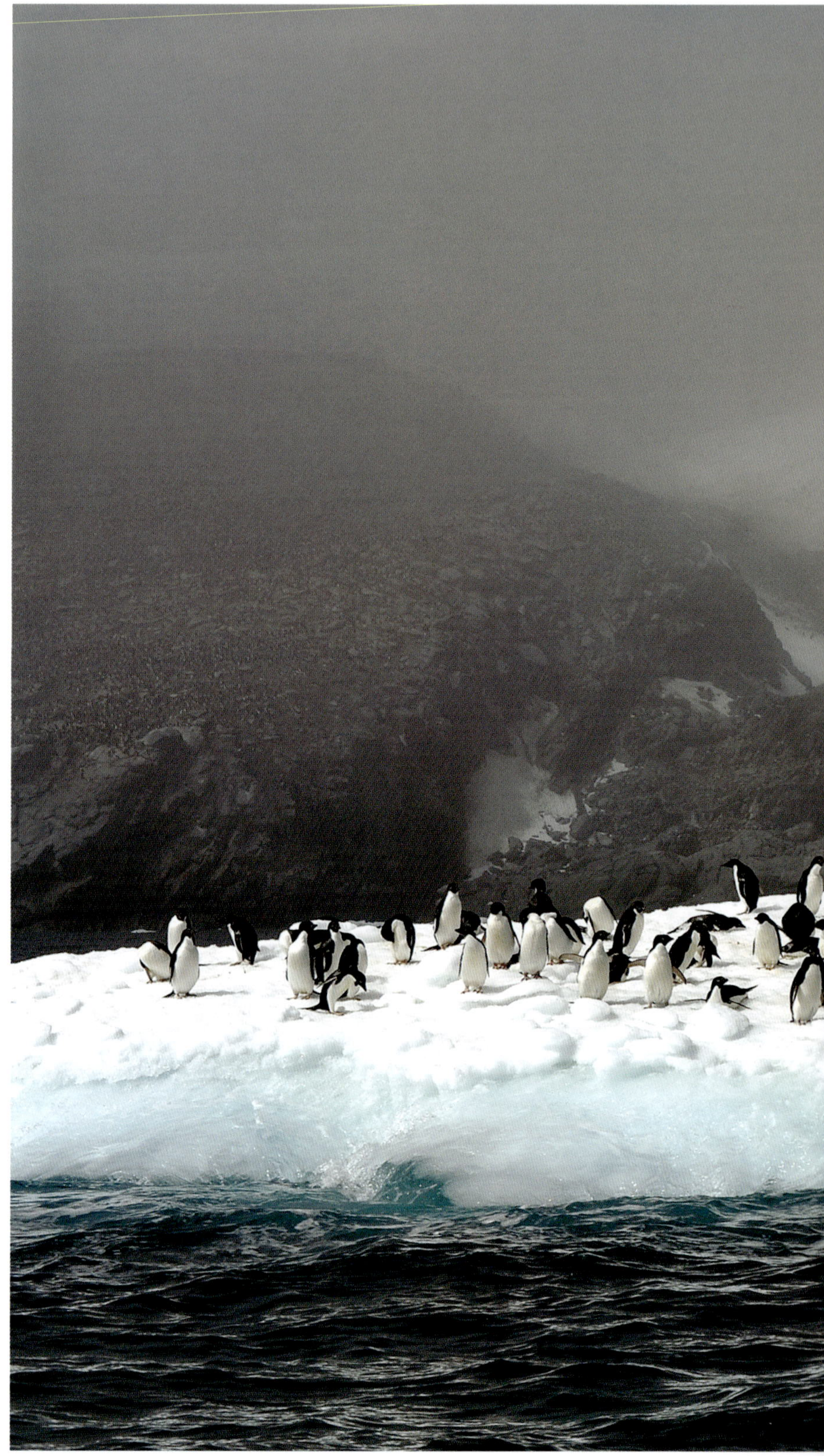

Dancing Adelies

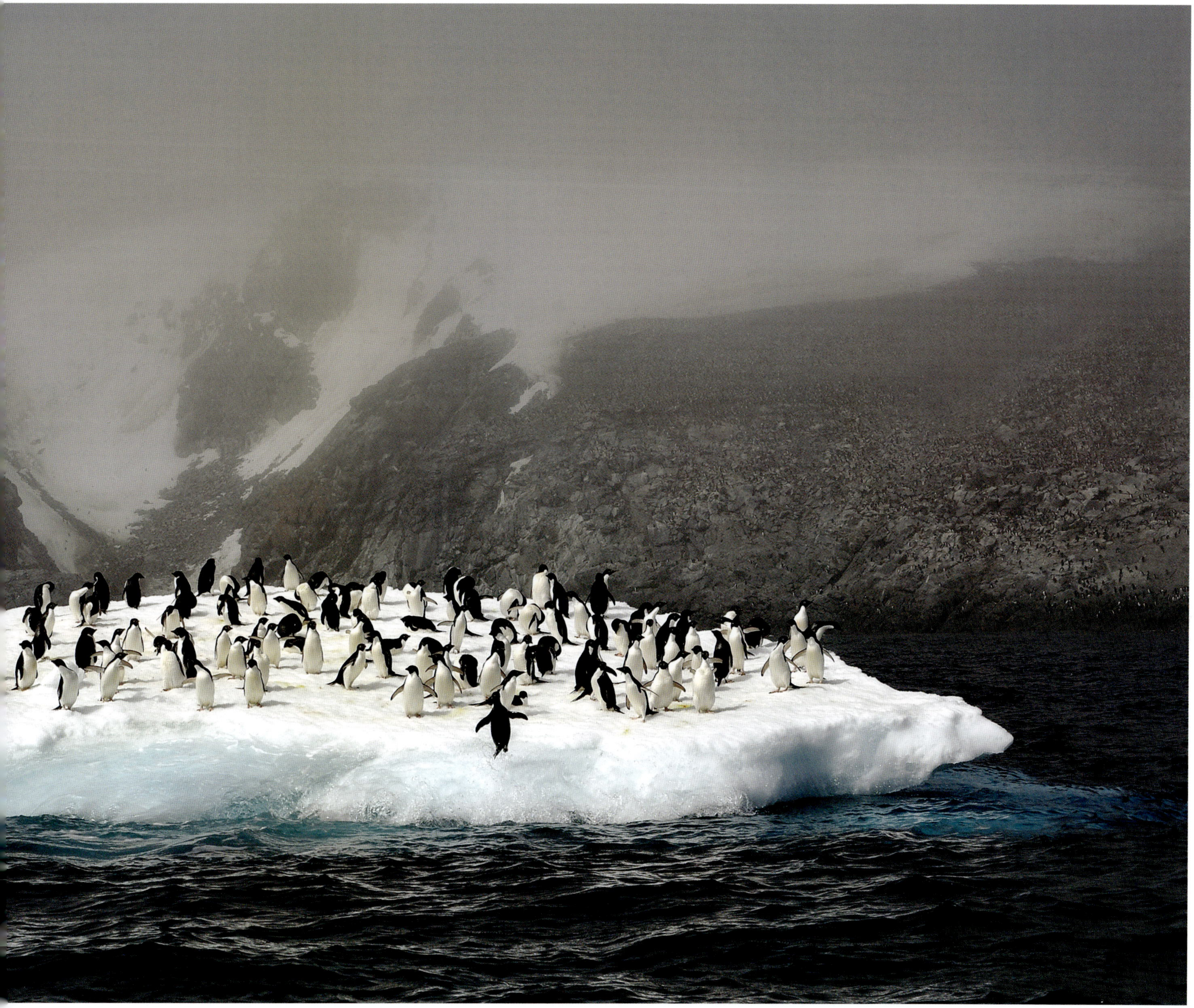

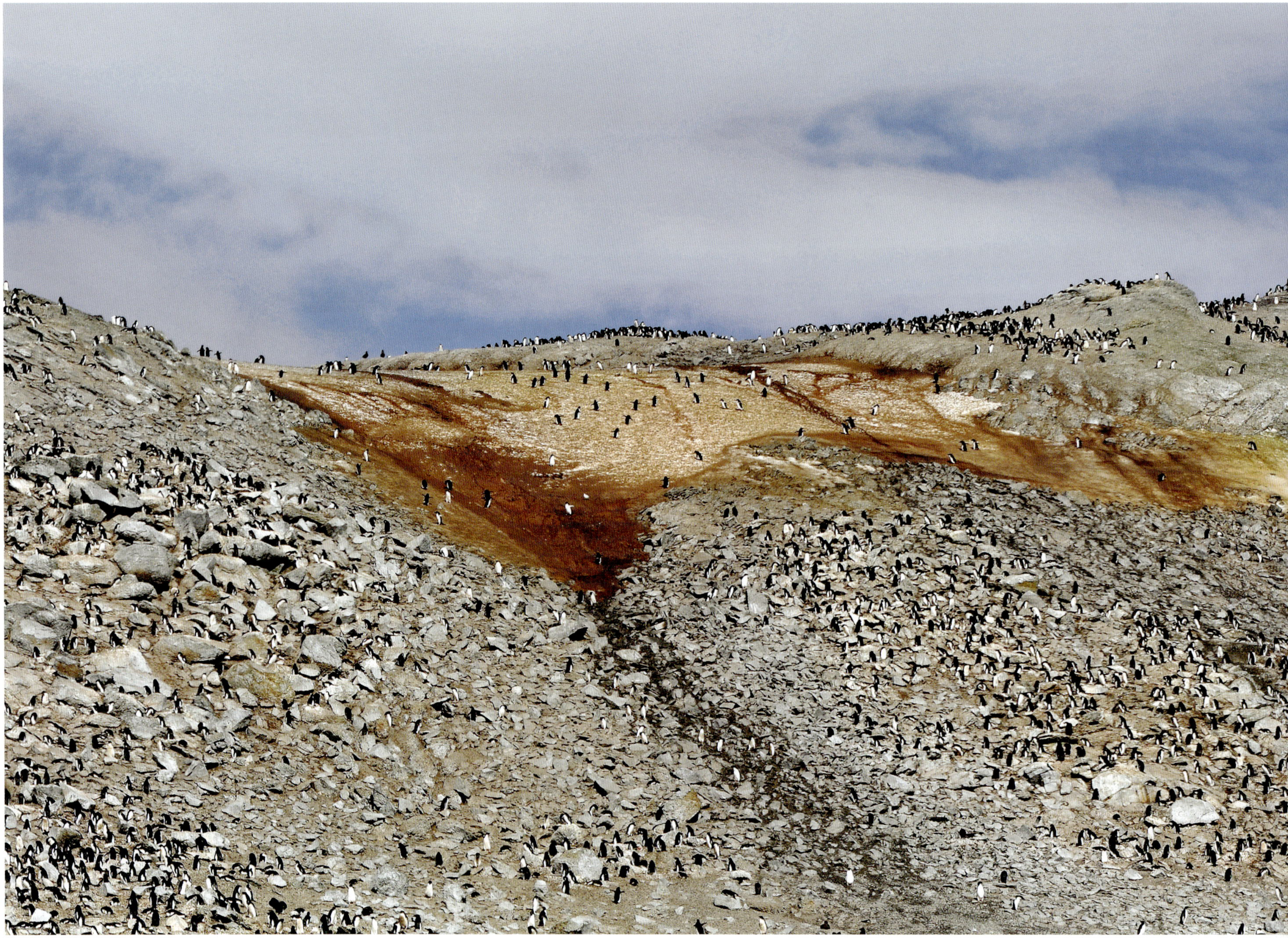

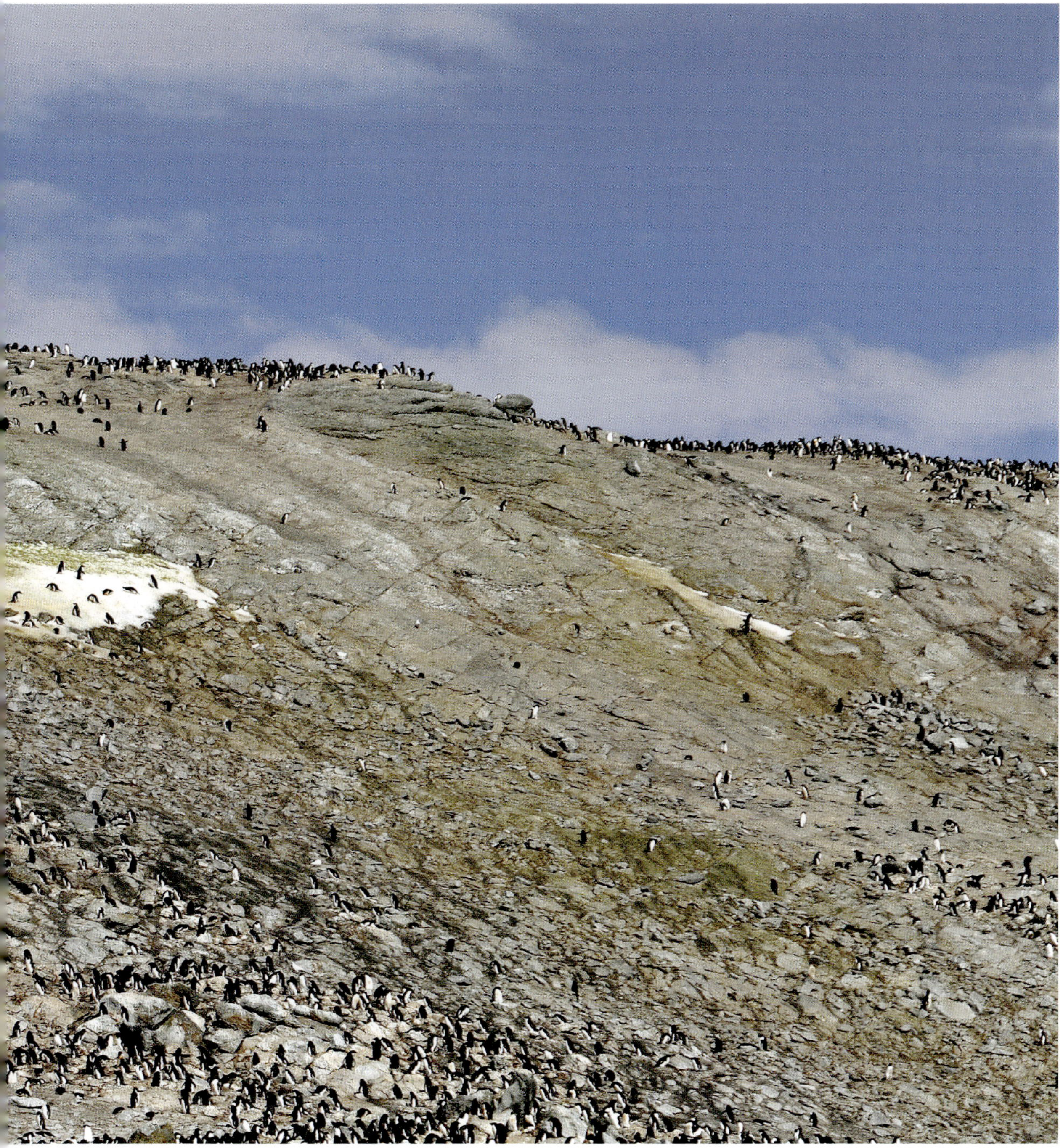

our direction and disappeared. I was now standing in the Zodiac nervously searching in every direction watching and waiting. Aware of something I could not see, the loud and excited braying of the fidgeting Adelies mounted. Moments later repeated splashes of flying water could be seen spraying up from behind the ice. And another penguin had fallen victim to the seal's hunger. It was an event that I would witness many times over the course of my trips. The first time it had a powerful effect on me and quickly brought a sense of reality to a seemingly idyllic scene. Meanwhile the fog became transparent and in the remaining mist there unveiled a steep and congested hillside. It was like looking at a huge factory floor where each penguin through their own eyes performed an assigned and meaningful task. To me however the noise and activity was mostly unintelligible, random and unfamiliar.

Antarctic summer

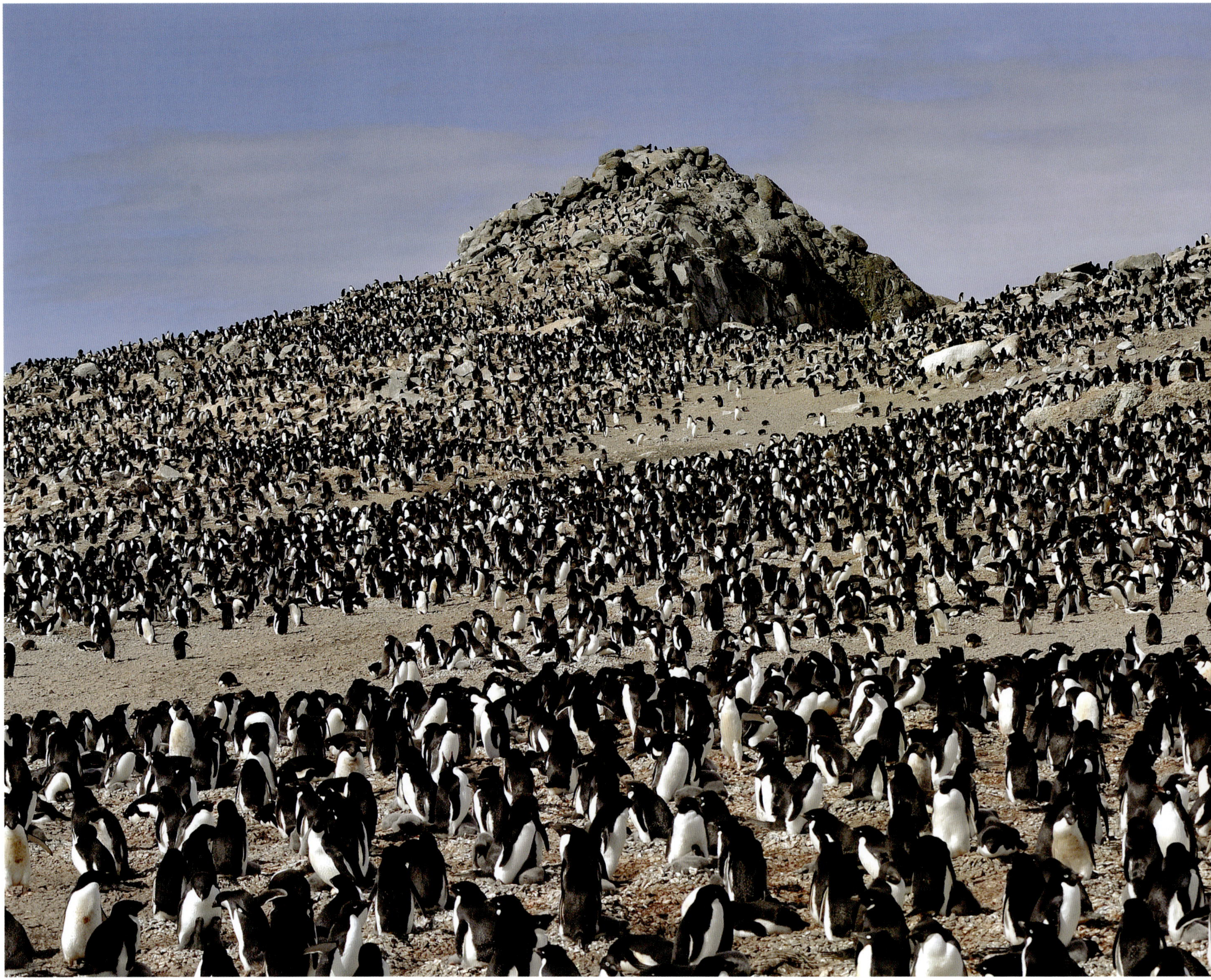

Adelie penguins. It was all acoustic.

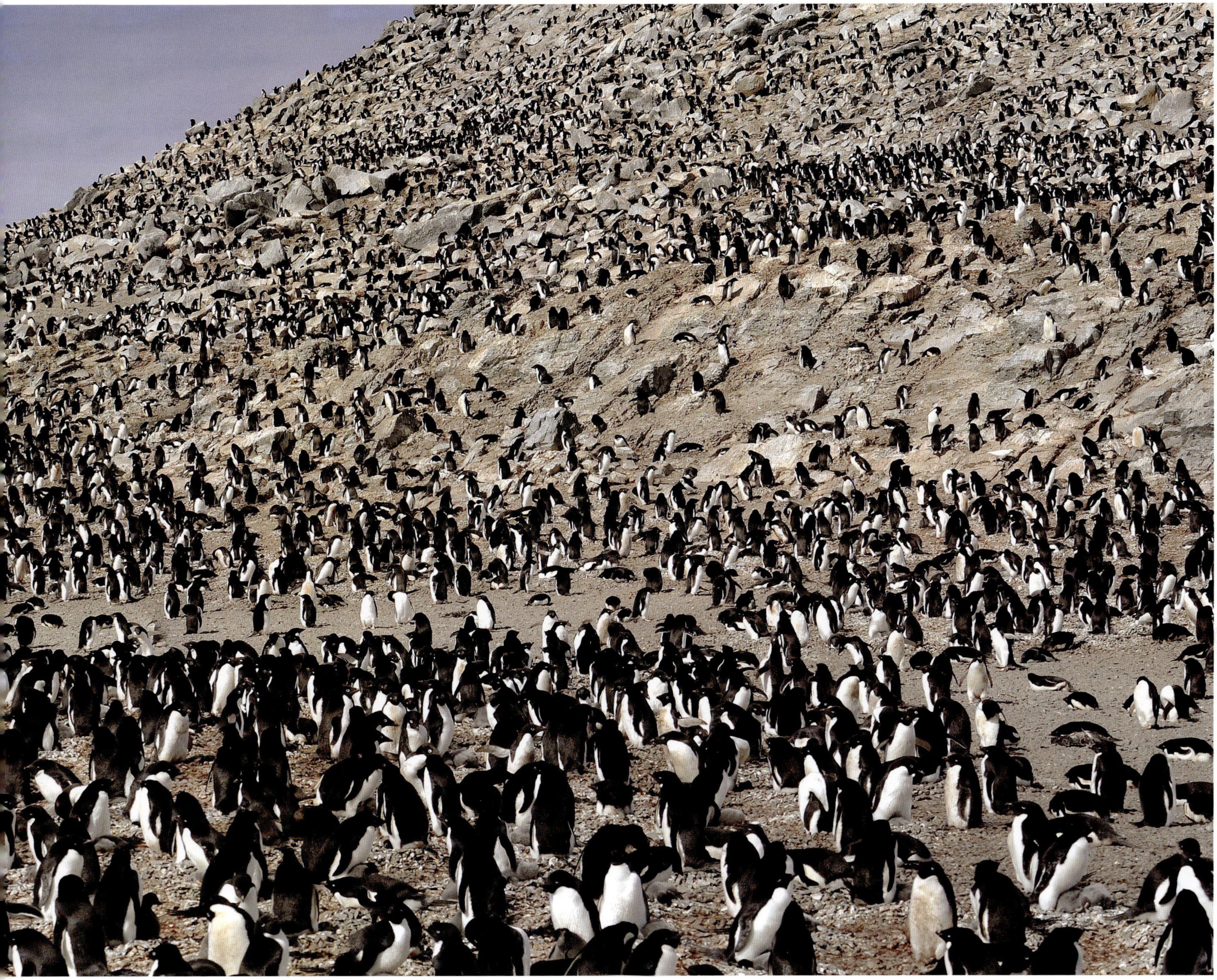

SITUATED AT APPROXIMATELY 63.6 degrees south latitude and 55.8 degrees west longitude this very small volcanic island off the tip of the Antarctic peninsula was home to twenty men. In the winter of 1903, trapped and surrounded by the crushing ice, their badly damaged and leaking ship the *Antarctic* sank leaving the men stranded. Some forty kilometres away was Paulet Island. It was their only hope. Now began their horrible fourteen day ordeal pushing their lifeboat over the shifting and unstable ice moving ever so slowly towards their goal. Winter was upon them and time was now a major factor in their survival. Hacking with a few axes they worked and shoved their way towards opening leads ferrying their goods again and again until they reached open water and finally after a six-hour row landed on the penguin filled shore of Paulet Island.

In his diary, Carl Skottsberg, a member of the shipwrecked crew, mentioned how distasteful and revolting it was to drink from their only source of water, a small green-coloured pond in the middle of the island. The waste from the thousands of penguins that lined the surrounding steep hills drained into it and threatened their health and just magnified their misery.

The following summer their eight month ordeal ended with rescue but not without heartbreak. In May, after three months of living in a cold stone hut whose walls were stuffed with rotting penguin skins, Wennersgaard, a Norwegian member of the expedition, passed away. Having suffered for some length his last days were spent by speaking only of his home and writing notes to his loved ones. He died sitting upright in his sleeping bag murmuring softly.

After wrapping him in his sleeping bag all the

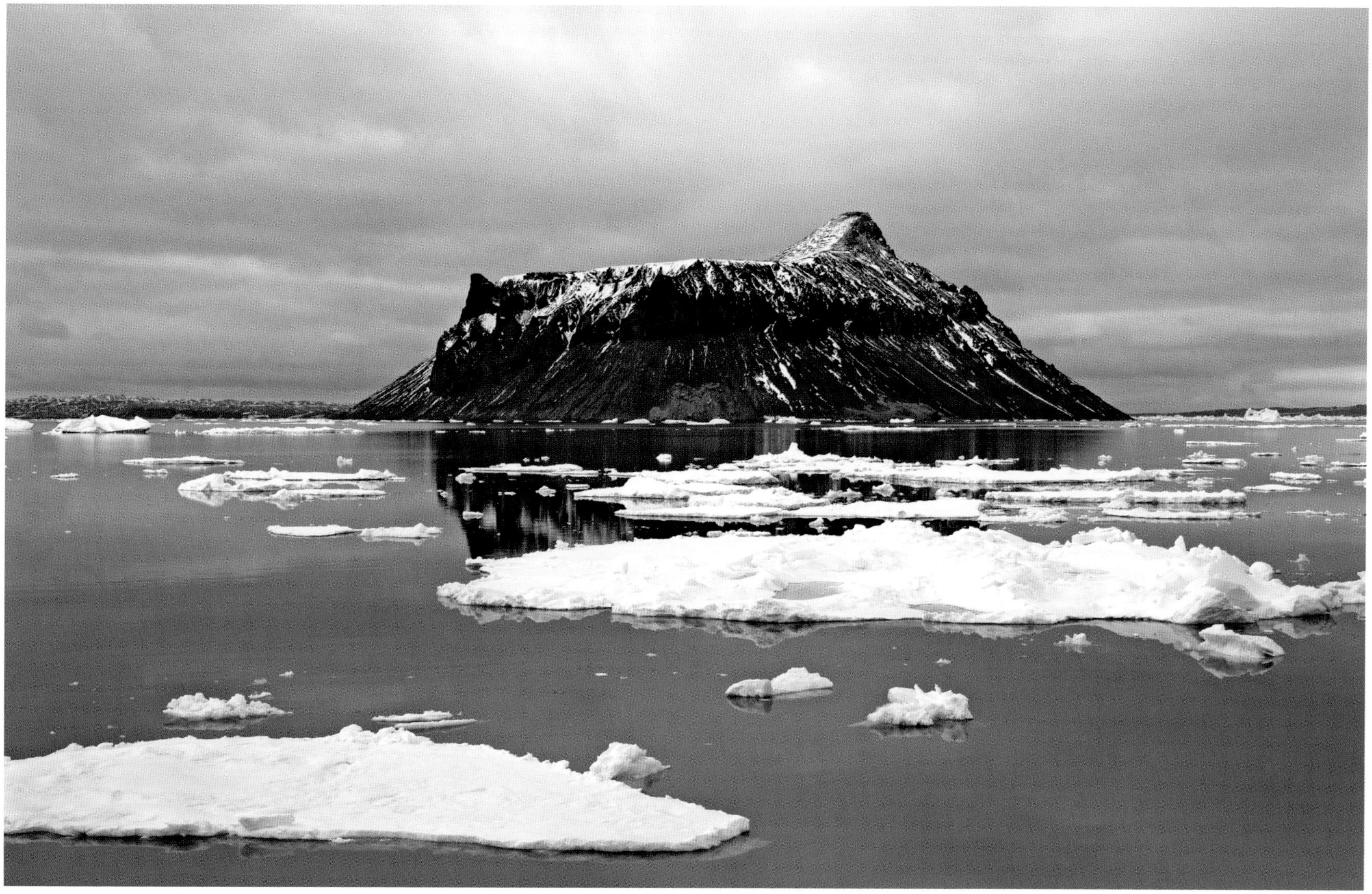

Paulet Island

men in a solemn procession carried the body to a distant snowbank and temporarily buried it until the summer when they could give him a proper burial among the rocks further along the shore.

The men at times thought he was suffering from consumption or pulmonary tuberculosis but today's diagnosis seems to conclude it was heart disease.

At last on the 10th of November 1904, while deep asleep, exhausted by the previous day's egg hunt, the men woke to an unusual sound. At first in disbelief no one reacted at what they might have heard. But with a second sharp blast from the ship their dreams of rescue turned into reality. Now a disorderly and disheveled group of nineteen men, half dressed and unkempt stared unrestrained at the ship. They could believe it. They would be home for Christmas.

While slowly pulling away from shore, Skottsberg in his last entry on board his rescue ship, stared back at the small stone shelter that was their home for eight months. In his diary he wrote with an unexpected sense of sadness and affection. One becomes mindful of his appreciation for life, no matter how unfavourable or adverse.

"The door is barred; I have passed over its threshold for the last time.

I am in a ship that moves. I can scarcely comprehend it. Slowly we glide away, away from the island, whose dark peak looms above us threateningly as before. But I cannot take my eyes away from it. I can hardly grieve that our prison doors have been opened, and yet it is with a sense of sadness and of regret that I see Paulet Island disappear behind the dazzling inland ice—disappear maybe forever.

Has it not been my home? "

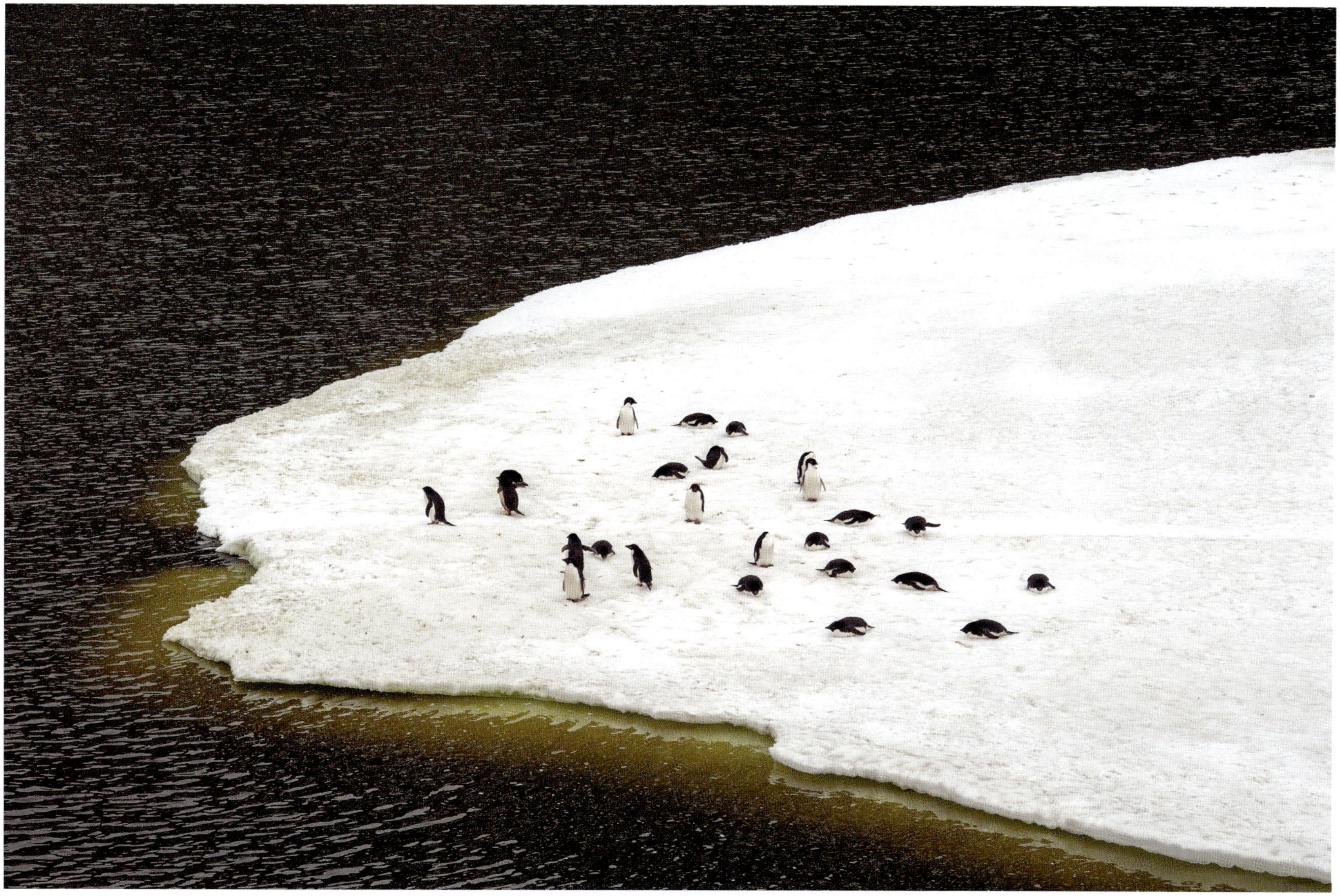

Pond on Paulet

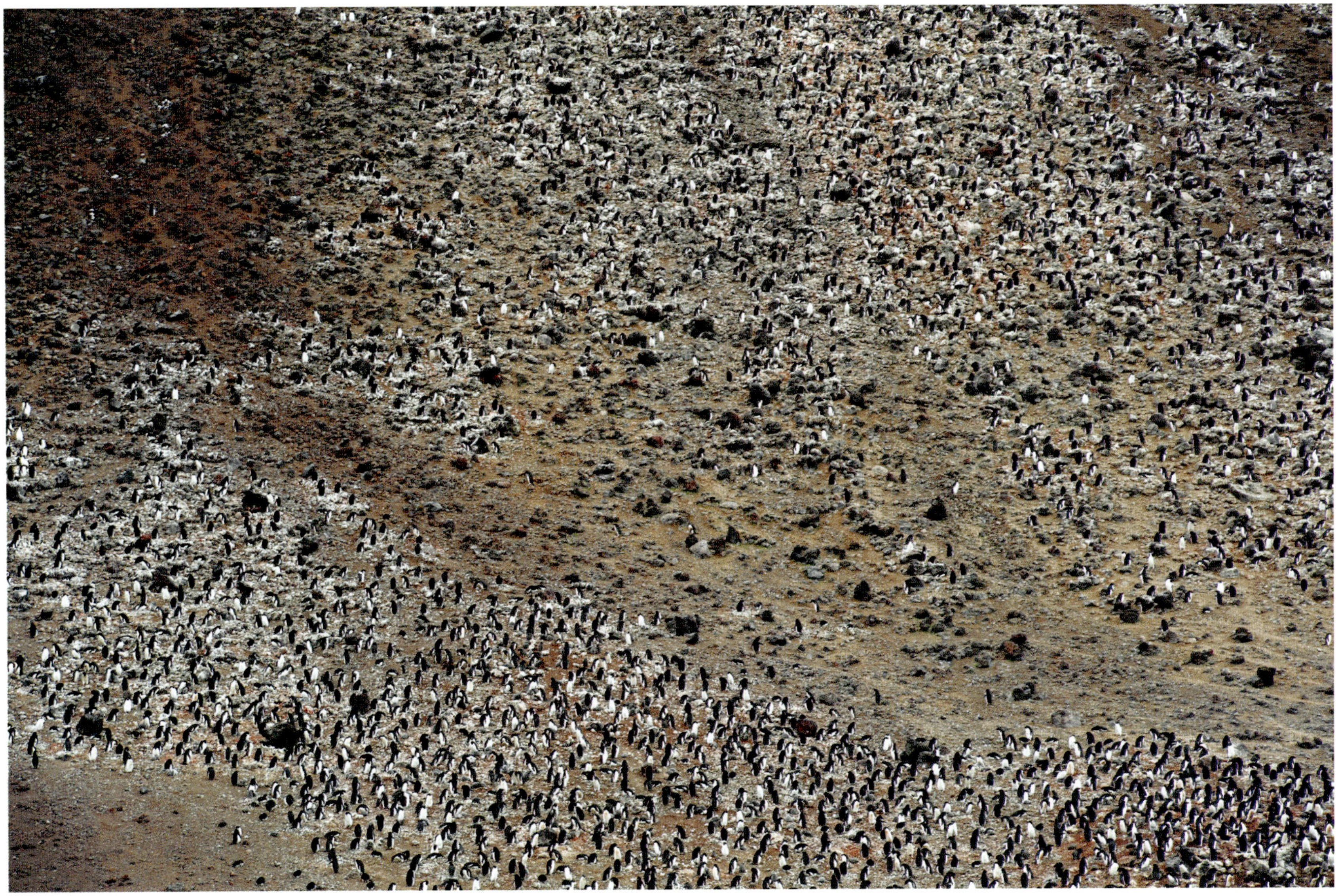

Paulet rookery

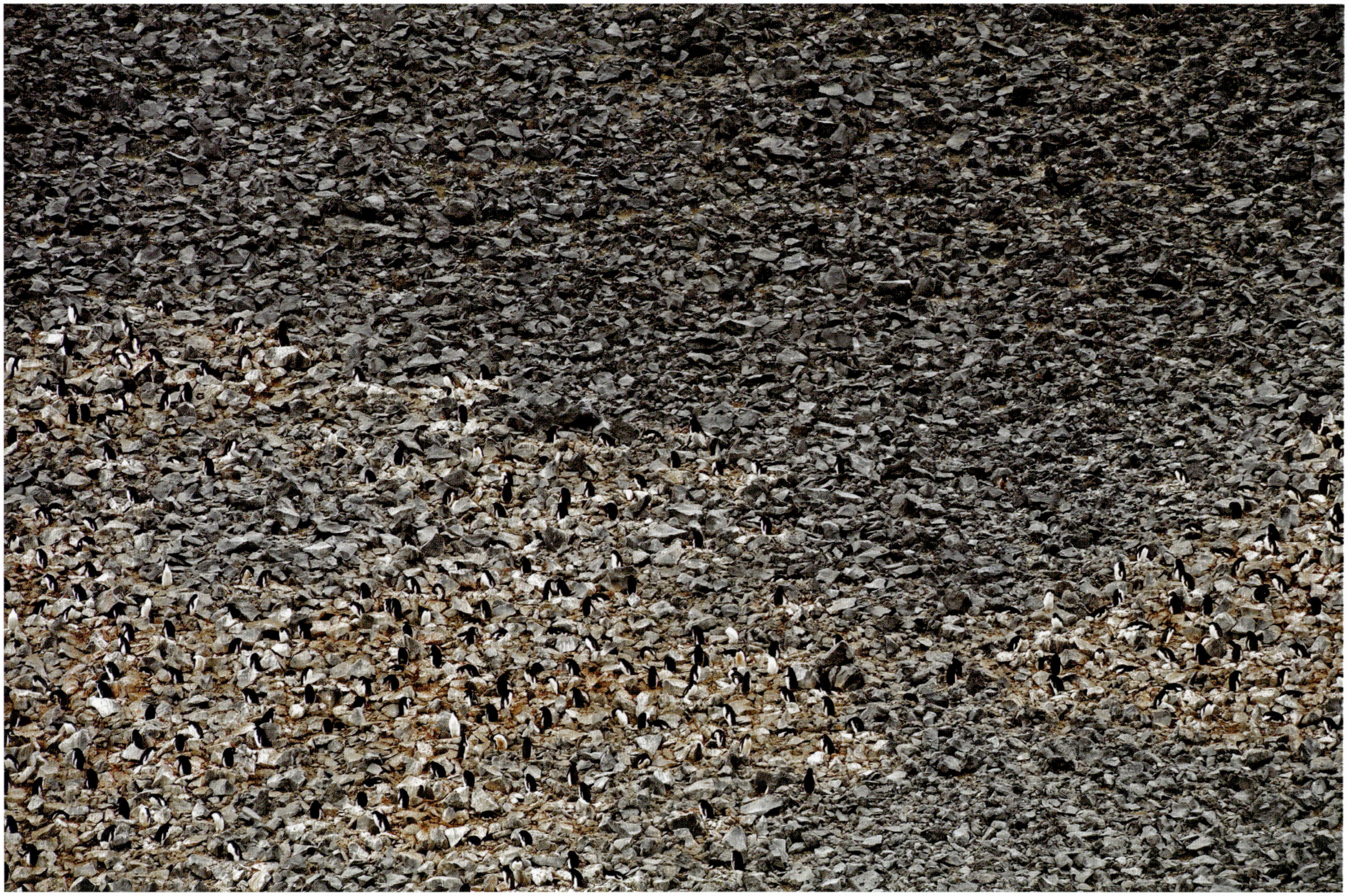

Frost shatter and rookery

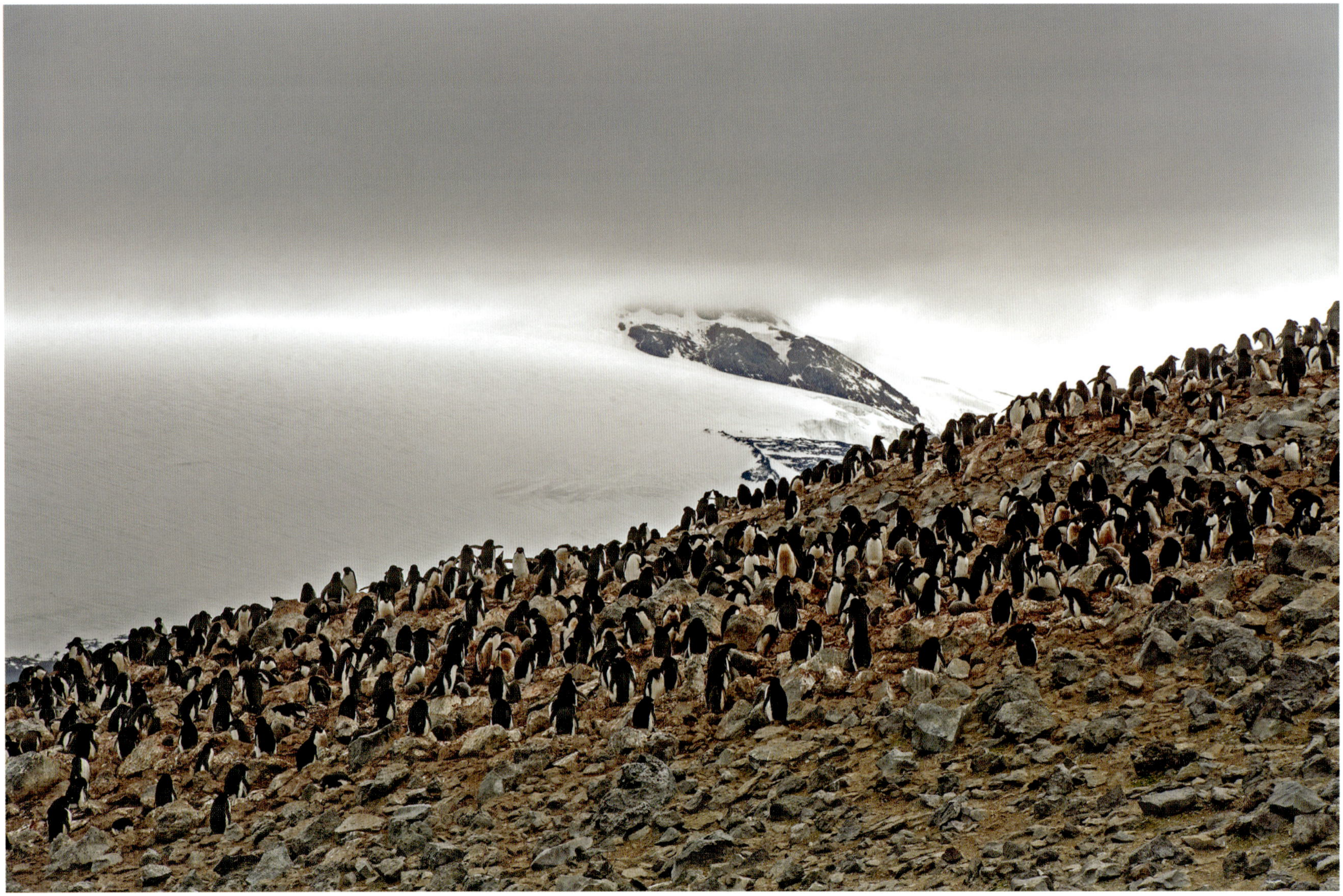

Glacial haze over Paulet Island

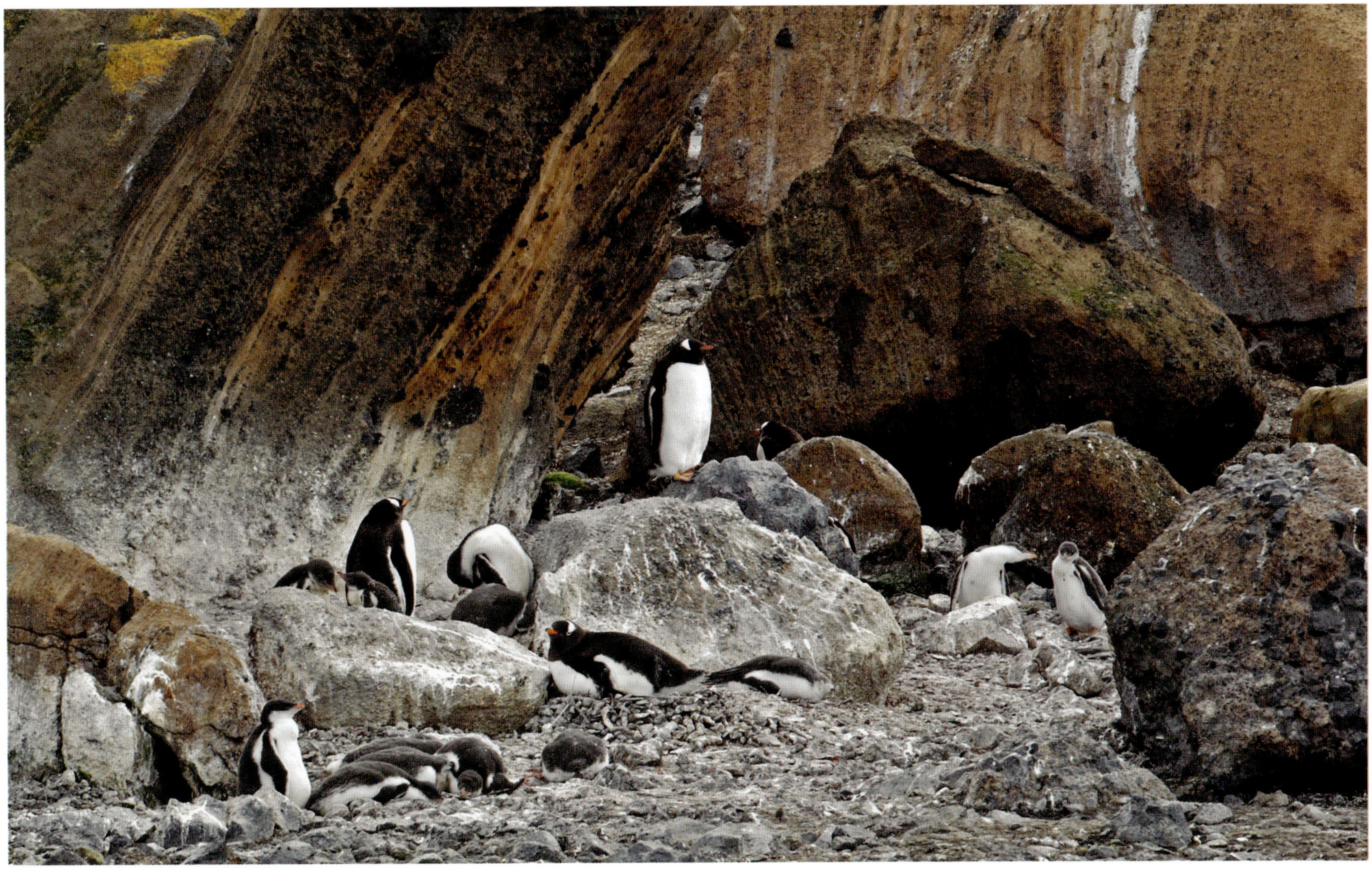

Nesting gentoo penguins

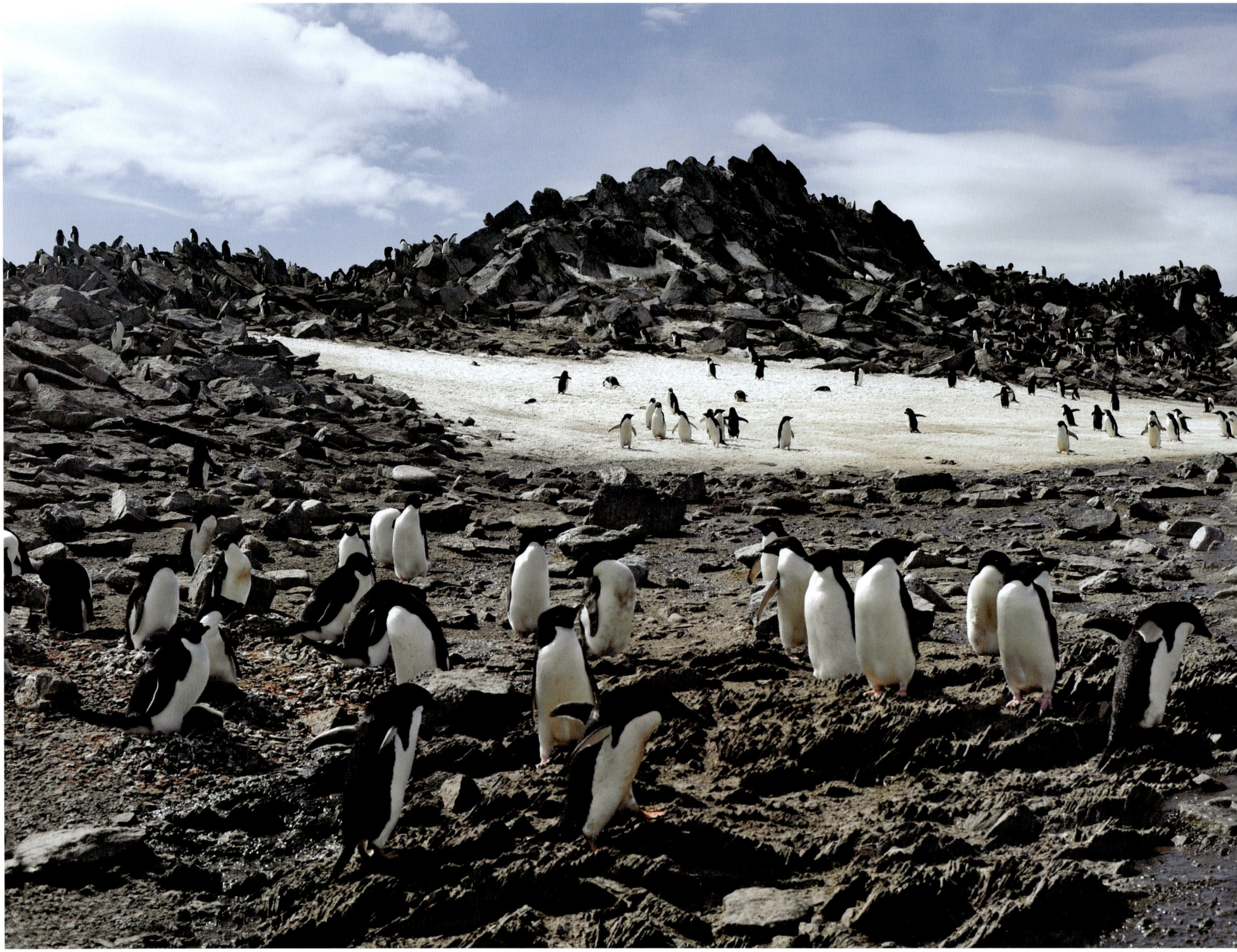

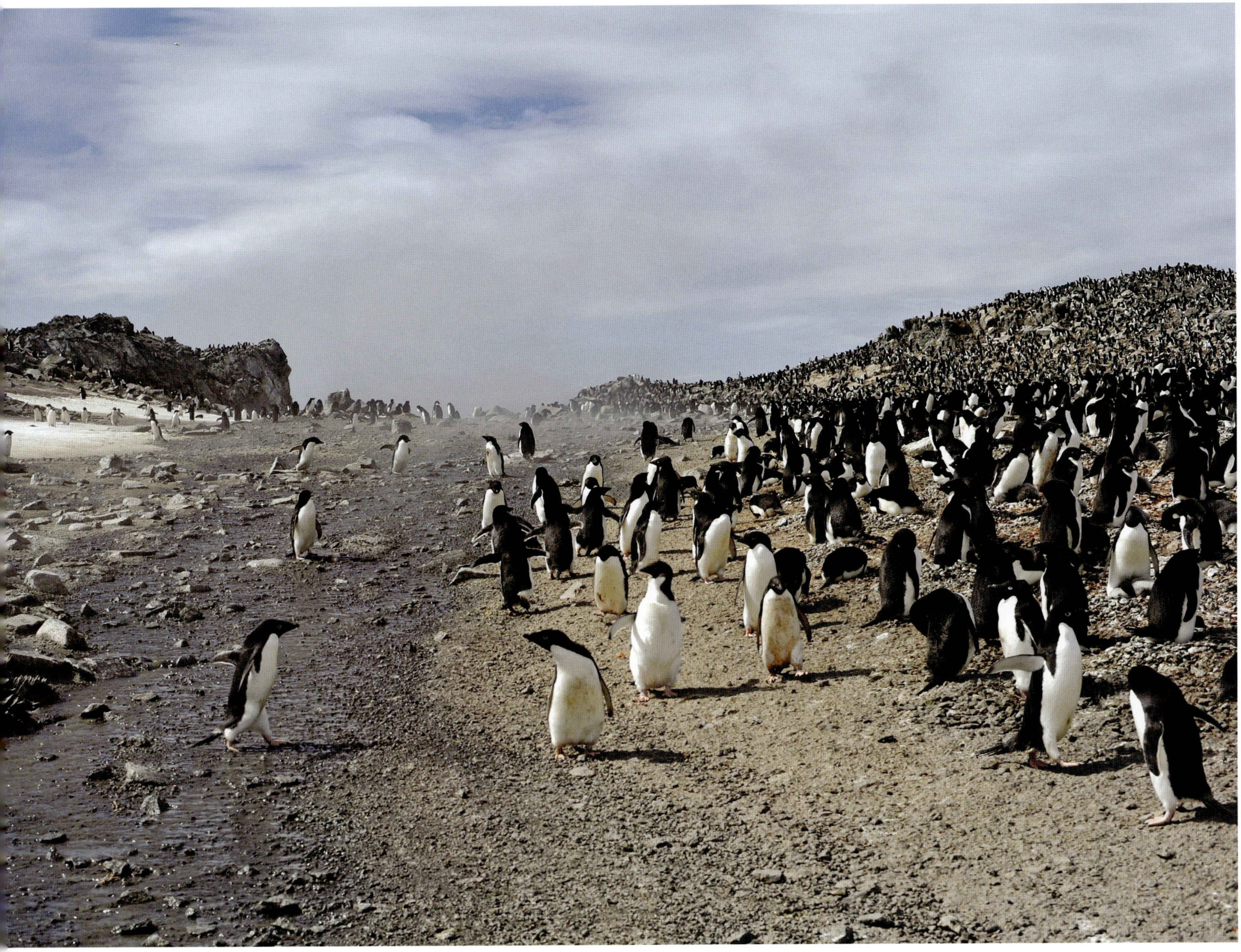

Heroina Island

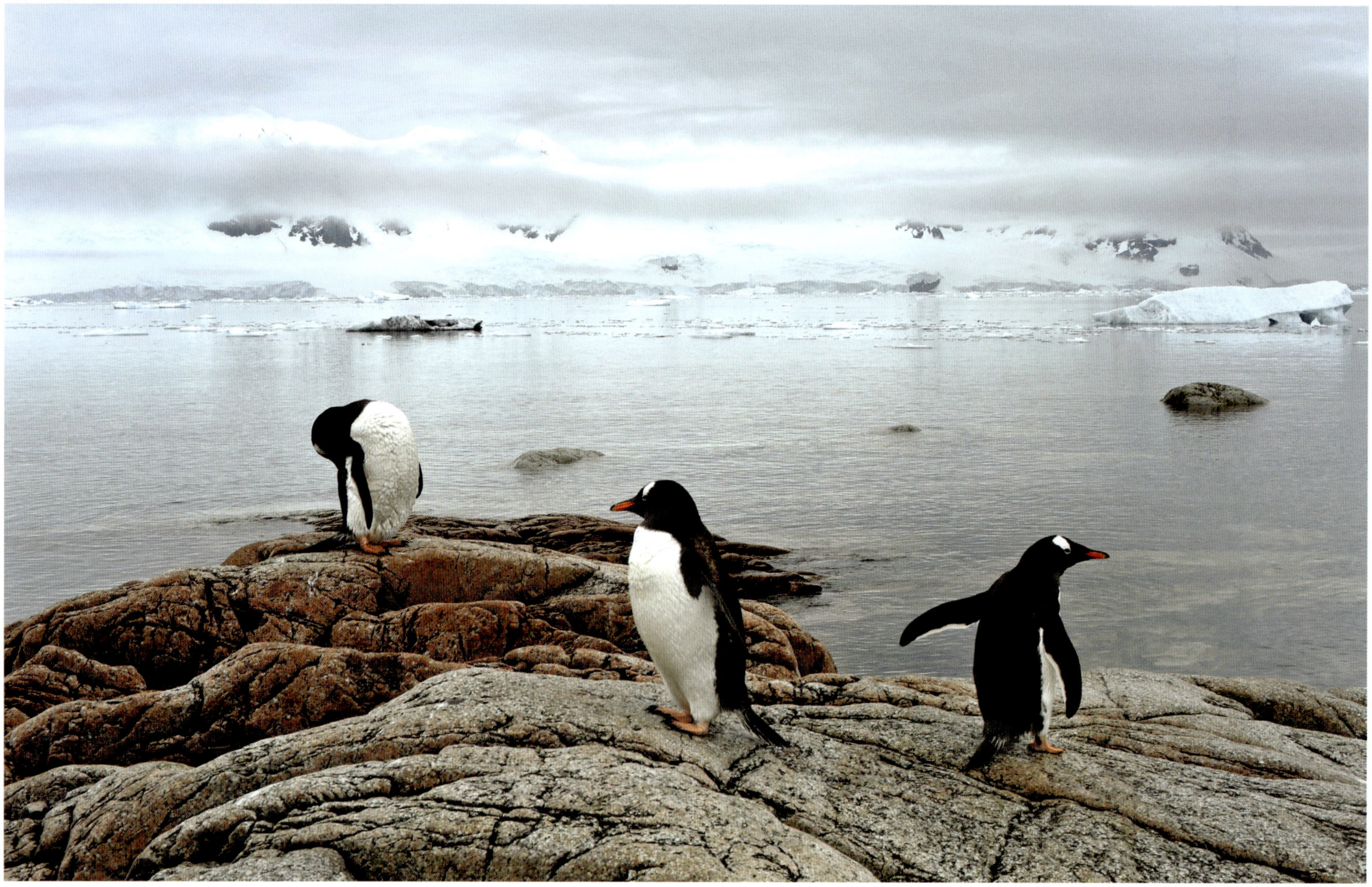

Wauwerman Islands

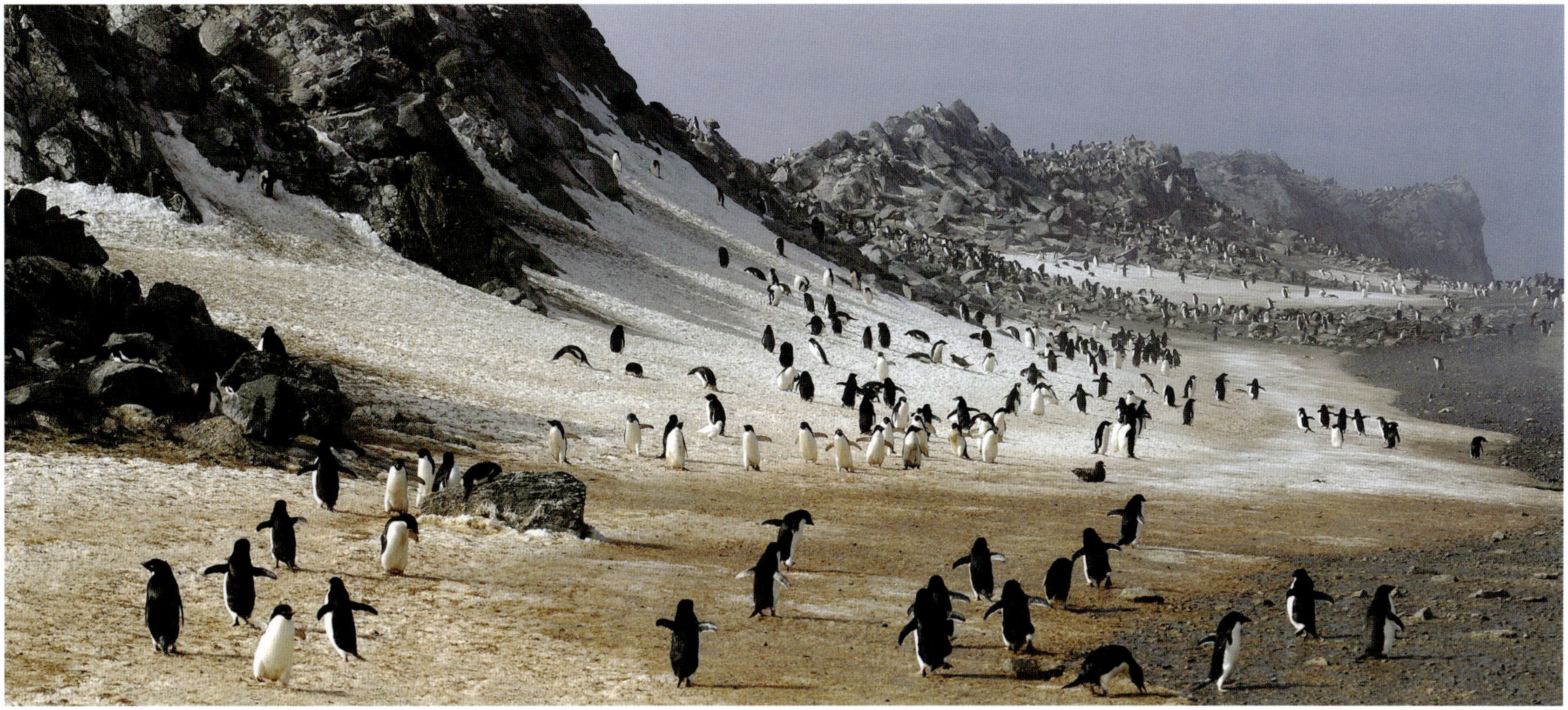

Penguin traffic

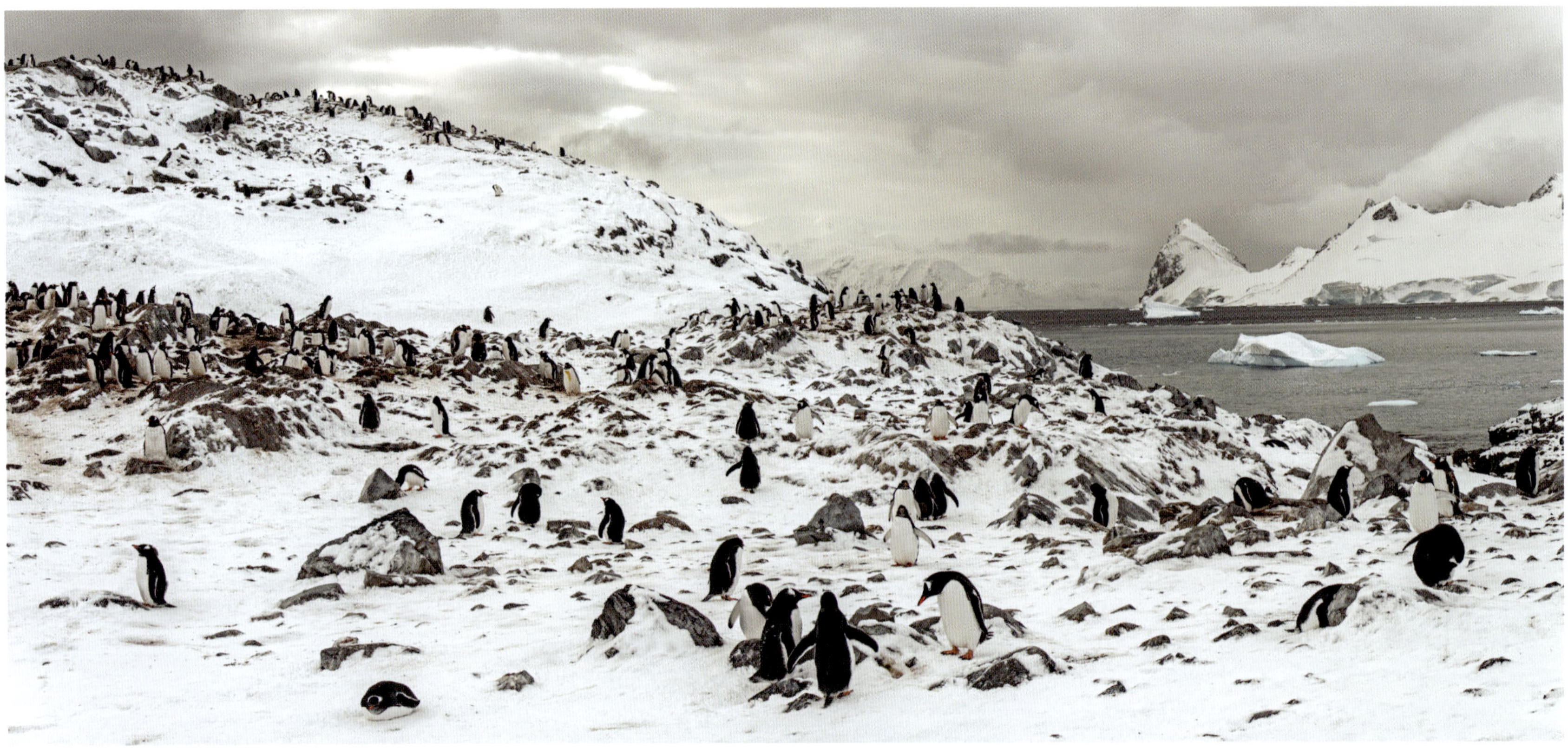

Evening Cuverville Island

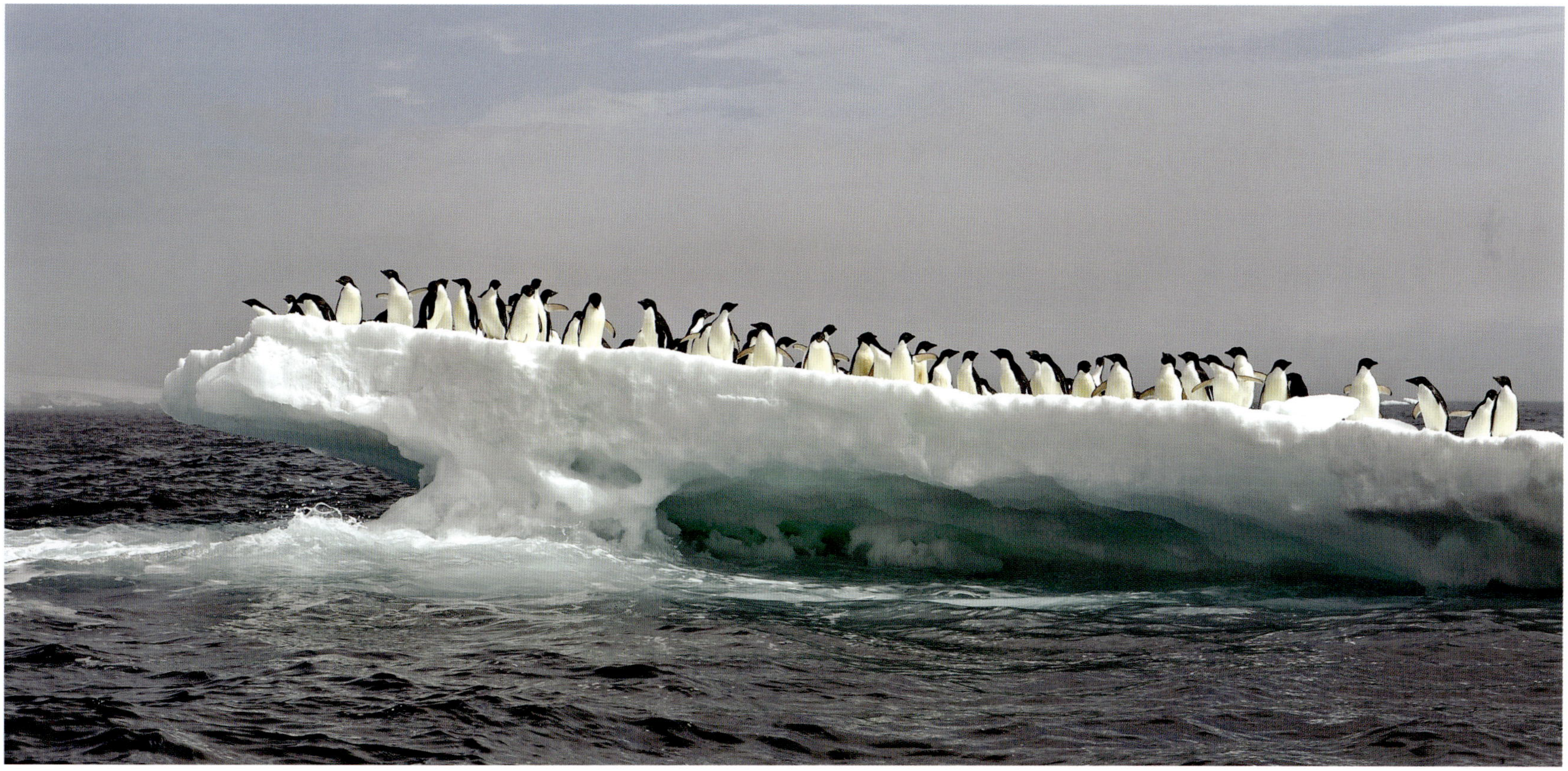

Shipwrecked

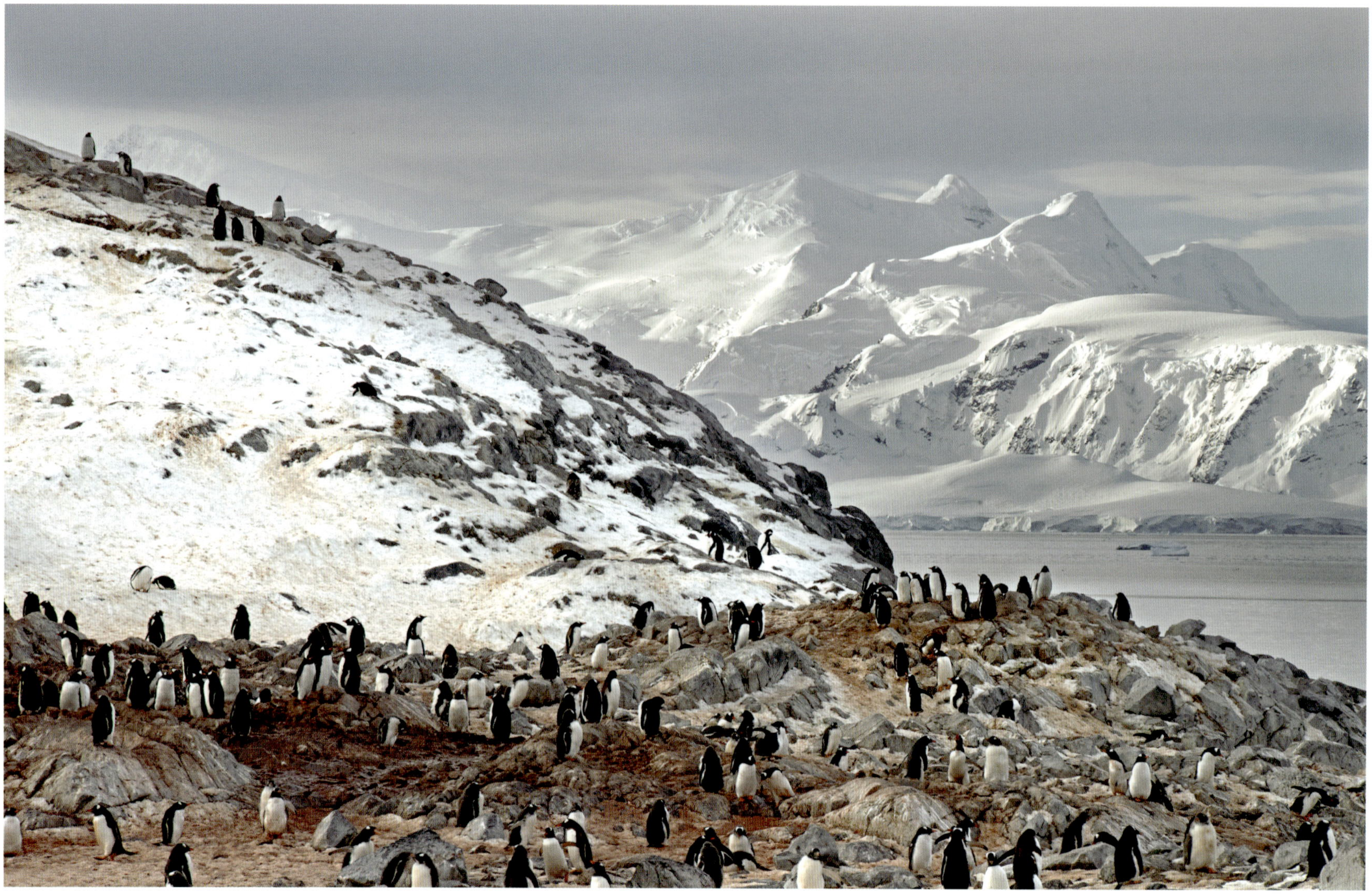

Gentoos and Arctowski Peninsula

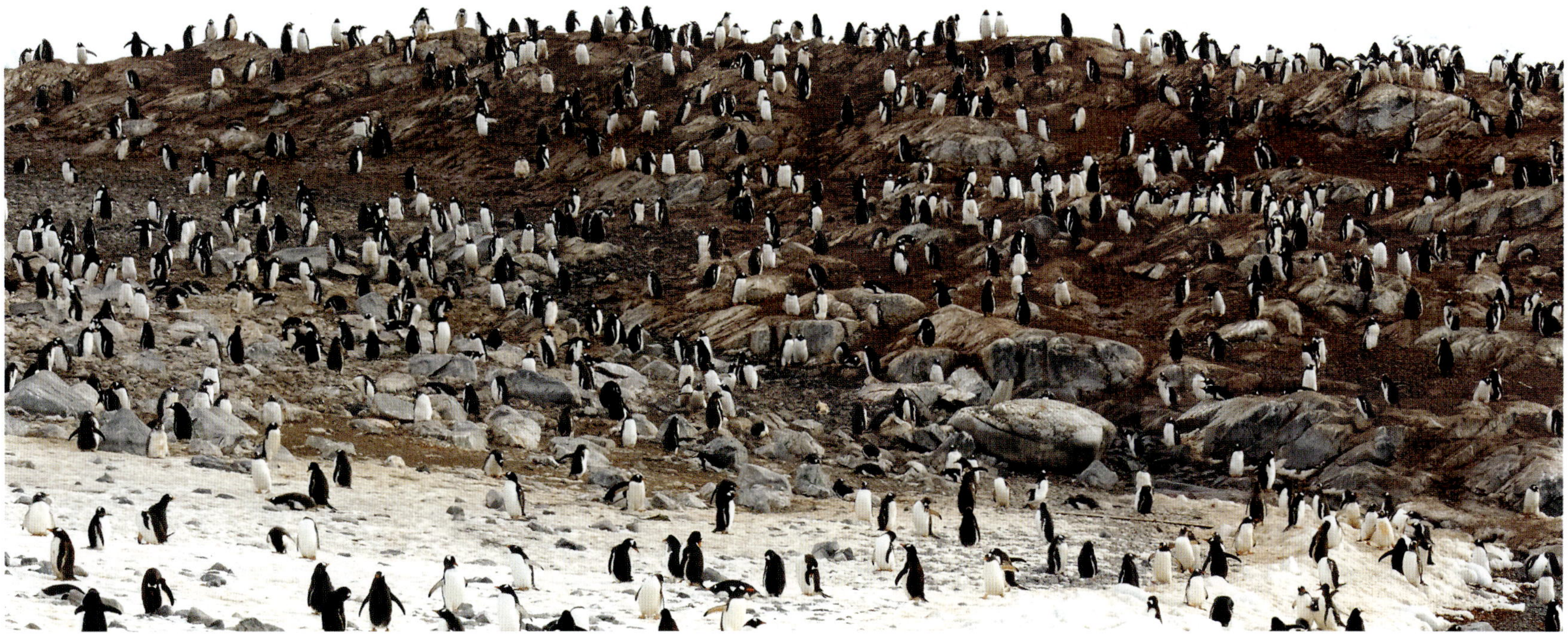

Gentoo colony

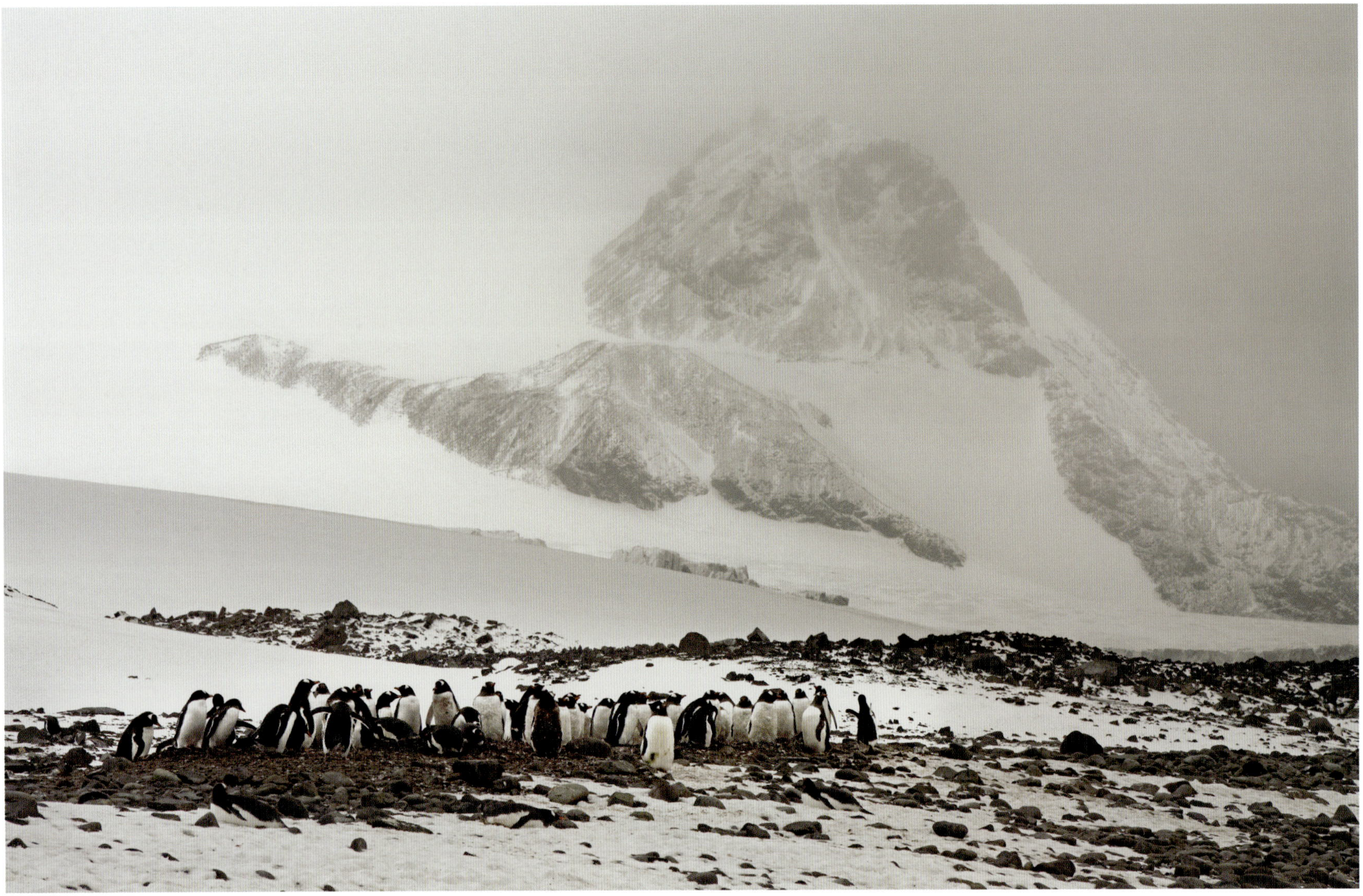

Gathering Yankee Harbour

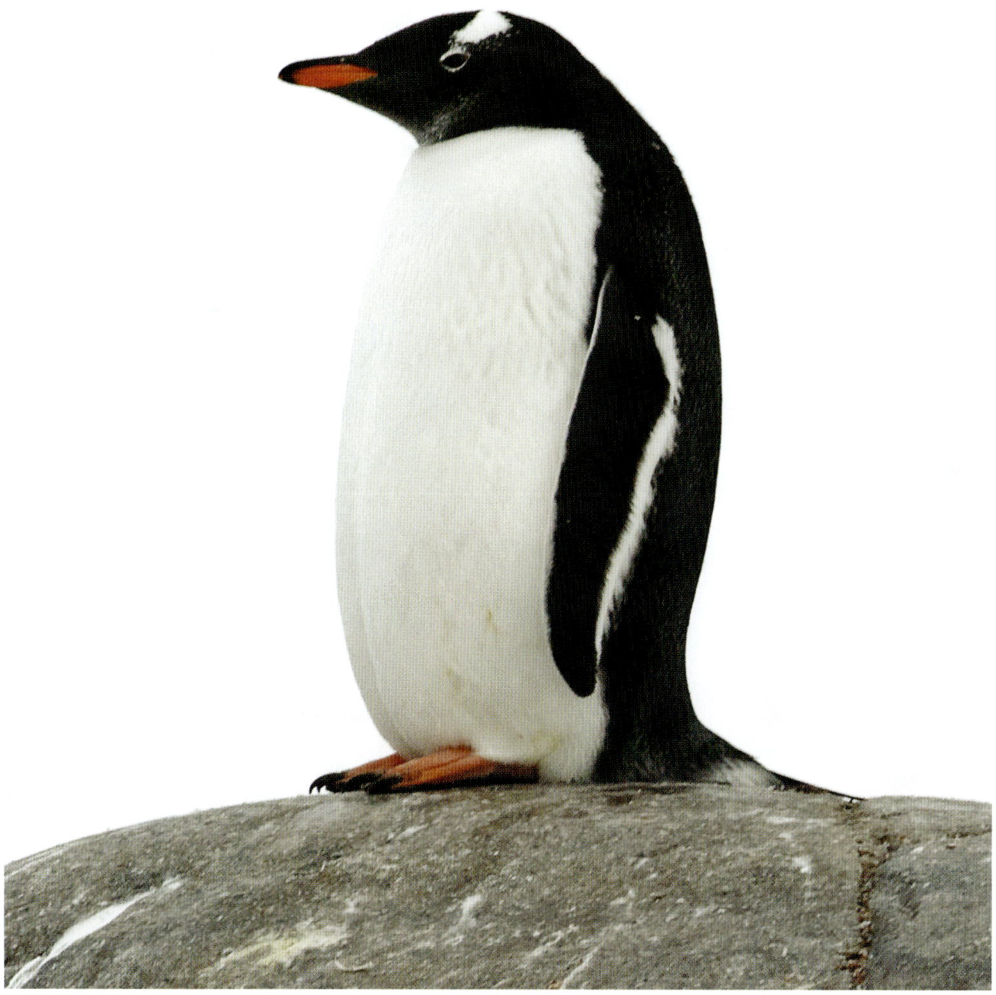

Among the boulders

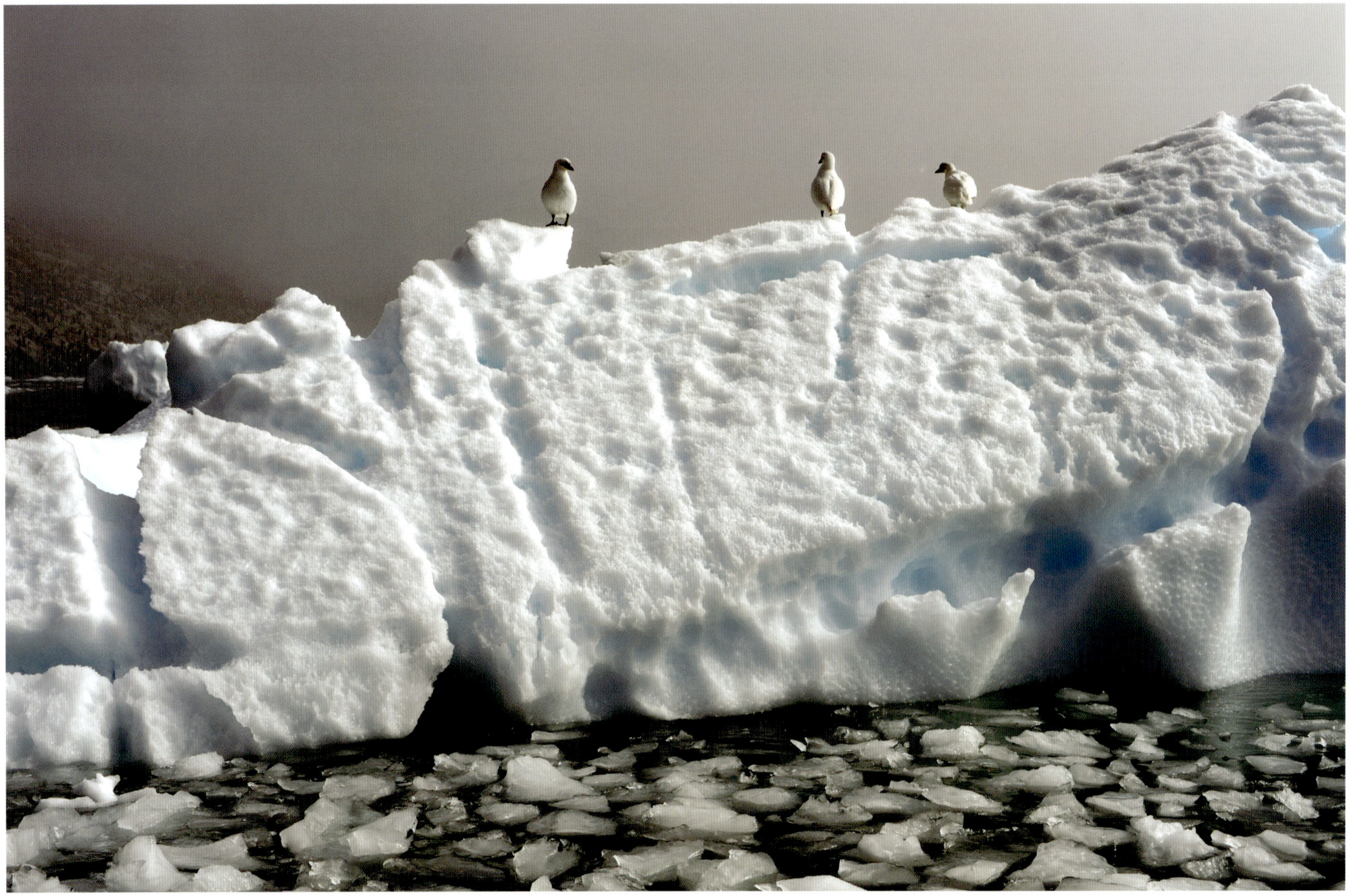

Sheathbills

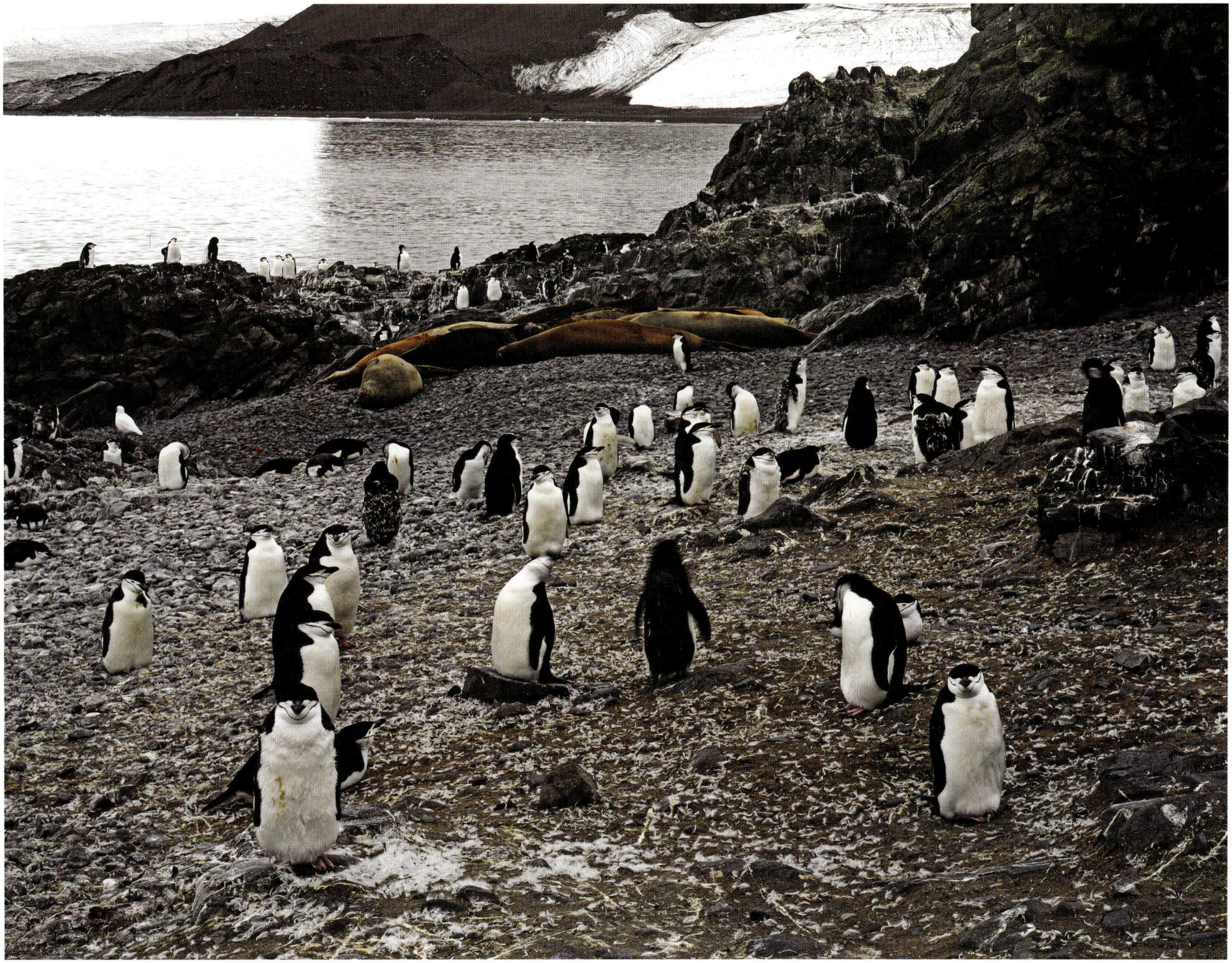

Molting penguins Hannah Point

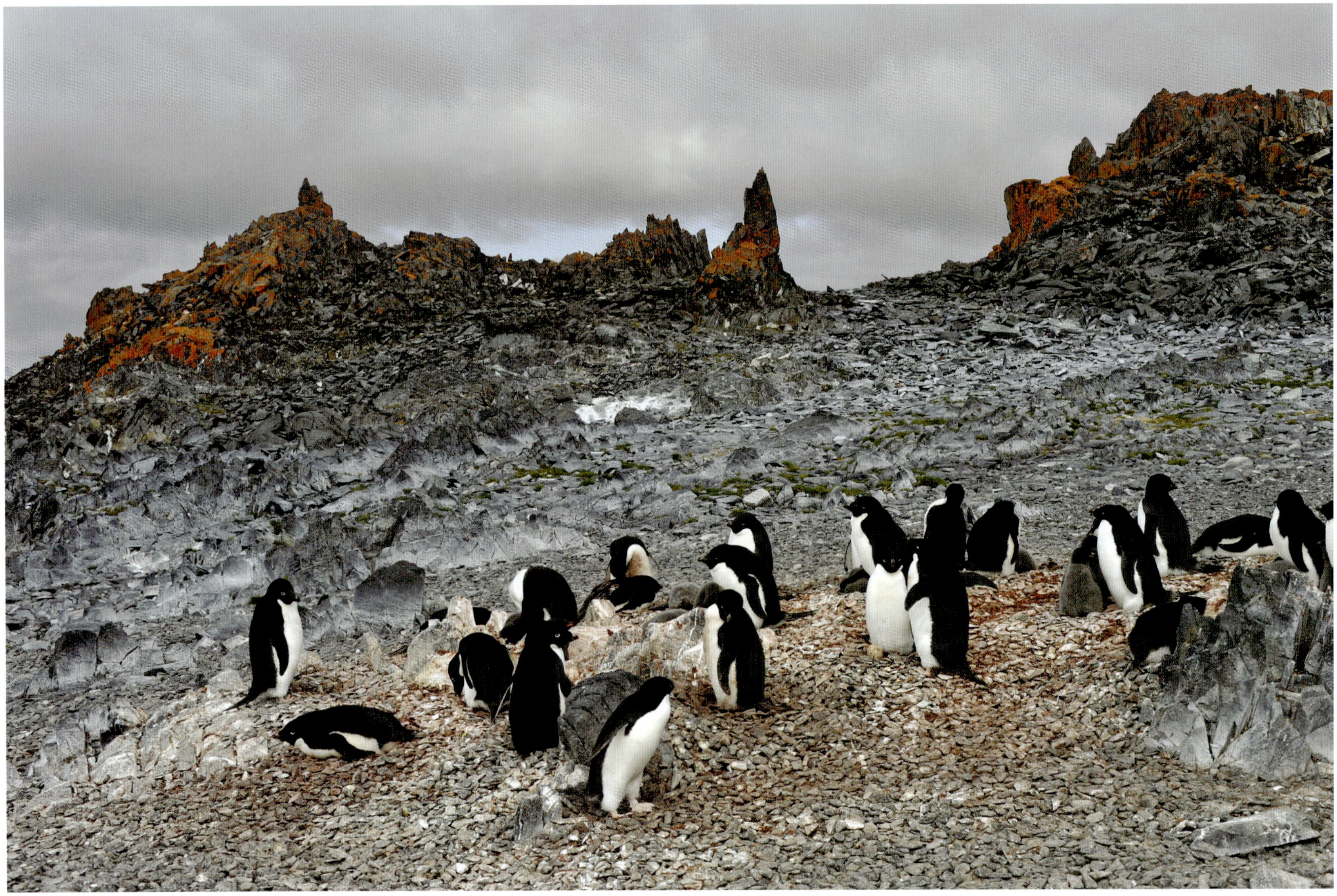

Outcast Islands Bismarck Strait

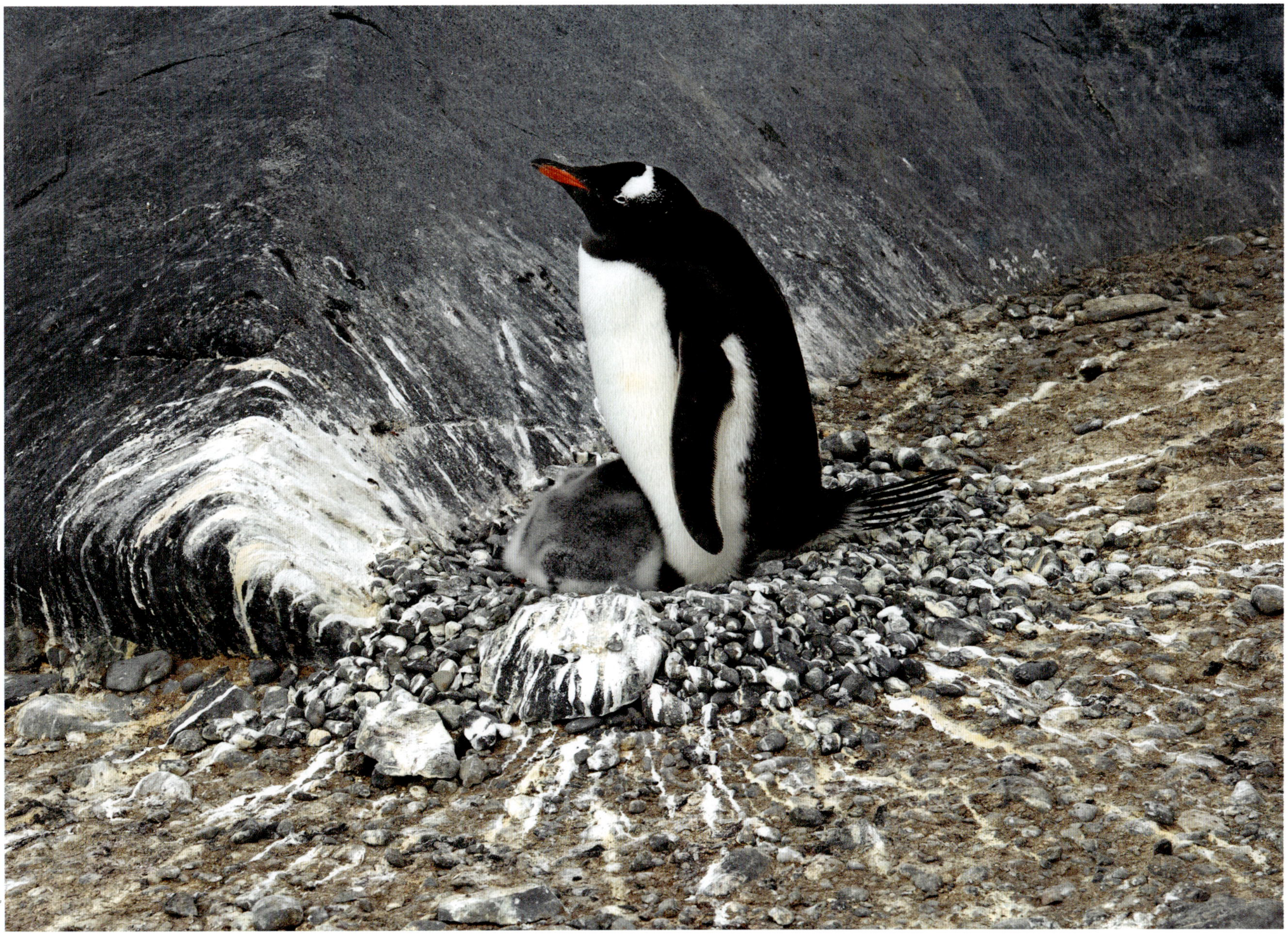

Gentoo and young surrounded by a stellate pattern

SCIENTISTS HAVE REPORTED that after the 1970 volcanic eruption there was a change in the local flora. Algae and moss carved out new niches amongst the dark and warming rock establishing a foothold on the porous ash. Rising heat from the recent underground activity now extends their growing season. Abrupt changes drives evolution and an insurance for survival.

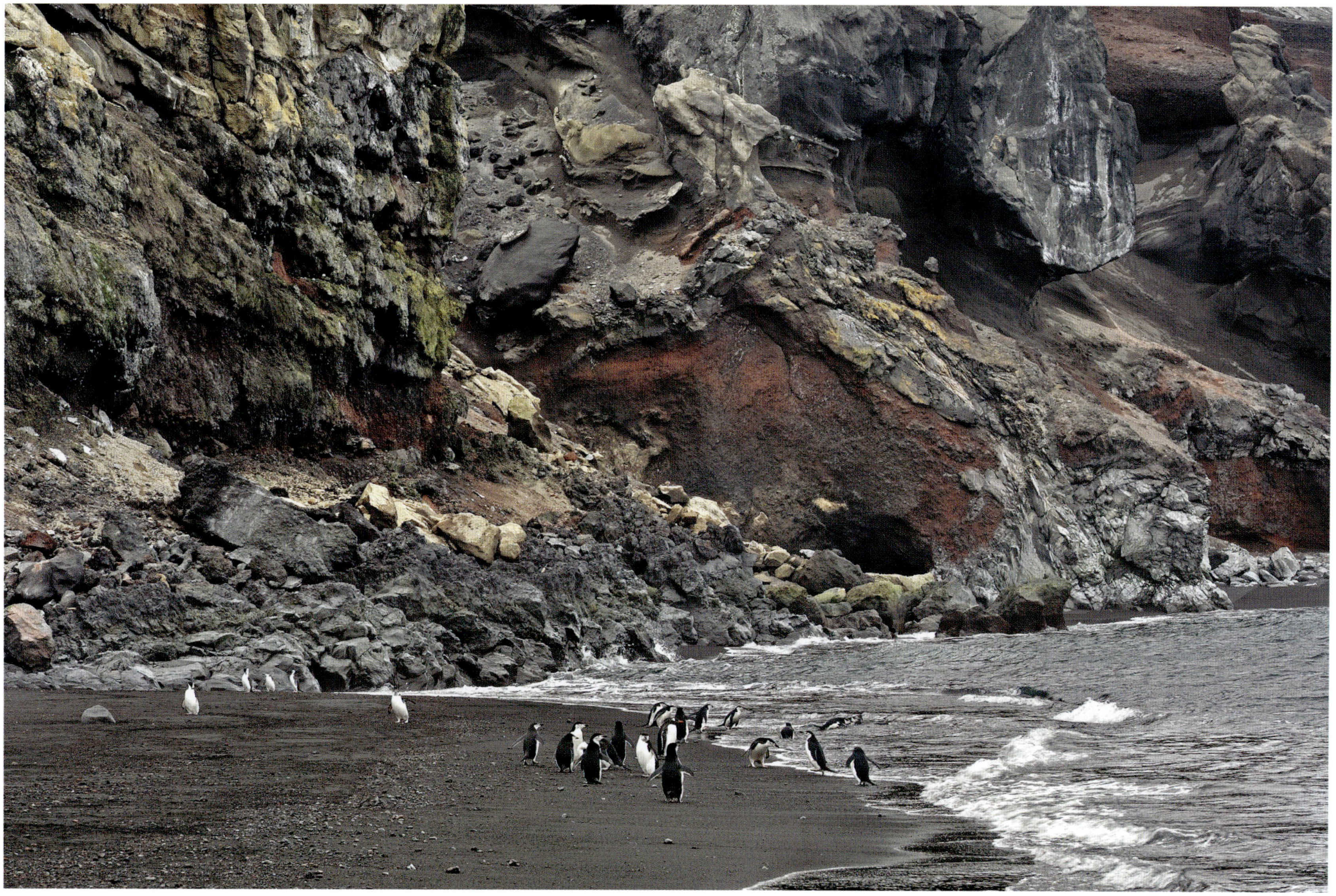

Volcanic beach Foster Bay Deception Island

EVEN AT A distance I could see this berg looked a little different and I was anxious to get closer. It wasn't the deeper blue found in numerous Antarctic icebergs but the slightly different sheen and texture. On the lower part it was polished by winds and waves giving the highly compressed ice a translucent glassy look. However its upper section was more textured like coarse cloth and dusted with snow that drifted and filled numerous cracks and crevices. I could see that this berg had rolled and changed its centre of gravity and waterline a few times. It had been glazed and subjected to much buffing giving it this appealing effect.

Below the waterline the light that was transmitted through the blue ice shone with a brilliant greenish cyan. It had a compelling effect against the darkened water. I was busy with my camera making sure I would capture its intense glow when I noticed a dark spot on its nearly vertical side. It was a lone Adelie penguin still climbing.

Now immediately the focus of my attention quickly changed from the berg to this minuscule and insignificant spot. It gave perspective to the iceberg's size but moreover it was the penguin's location that was so surprising and unexpected. Somewhat baffling. It raised obvious questions. Why was it so high on an icy and slippery slope? How did it manage to get there? Was it chased from the water? Was it still escaping, struggling upwards? How high will it climb? How will it get down?

The berg was even more vertical on the far side. Would it try and climb down or just slide or will it wait days hoping for a shift in the ice? The scene drifted by too quickly and I wished I could have lingered. The encounter remained with me for the whole day and my mind kept wondering about it all.

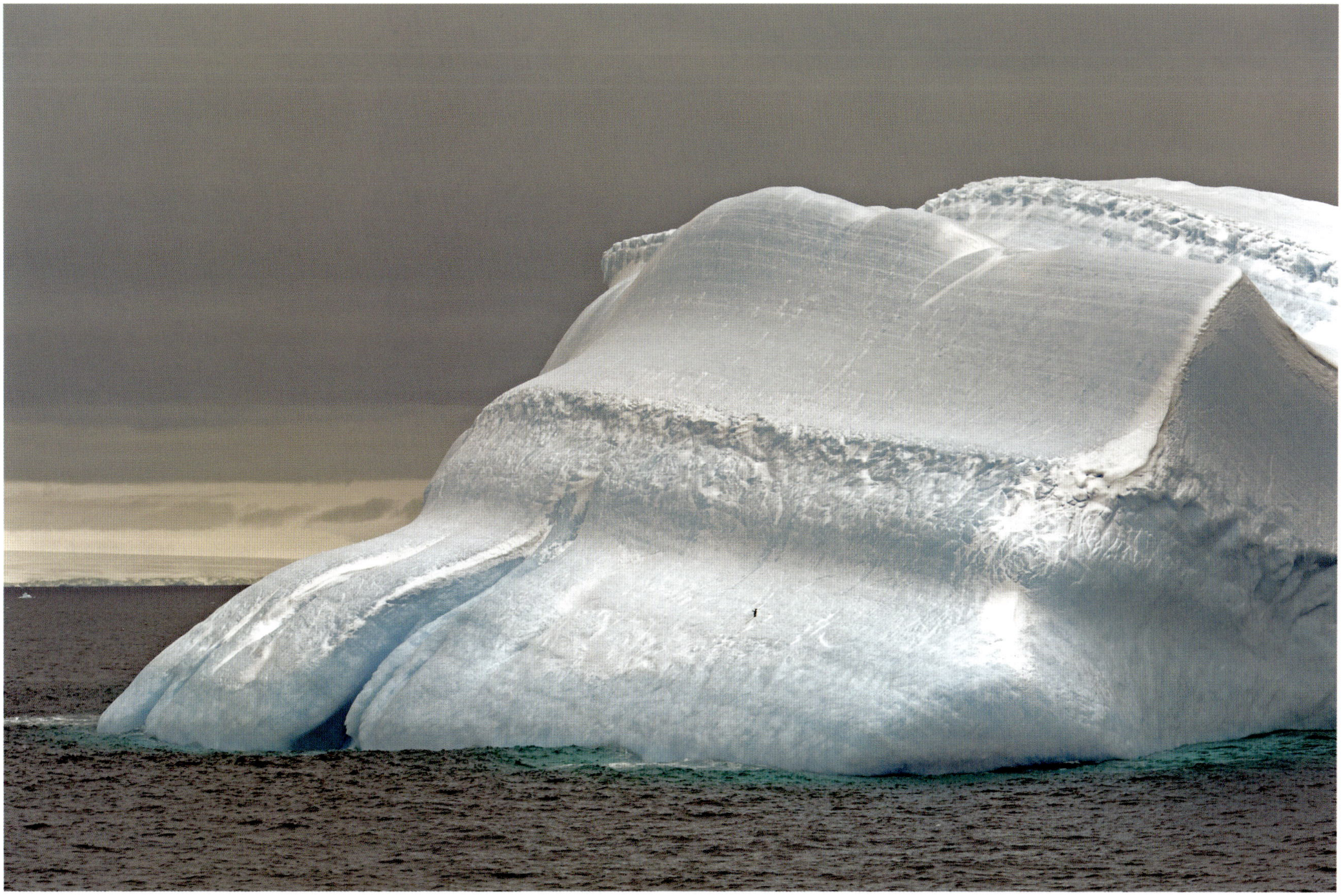

Lone penguin

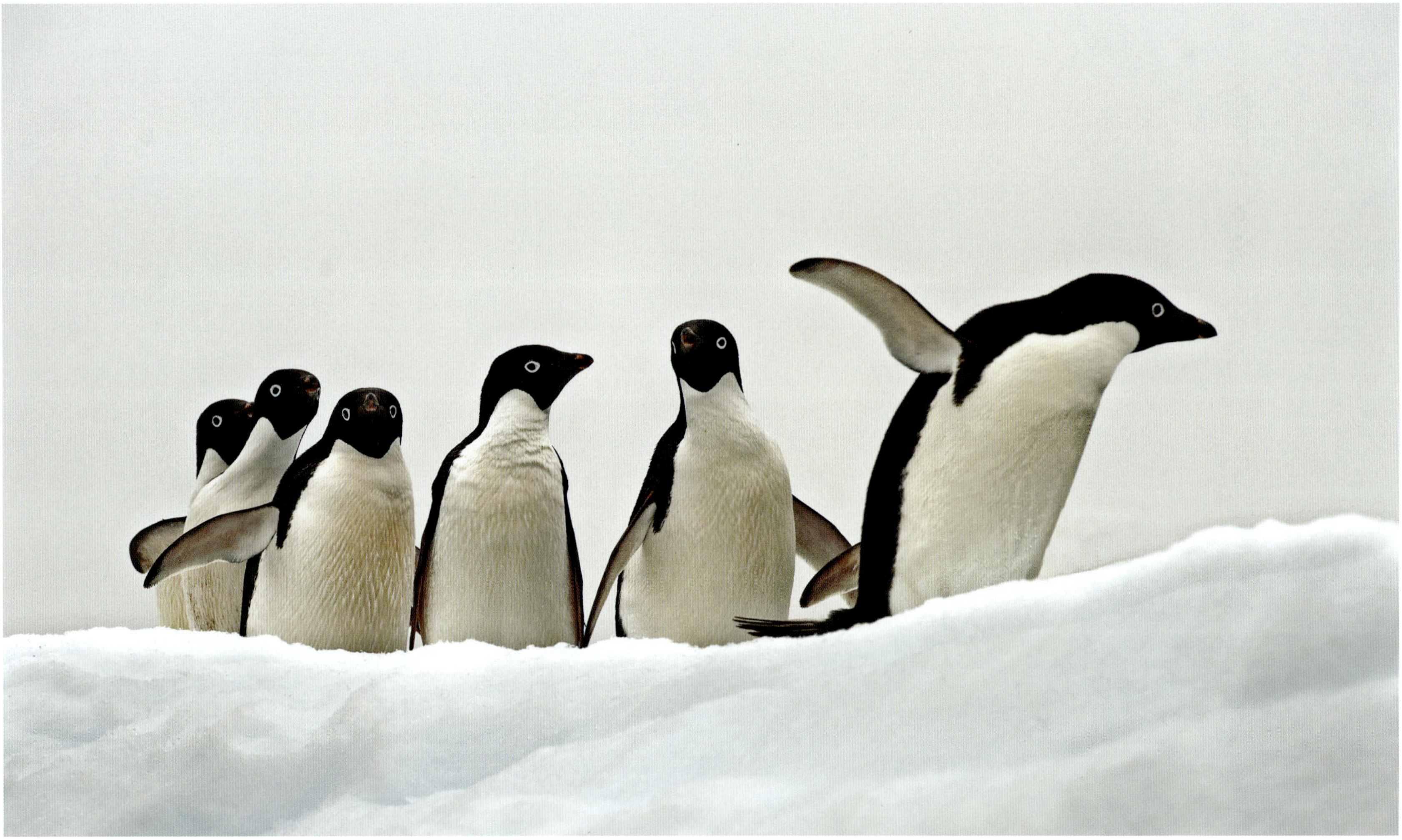

I HAD NOT noticed the climbing Adelie penguin (at right) till I reviewed my pictures back home. I was so stuck to my viewfinder concentrating on getting a strong composition to draw attention to the smooth washed out surface of this tilted iceberg that the little black speck eluded me.

Adelie eyes

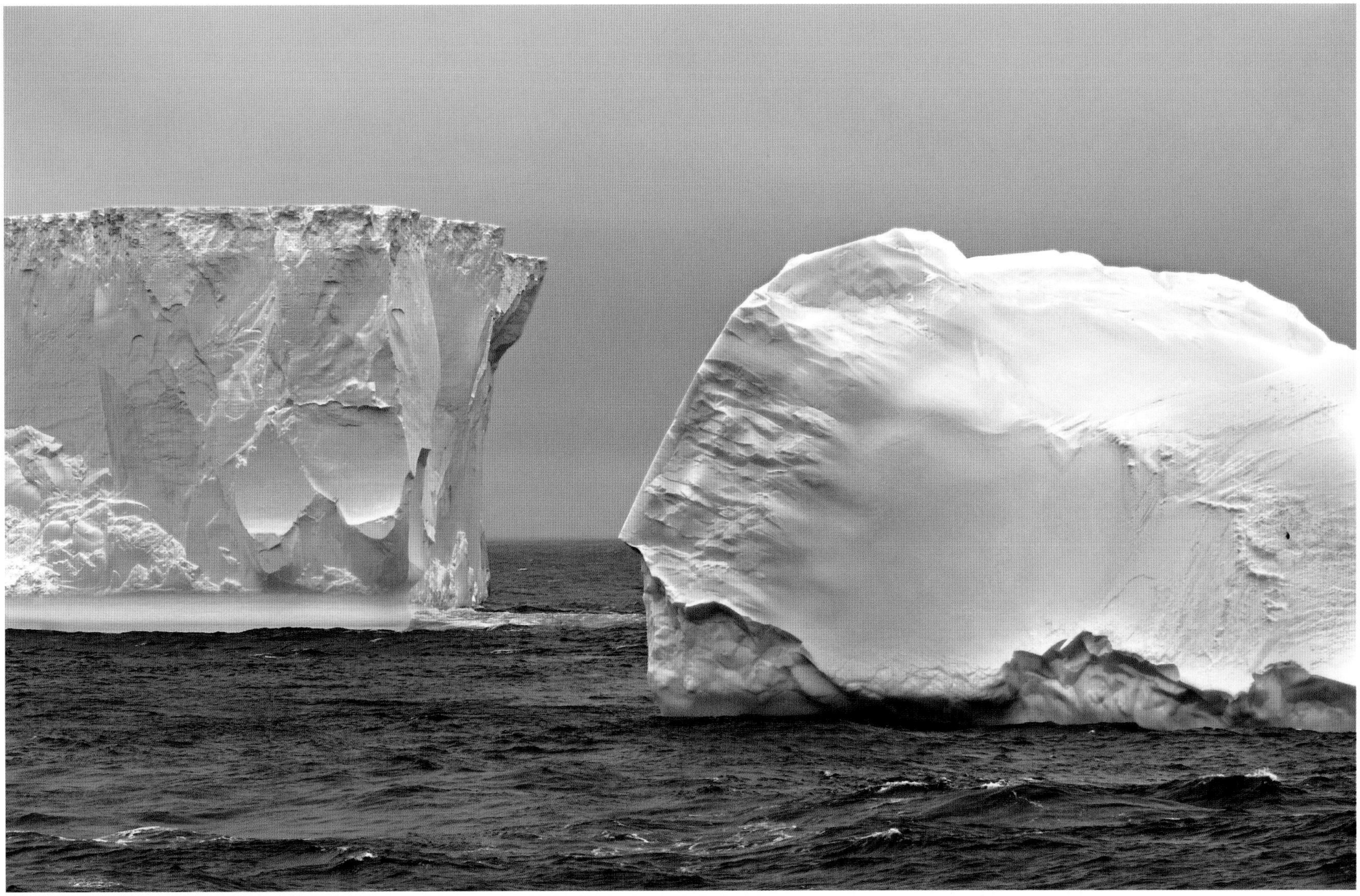

Climbing Adelie

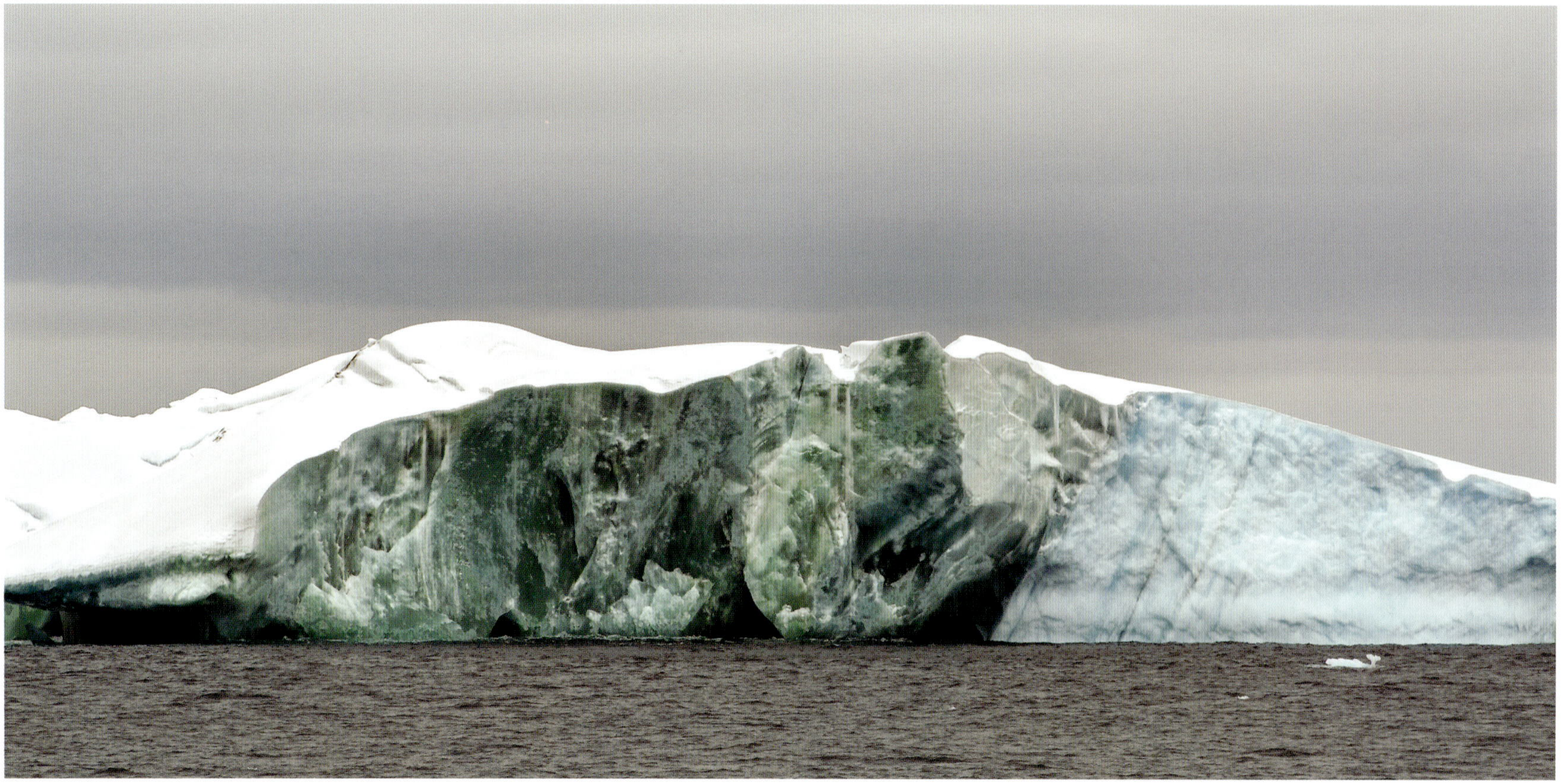

Ice captured algae

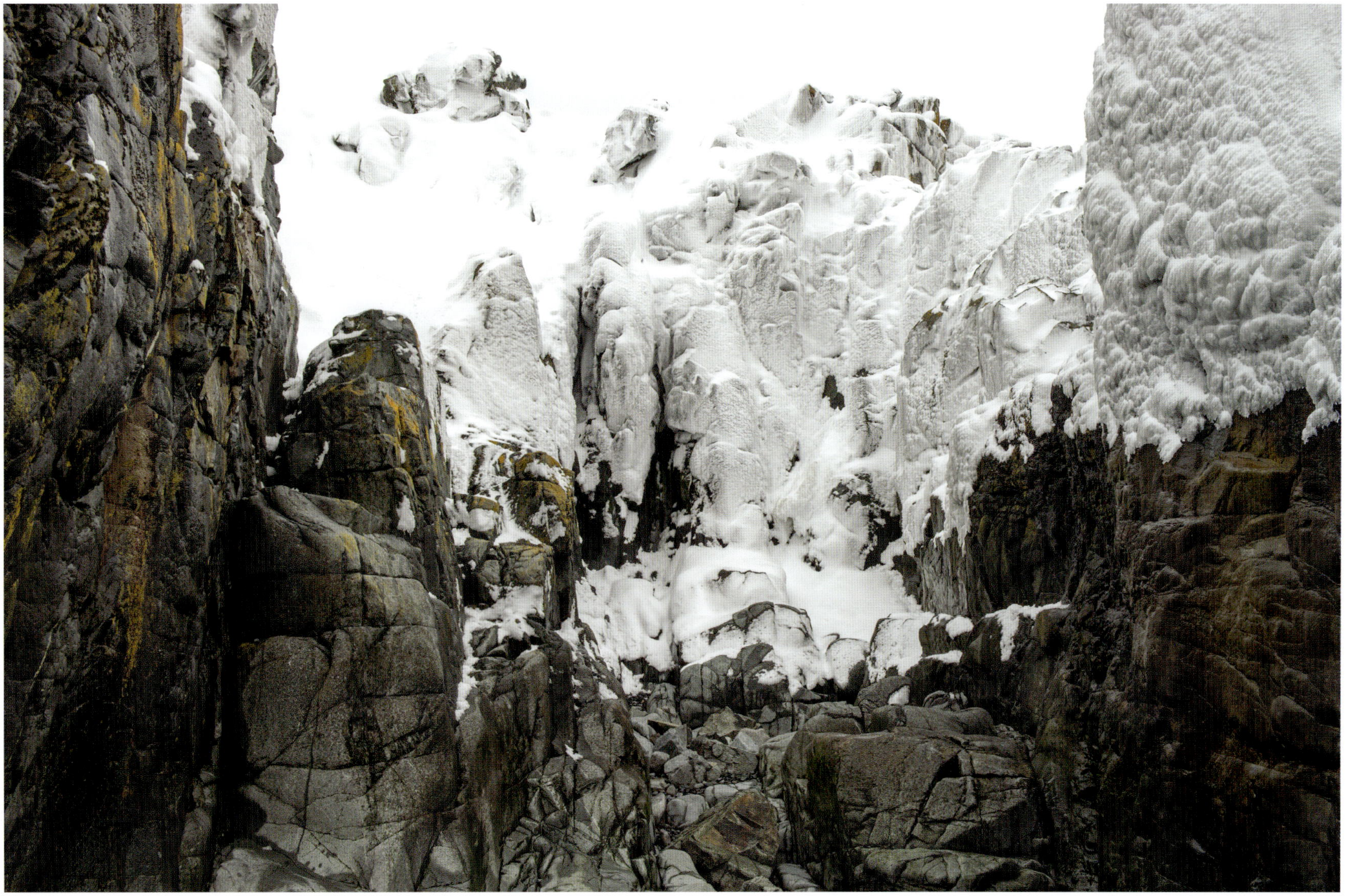

Cliff detail Half Moon Island

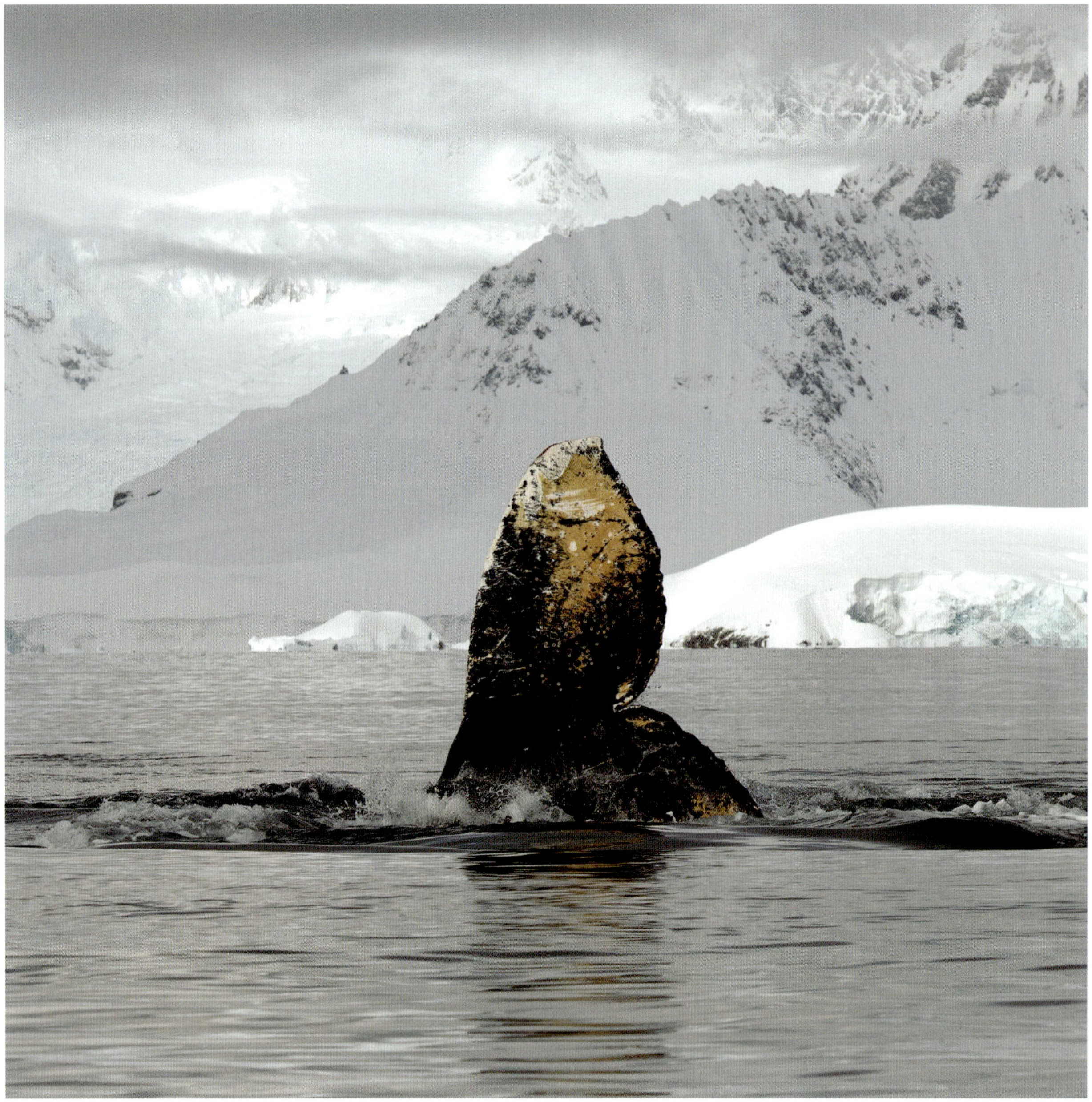

Stained fluke

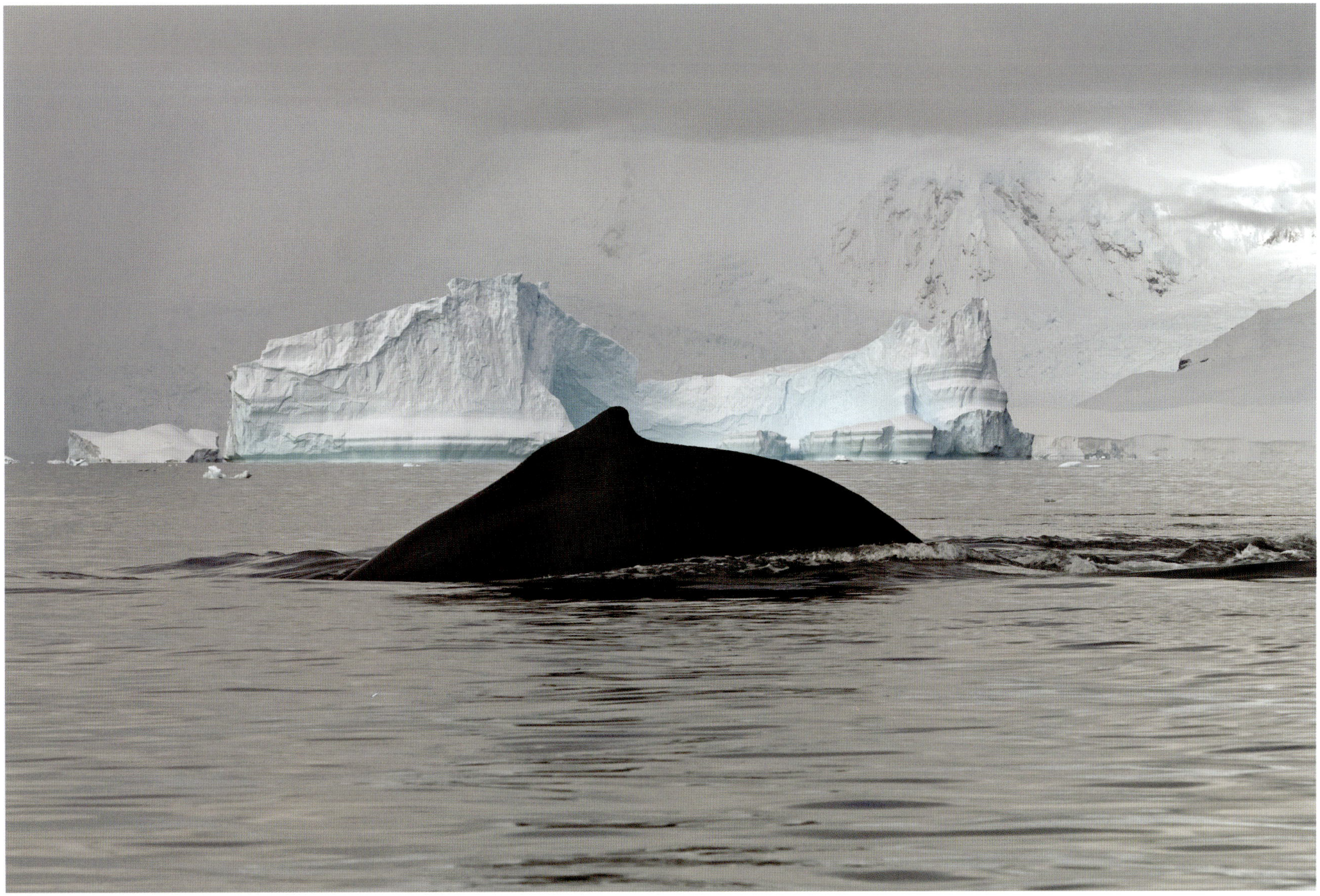

Whale breaching

THE INFLUENCE WATER has on manipulating the visual effects of light are astonishing. No matter its state, whether solid, liquid or gas there will be a pronounced change in outcome. No more is this phenomenon apparent and evident than here in the moisture laden air of the Antarctica peninsula.

Ninety-six percent of Antarctica's 14,000,000 square kilometres is ice and with the winter the freezing ocean doubles the existing ice surface making it the largest accumulation of solid water anywhere on the planet. With so much ice comes a variety of frozen liquid. From microscopic ice crystals that rain down from a cloudless sky, to enormous cascading glaciers stretching miles across, to icebergs as big as some counties—these all contribute in their own specific fashions. Refraction, reflection, absorption and diffusion are just a few of the ways light is manipulated into a kaleidoscope of visual effects.

A good example of such an occurrence is here in this image where one cannot be but mesmerized at the pure colours that seem to emanate from deep within the ice. While fresh snow fluffed up with air scatters light in a display of white, the ice in this stranded berg is so compressed as to absorb the reds and yellows and show only the shorter ones, cyan and blue. Even the water filled sky has a touch of magenta.

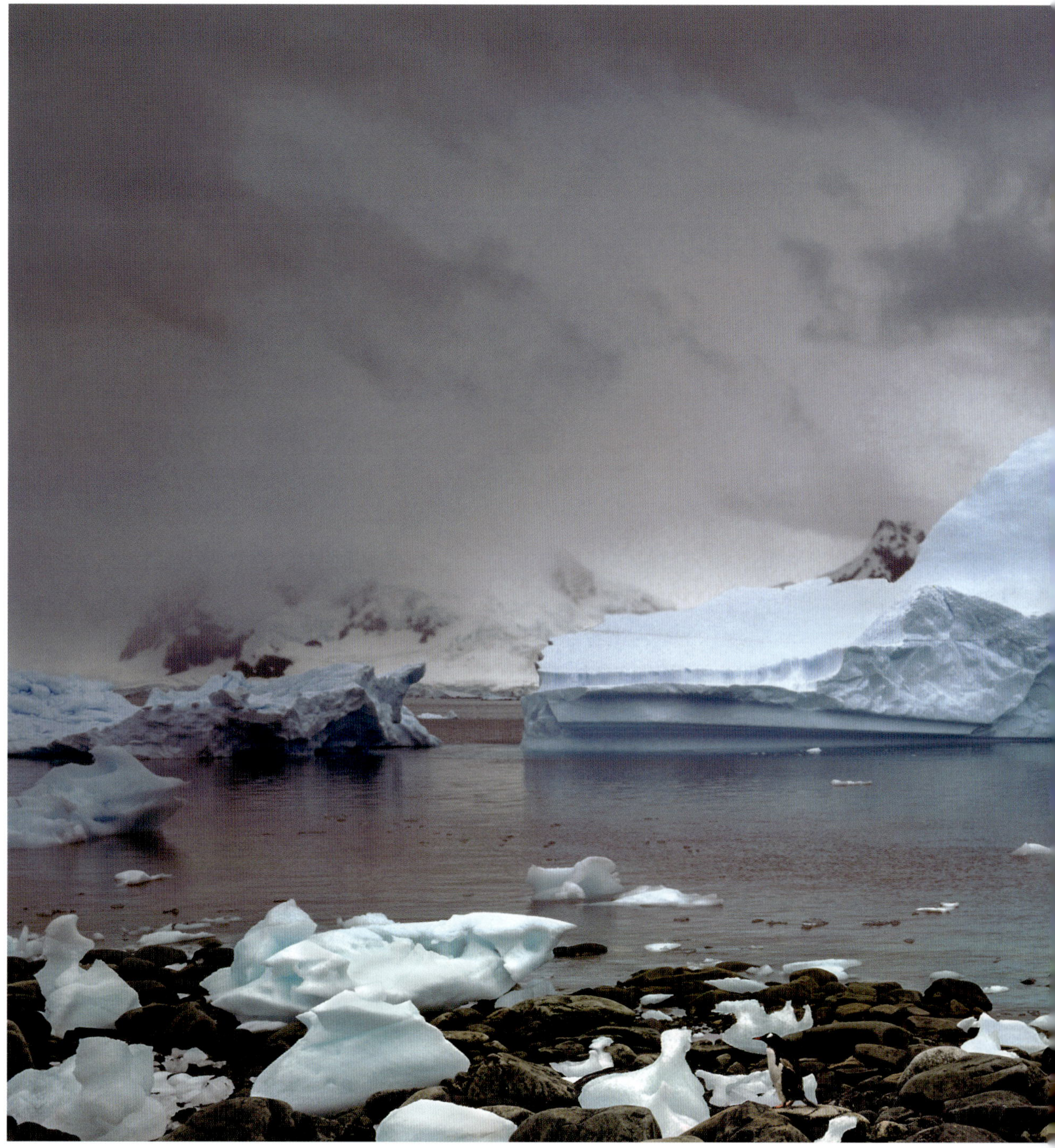

Iceberg and magenta sky

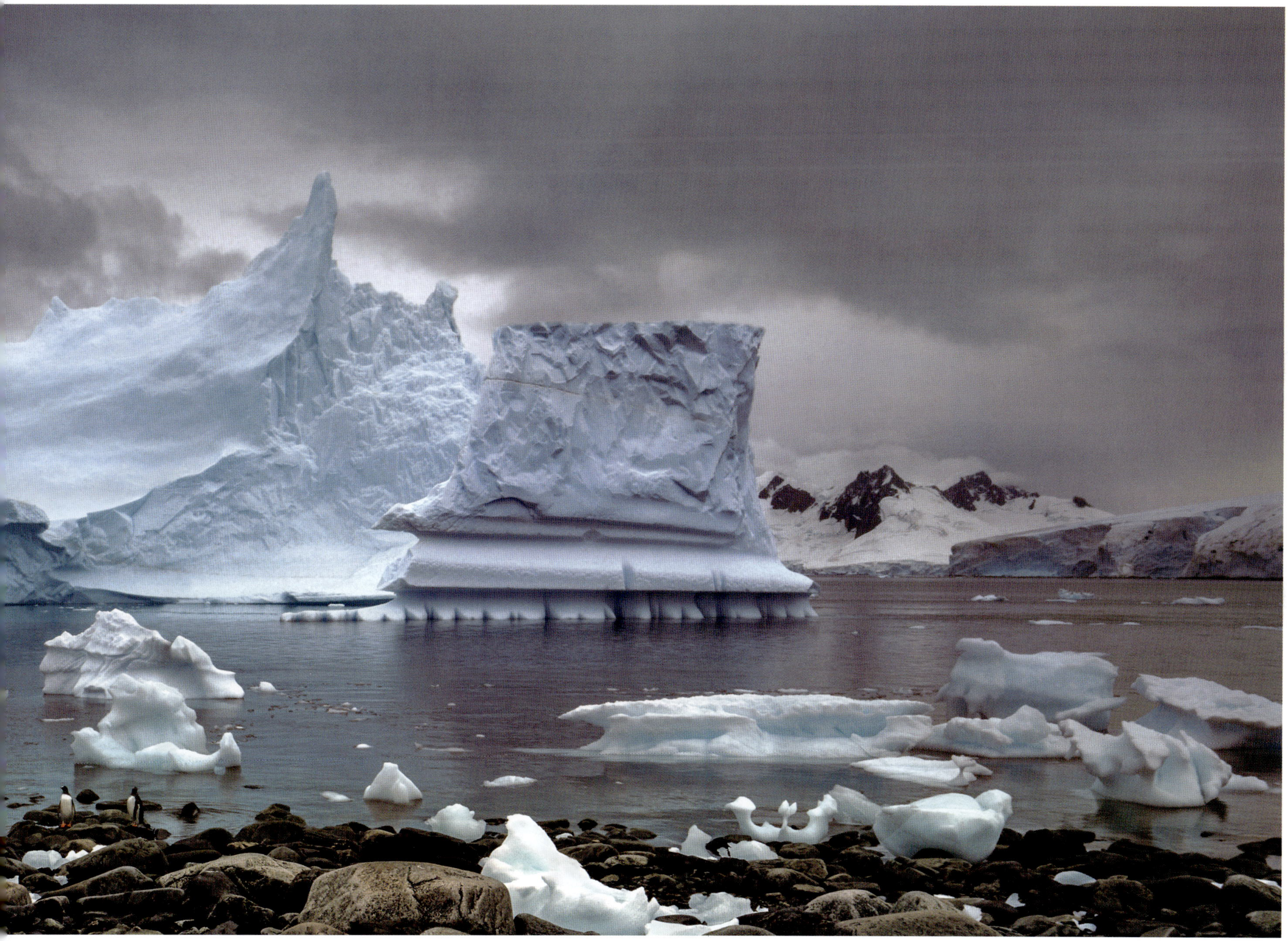

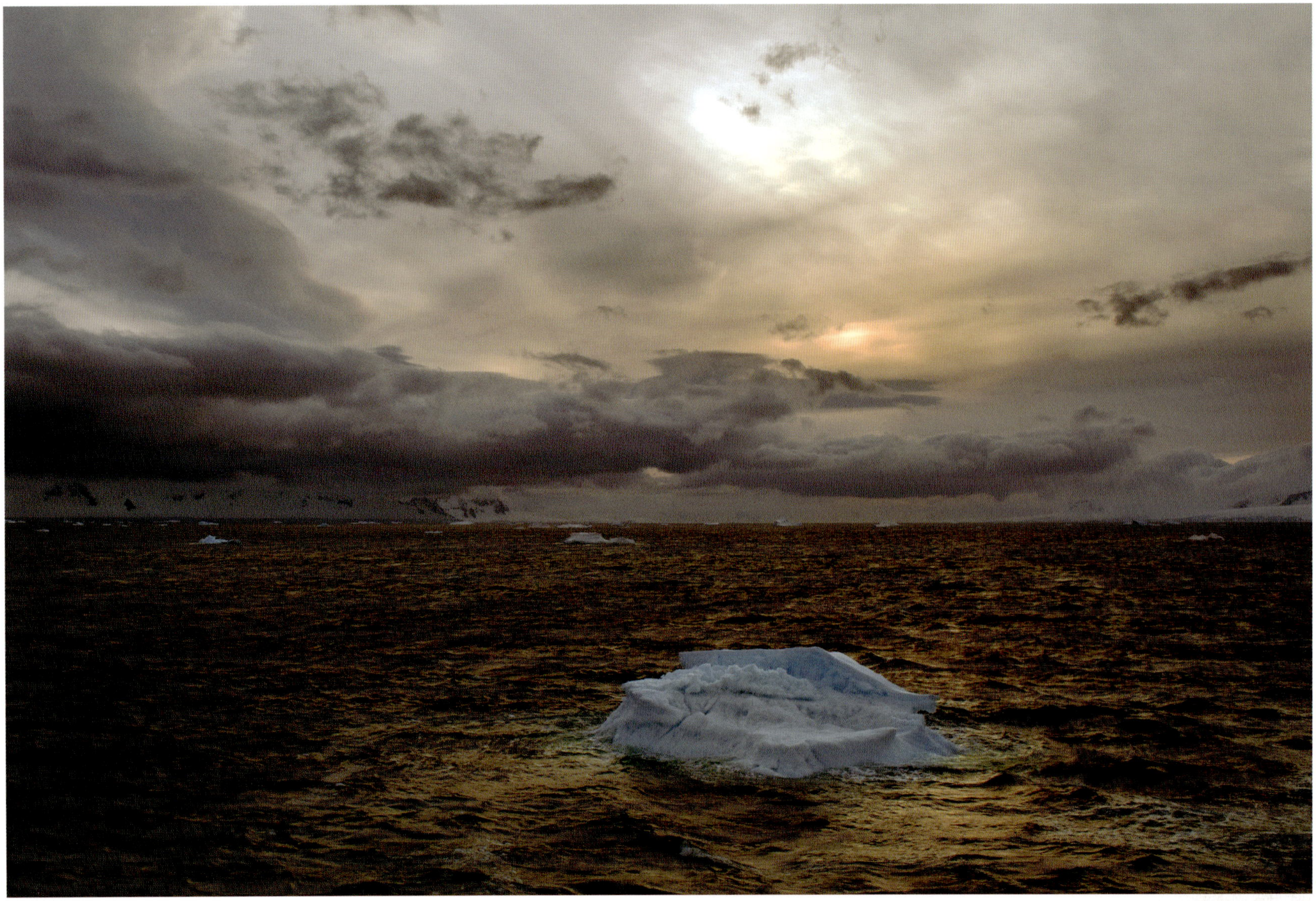

Blue growler

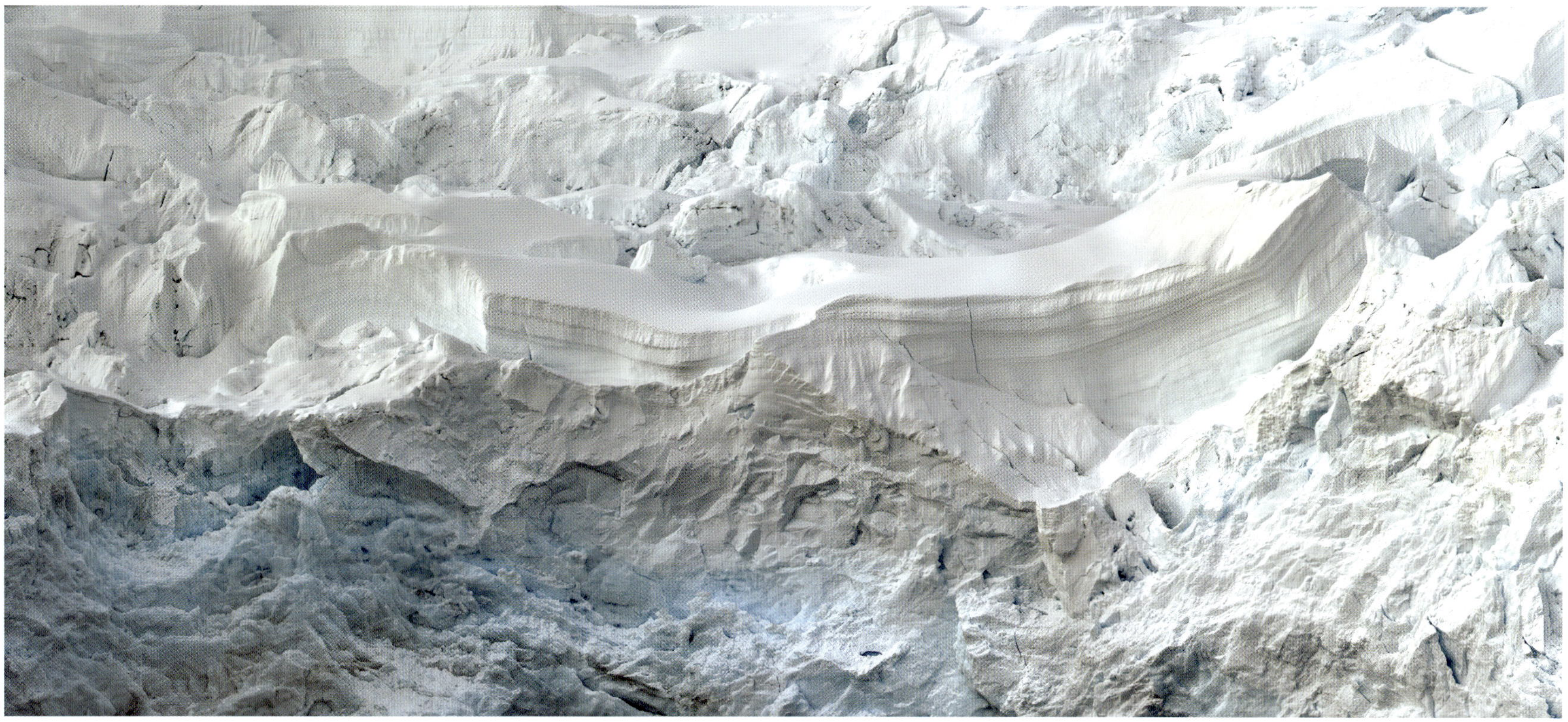

Glacier front detail

IN THE LAST twenty five years I have travelled in the Arctic and Antarctic and had many opportunities to take pictures of icebergs. Initially they were compelling because of their grandeur and novelty. I took many. The sky was usually blue, the ice white, and the water dark. However after many pictures they all began to look similar and commonplace. Granted each berg was different in shape and form and some were very attractive in their own right. However I had hoped for bergs that were enriched by mood and composition.

You can imagine my excitement years later when drifting slowly through the fog phantom shapes moved towards me. Diffused outlines and soft delicate profiles loomed on all sides. The air was perfectly still. Only our ripples disturbed the glassy water.

This was the light and opportunity I had been hoping for. Initially I thought that creating a strong composition and exposing for the muted light was my only challenge. I had forgotten about the waves and ripples of the boat that would creep into the image caused by the changes in direction or of our slowing down. Approaches were therefore planned and purposeful. There could be no second take.

And now there came both mist and snow,
And it grew wondrous cold:
And ice, mast-high, came floating by,
As green as emerald.

Samuel Taylor Coleridge

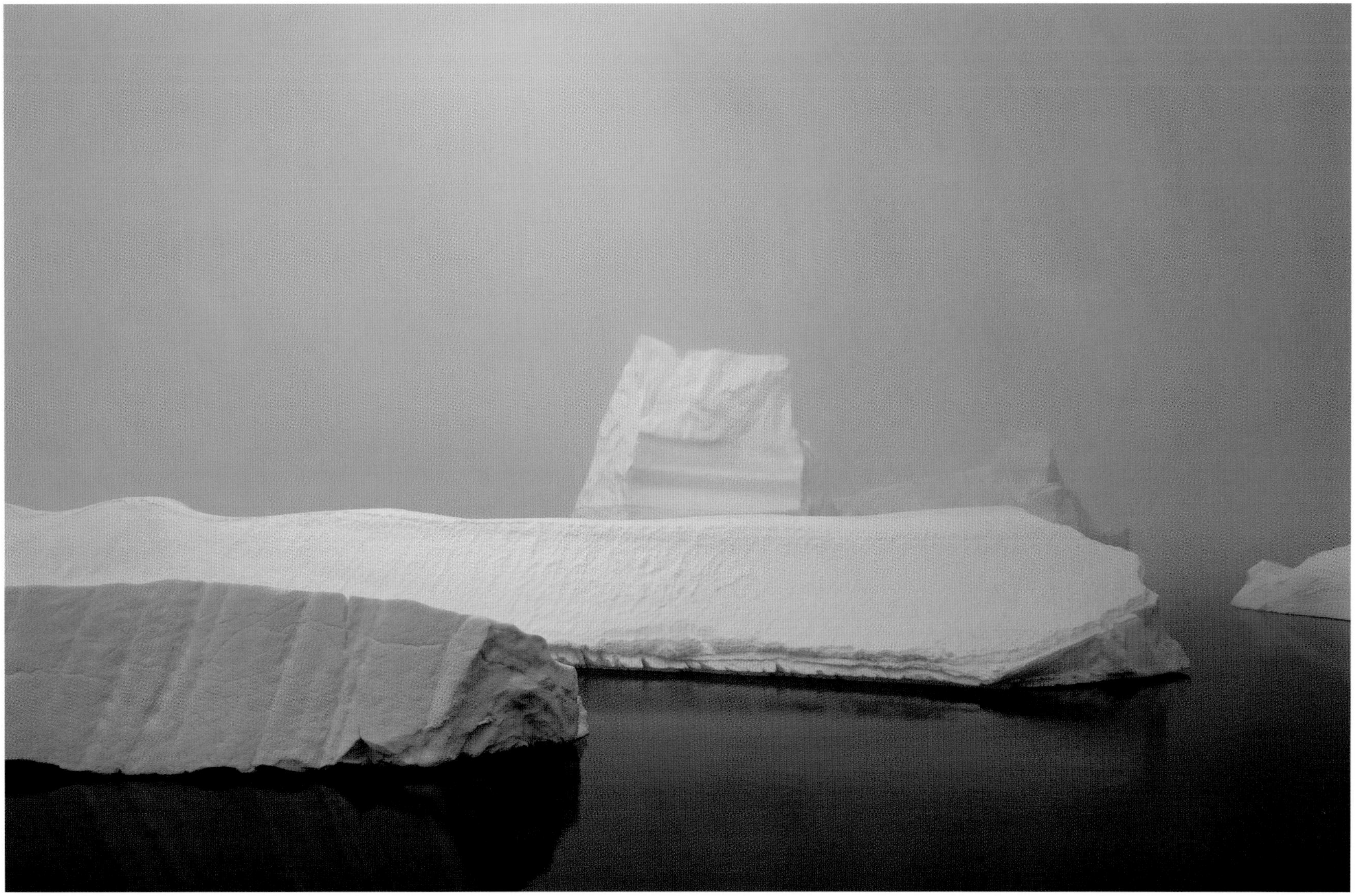

Water and light

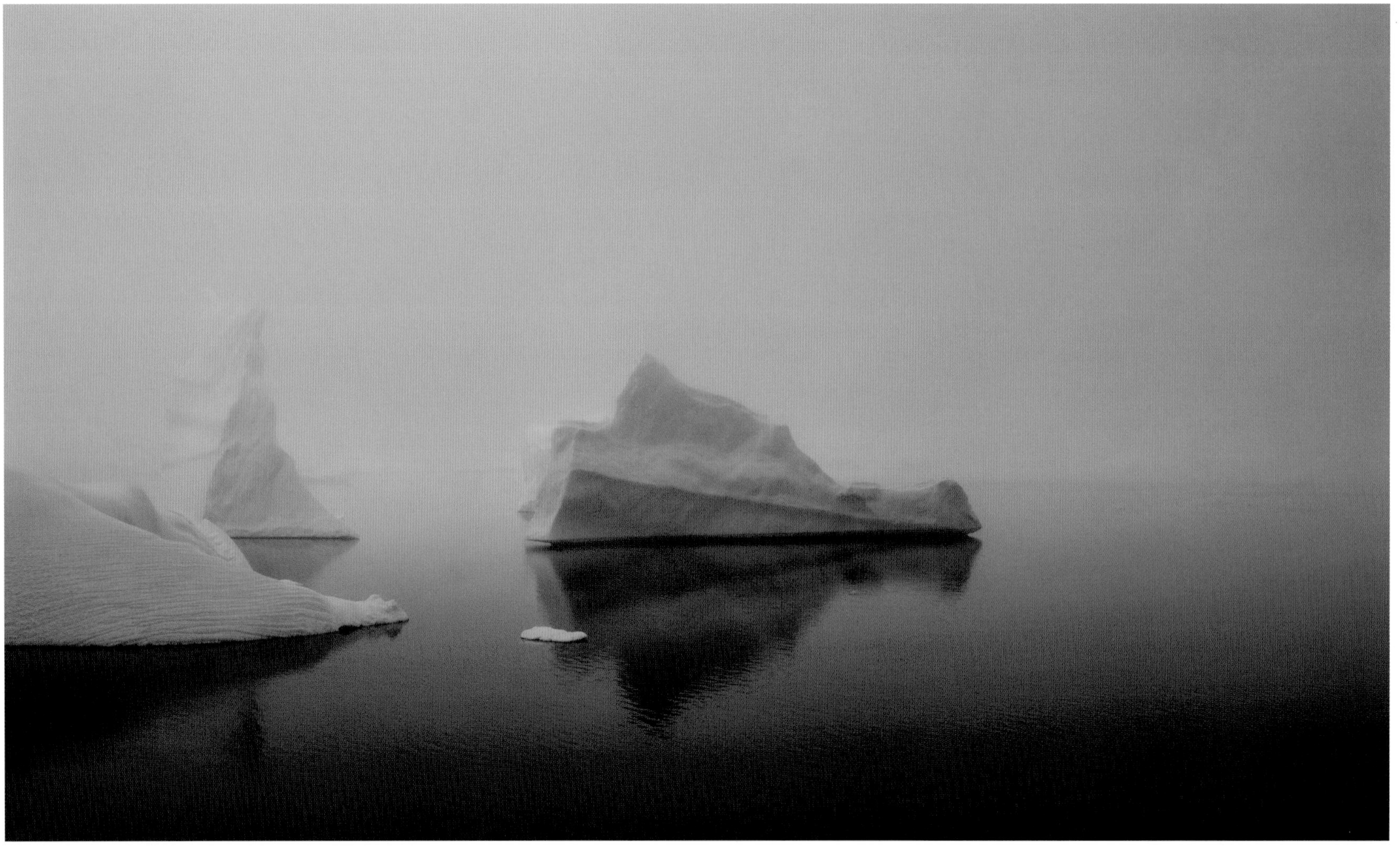

Changing light

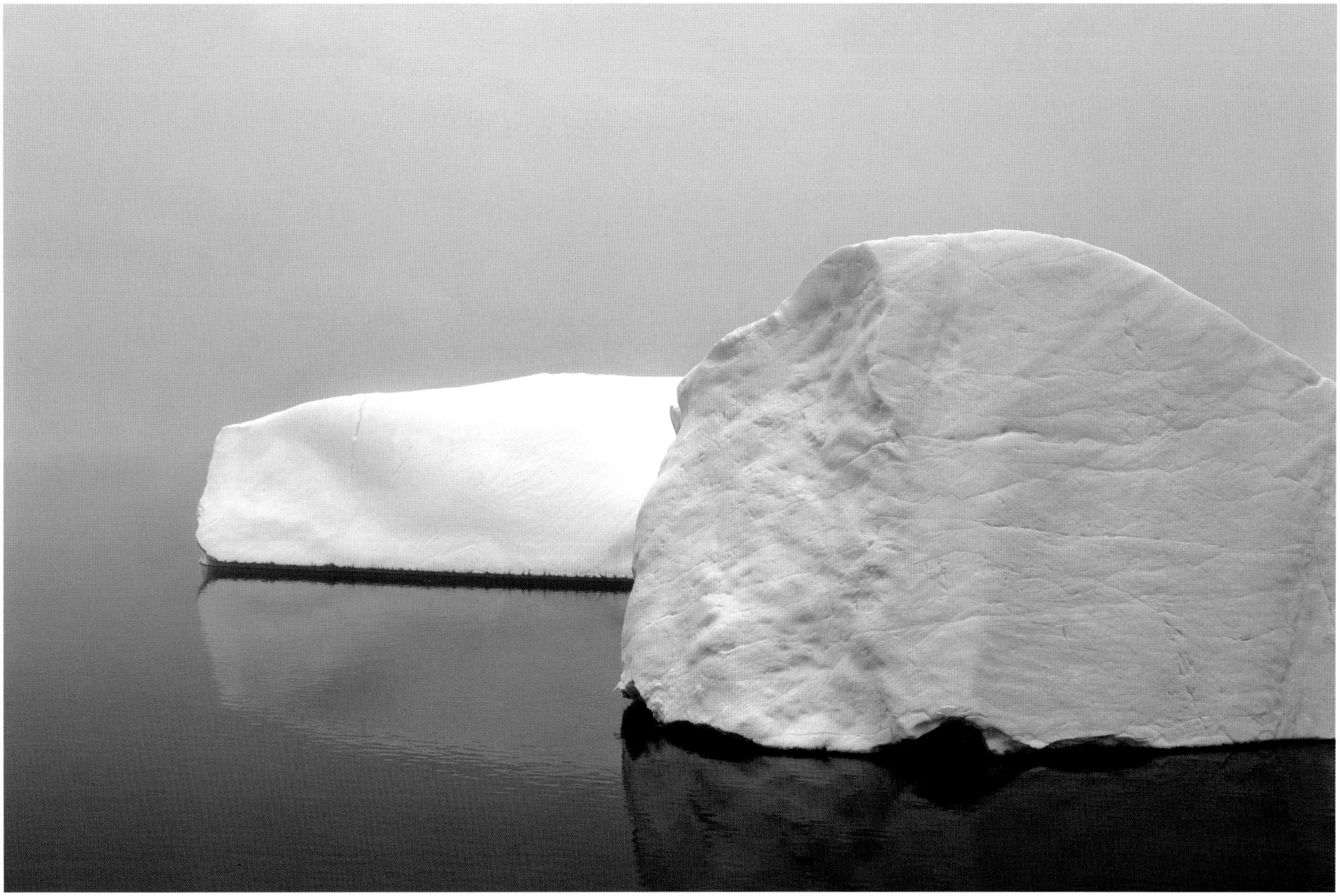

Two bergs

ANTARCTICA IS KNOWN for its tabular icebergs and here was a fine example.

I had been patiently waiting and watching this nondescript massive slab of ice slide by when a white light filtered by the mist spread over the horizon. Immediately it affected a change and the broad side of the berg glowed a brilliant white and the side in shadow was diffused with a transparent blue. A white sliver blaze spread over the berg and separated it from the darker sky.

The iceberg now had form and texture. I took a reading and made the picture.

Instinctively I examined the histogram on my camera and looked up to notice that the fog was thinning. Over the dark water the horizon took on a purplish hue and above the sky brightened. I took another picture. A minute passed when the swirling cloud and fog suddenly revealed a gloomy mountainous shape high above the horizon. Surprised with the height and proximity of this massif I took yet another picture. More satisfied with this image I now held the camera against my chest and enjoyed the varying light contemplating this eerie scene. I felt that I was done and nothing more could be added to this primeval spectacle when unexpectedly from out of the sterile mist appeared the ghost of sailors past, an albatross. It inspected the edge of the berg, dropped low and disappeared.

It is interesting how the Antarctic and the albatross have become amongst other things associated with Samuel Taylor Coleridge's poem, T*he Rime of the Ancient Mariner*. The famed albatross in the poem was well known to science but the Antarctic continent would not be discovered for another twenty years.

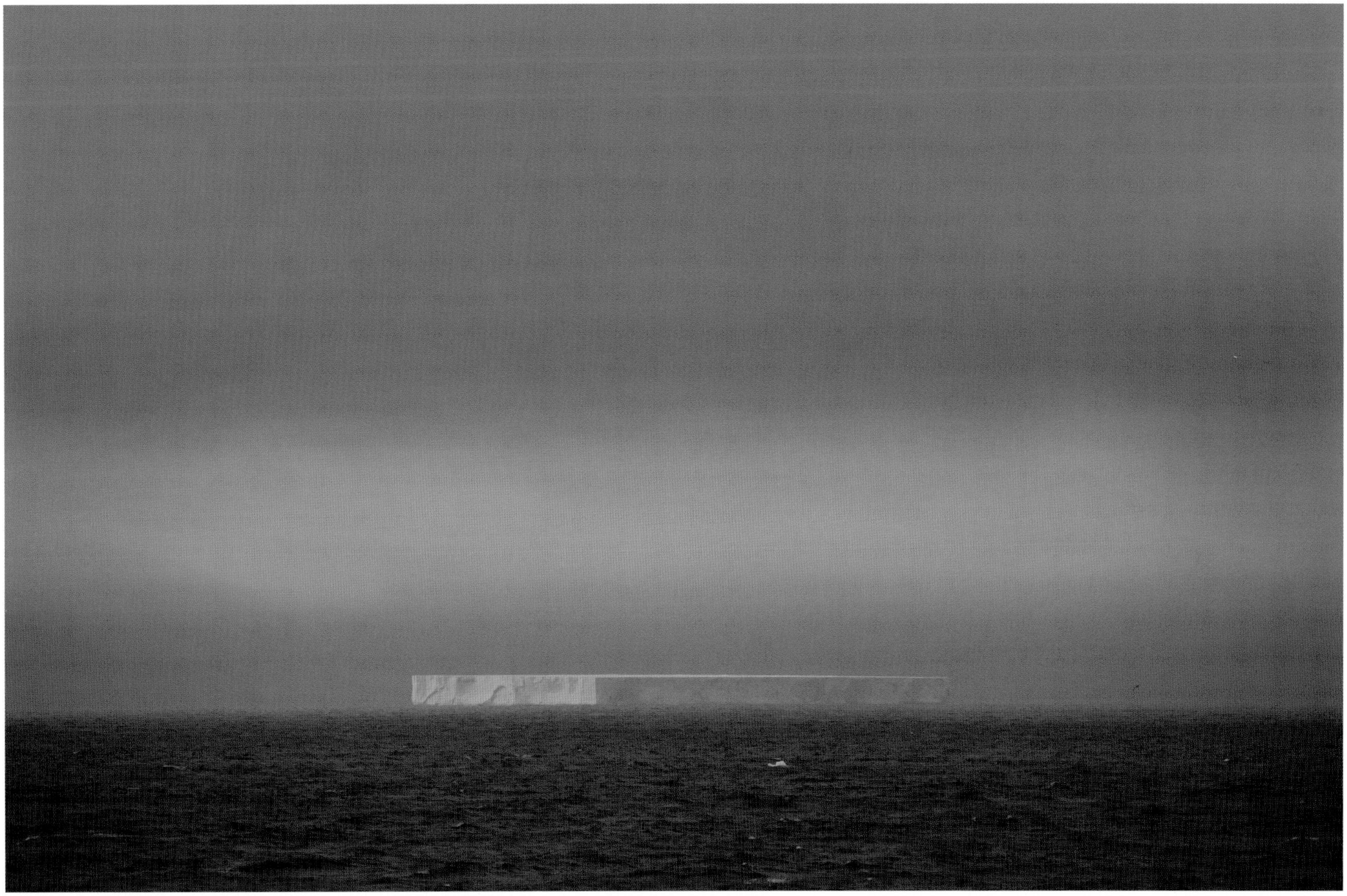

And out of the mist… the albatross

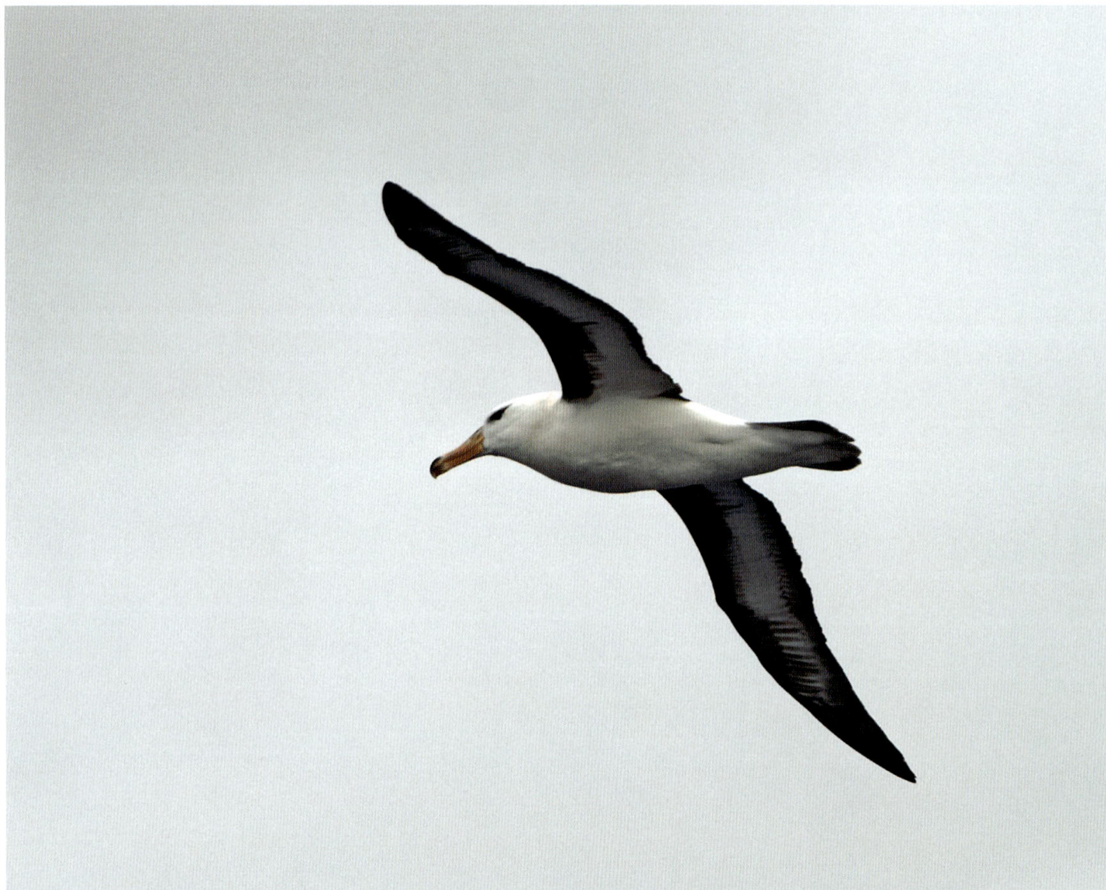

Black browed albatross

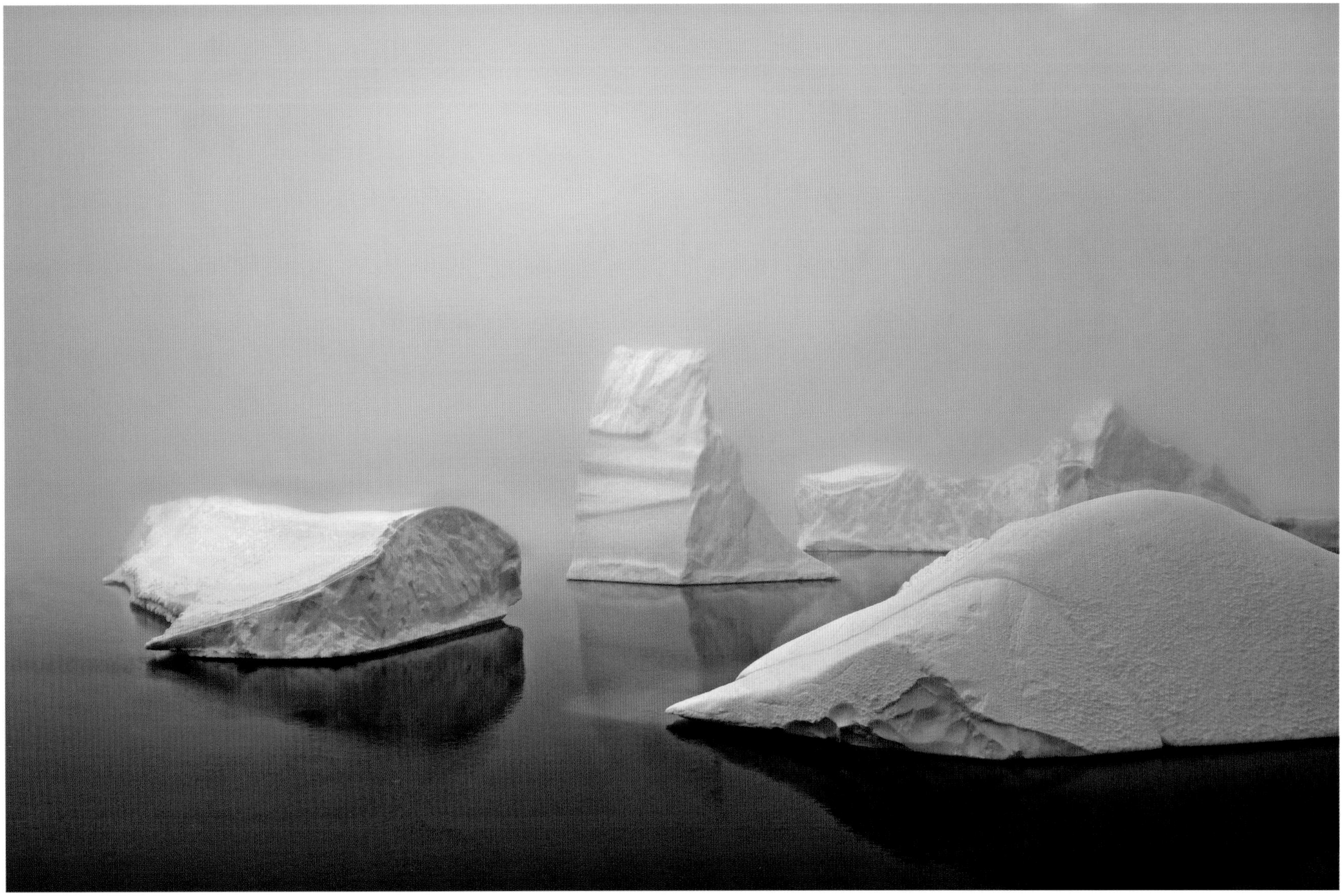

Misty horizon

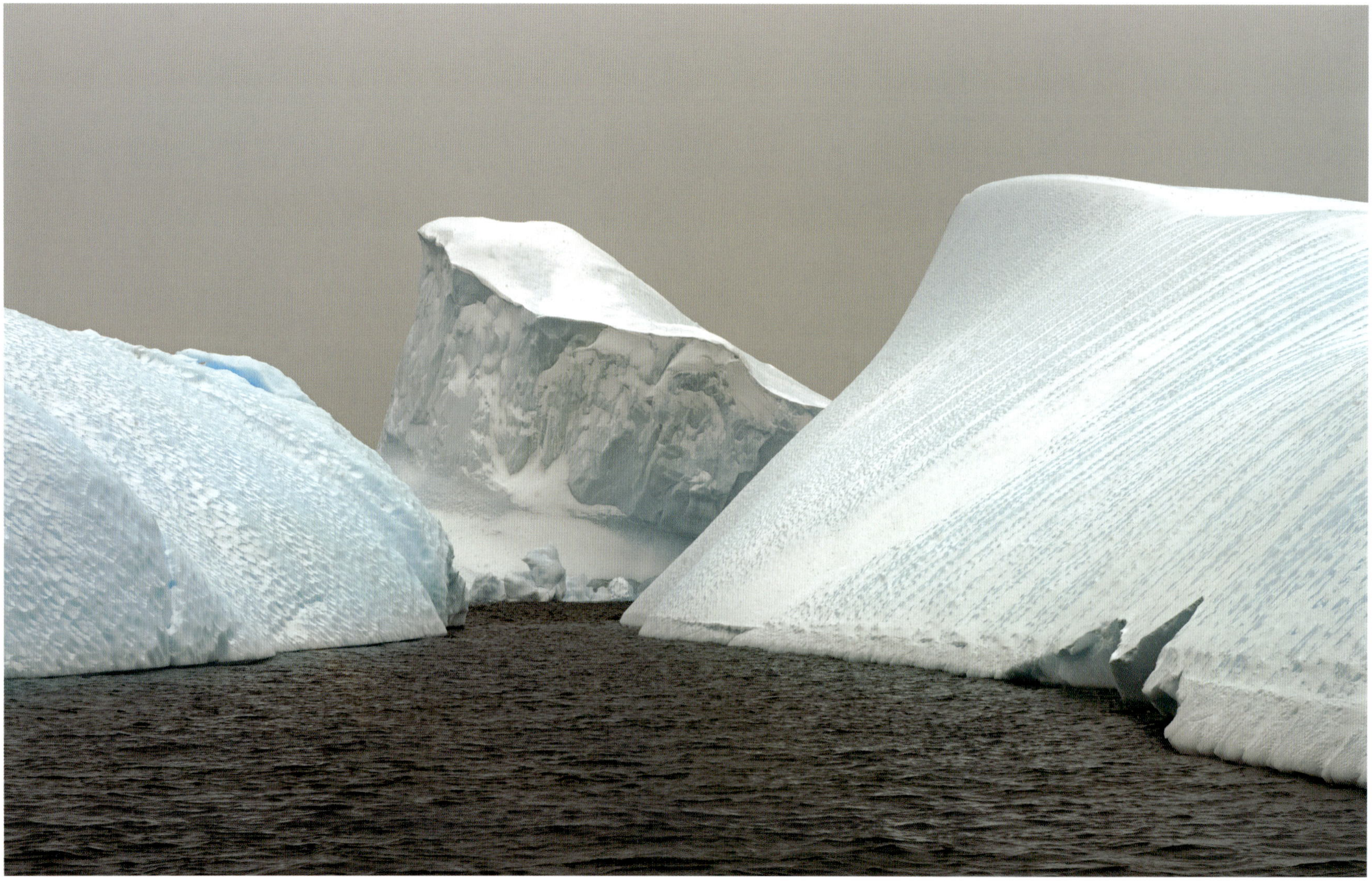

Scene 3

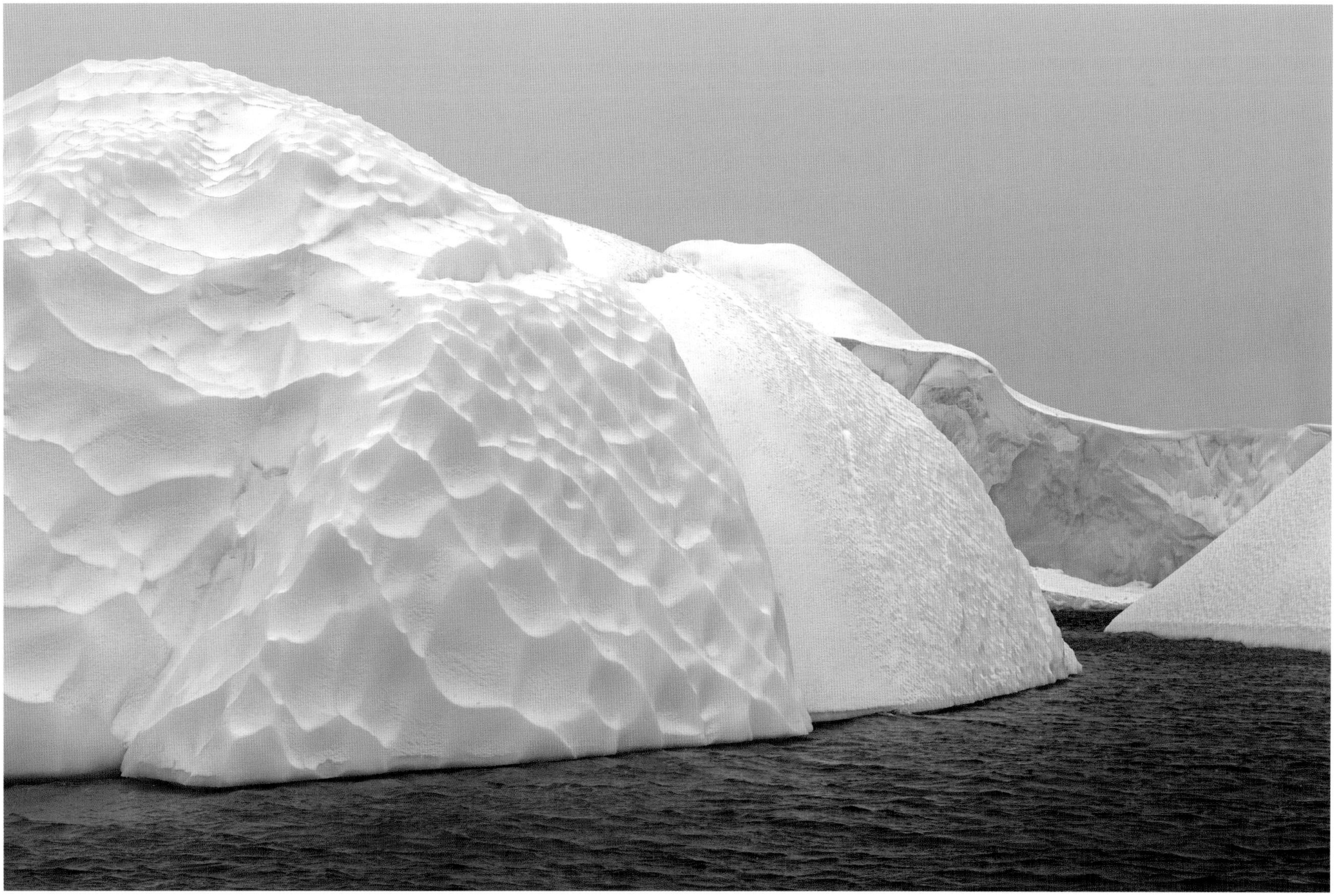

Scene 4

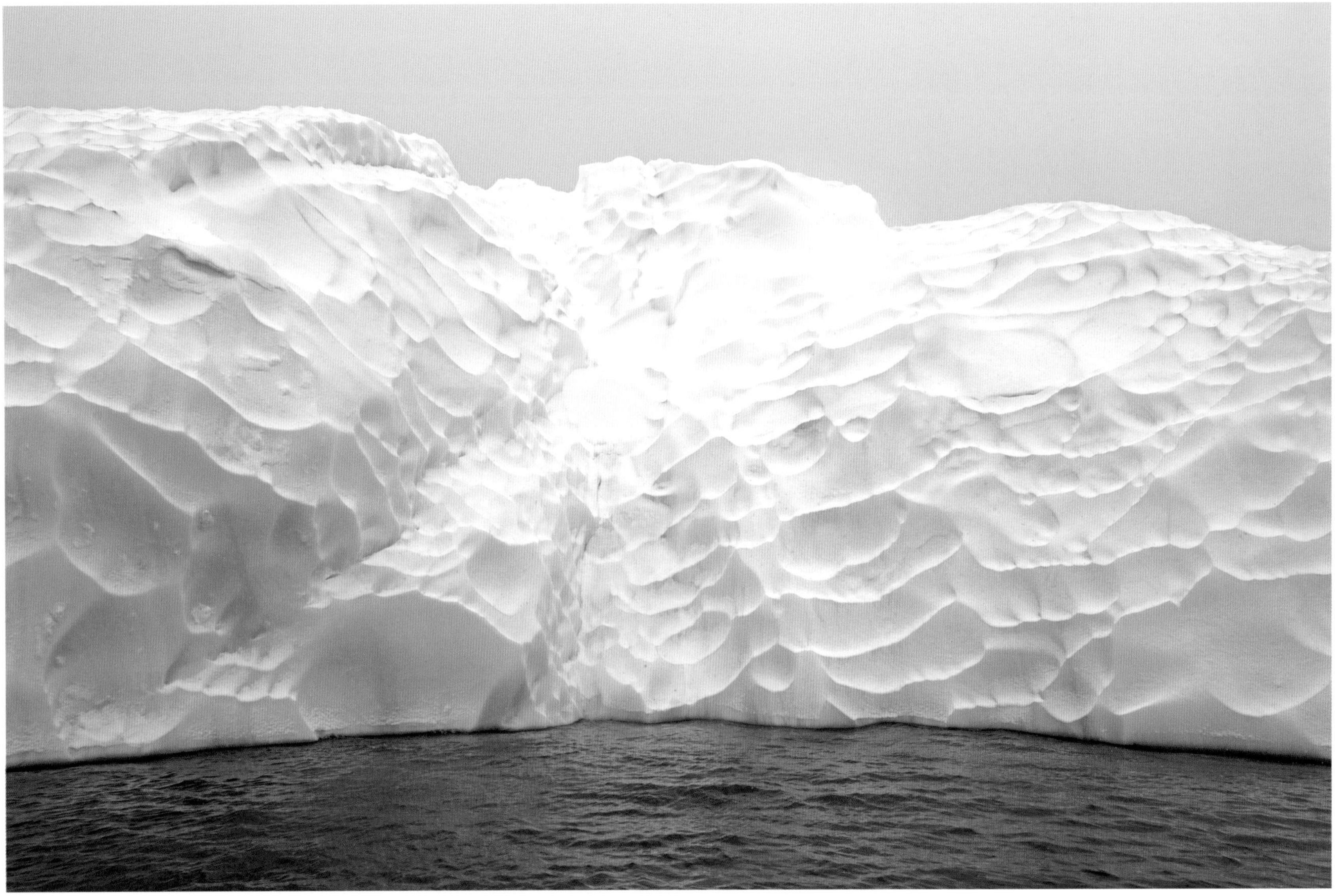

Scene 5

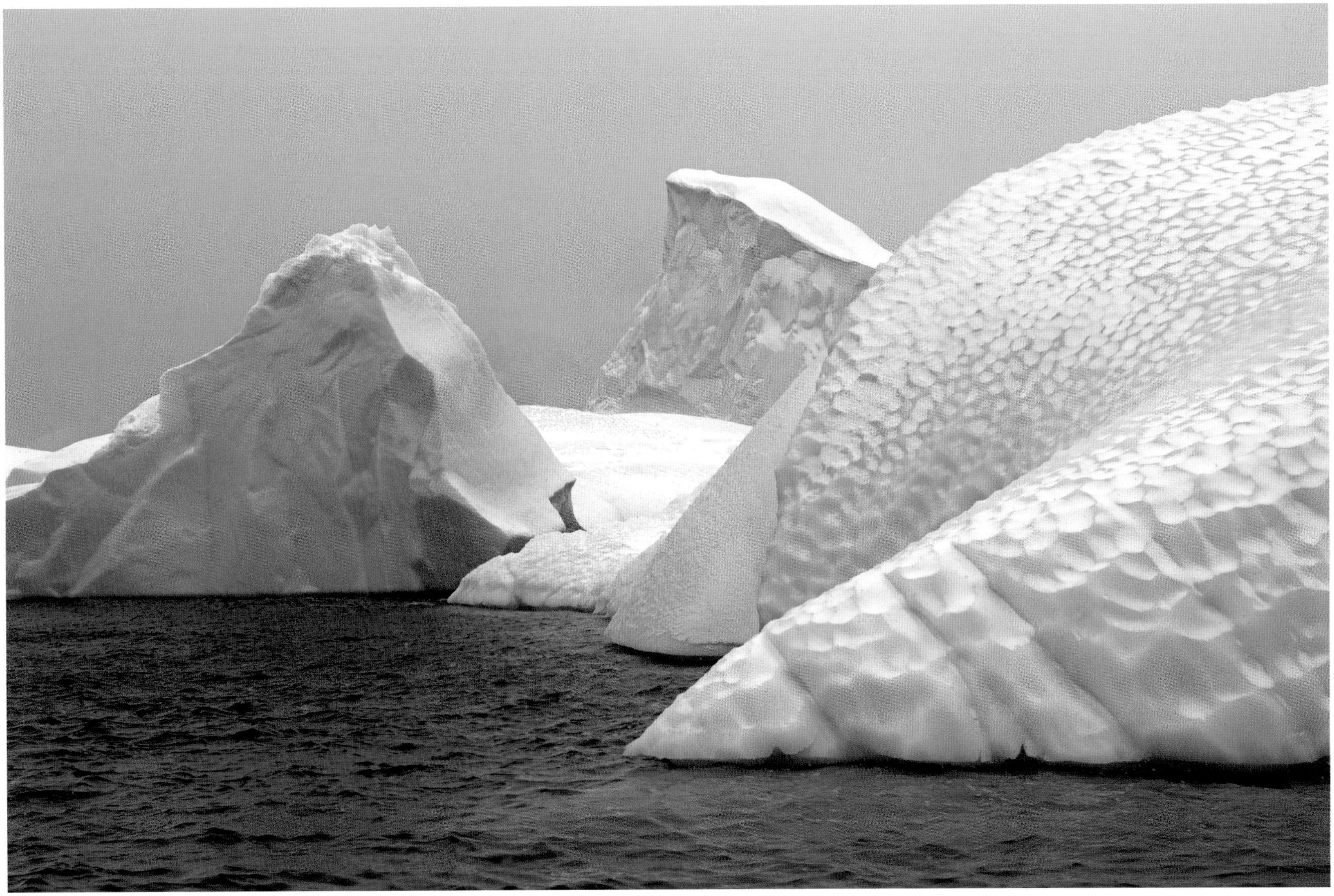

Scene 6

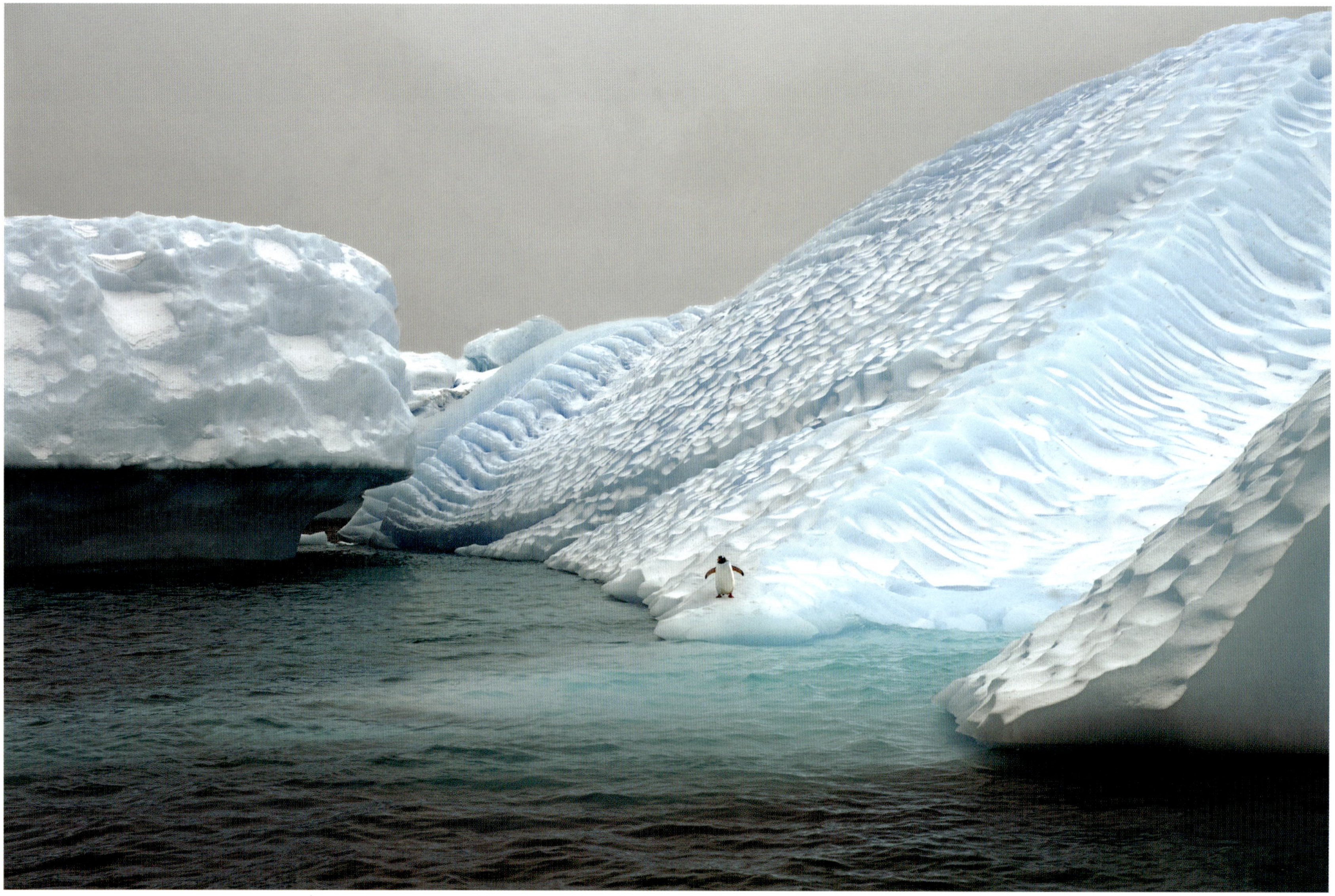

Scene 7

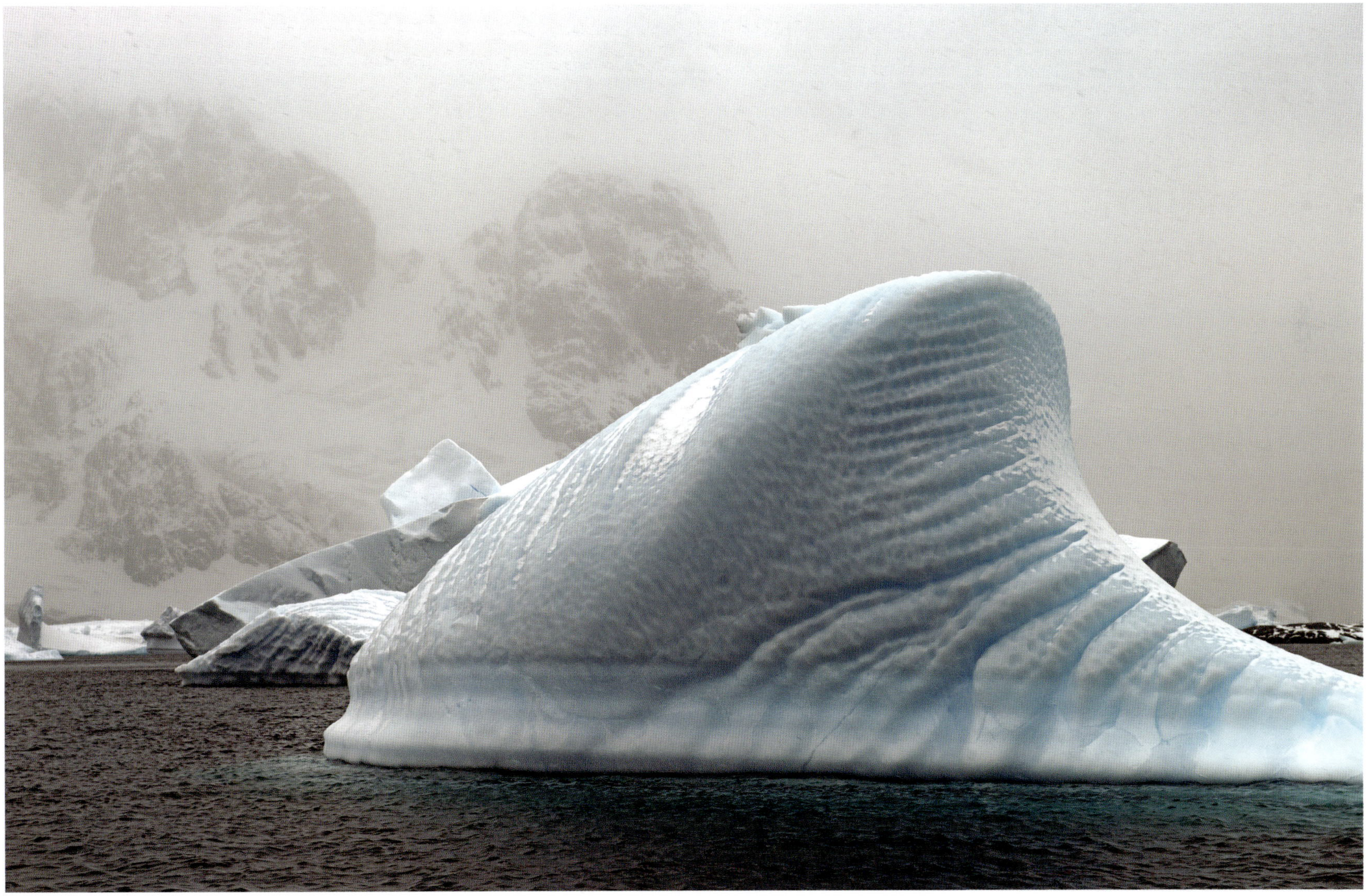

Scene 8

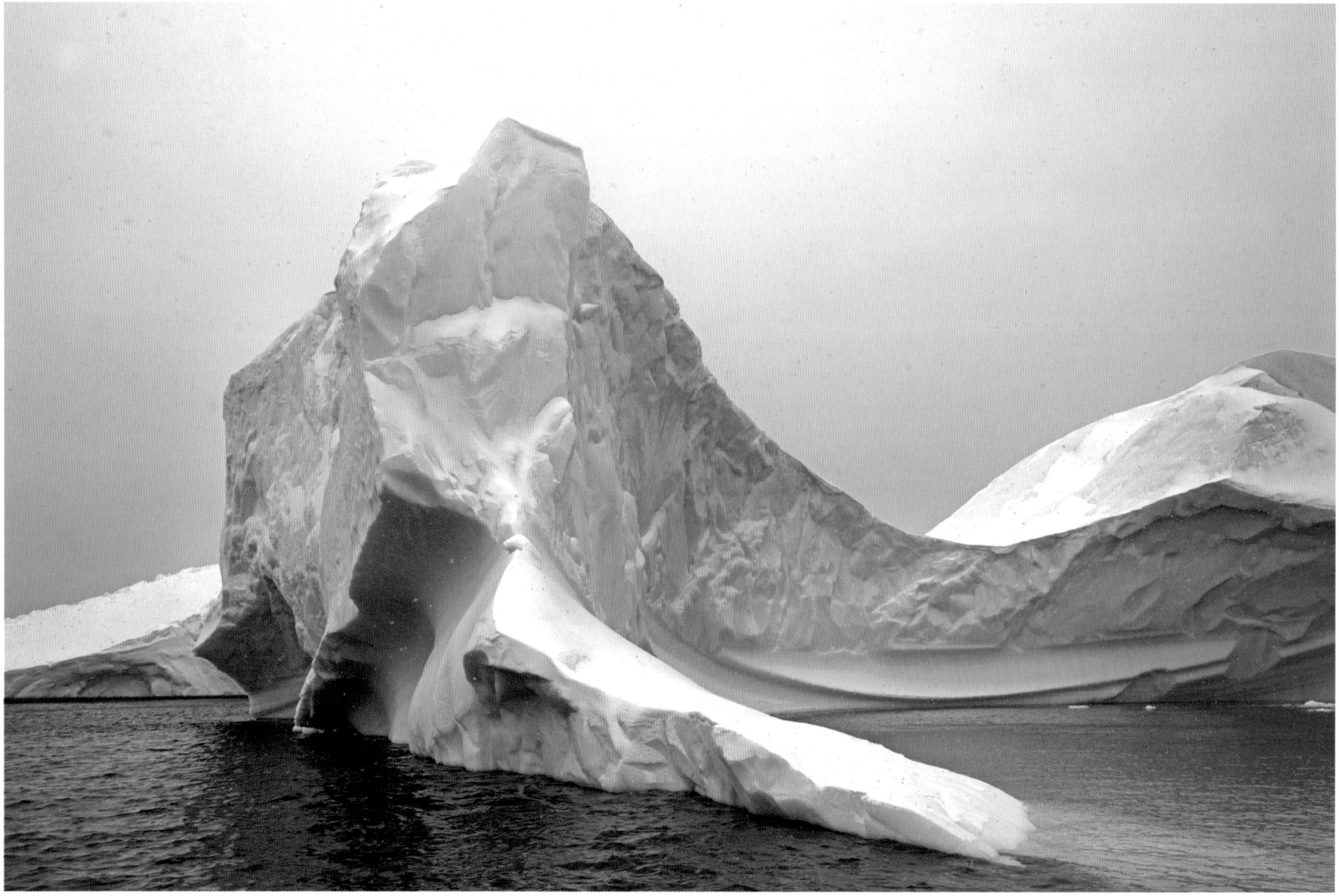

Scene 9

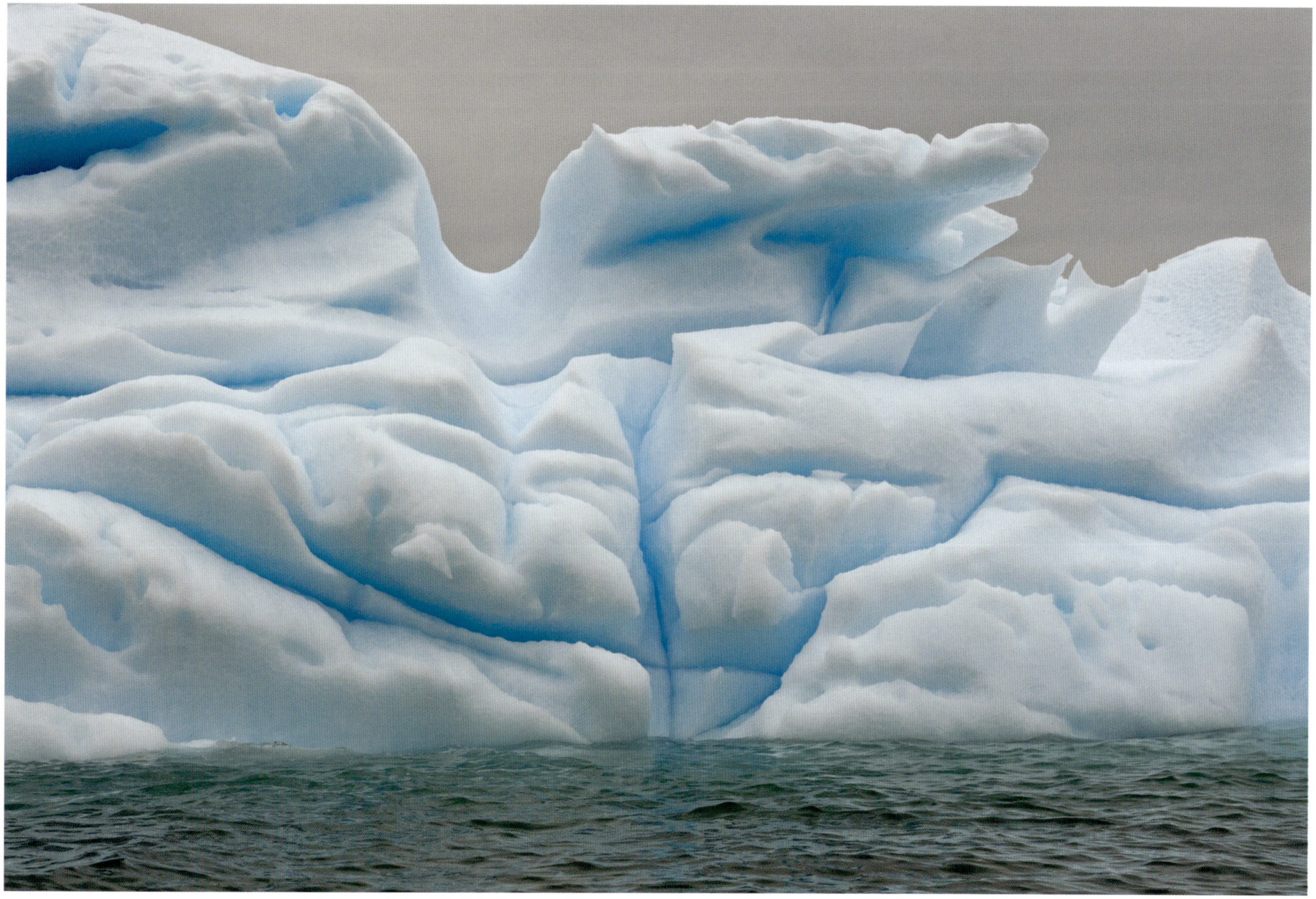

Scene 10

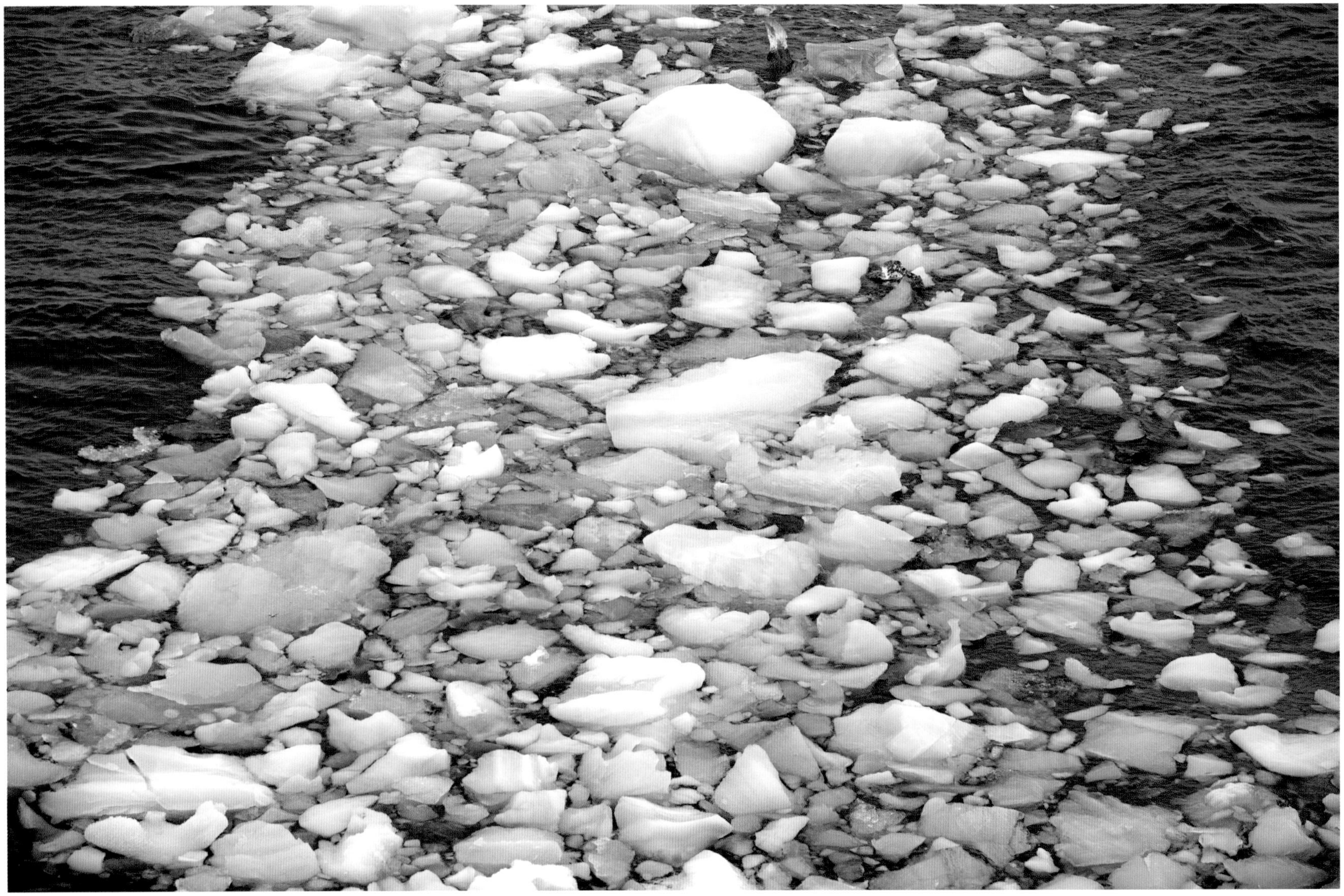

Brash ice

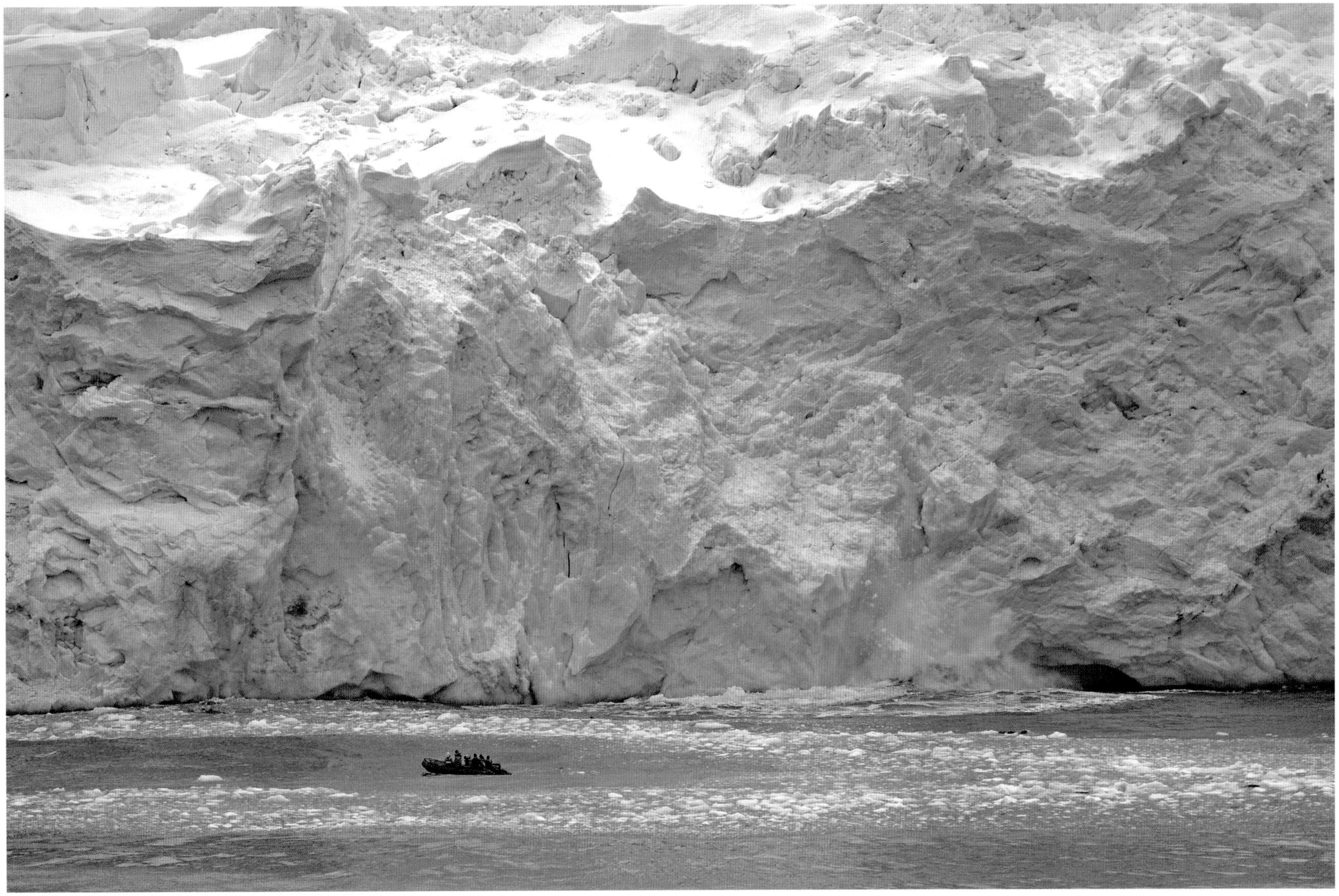

Zodiac and calving glacier

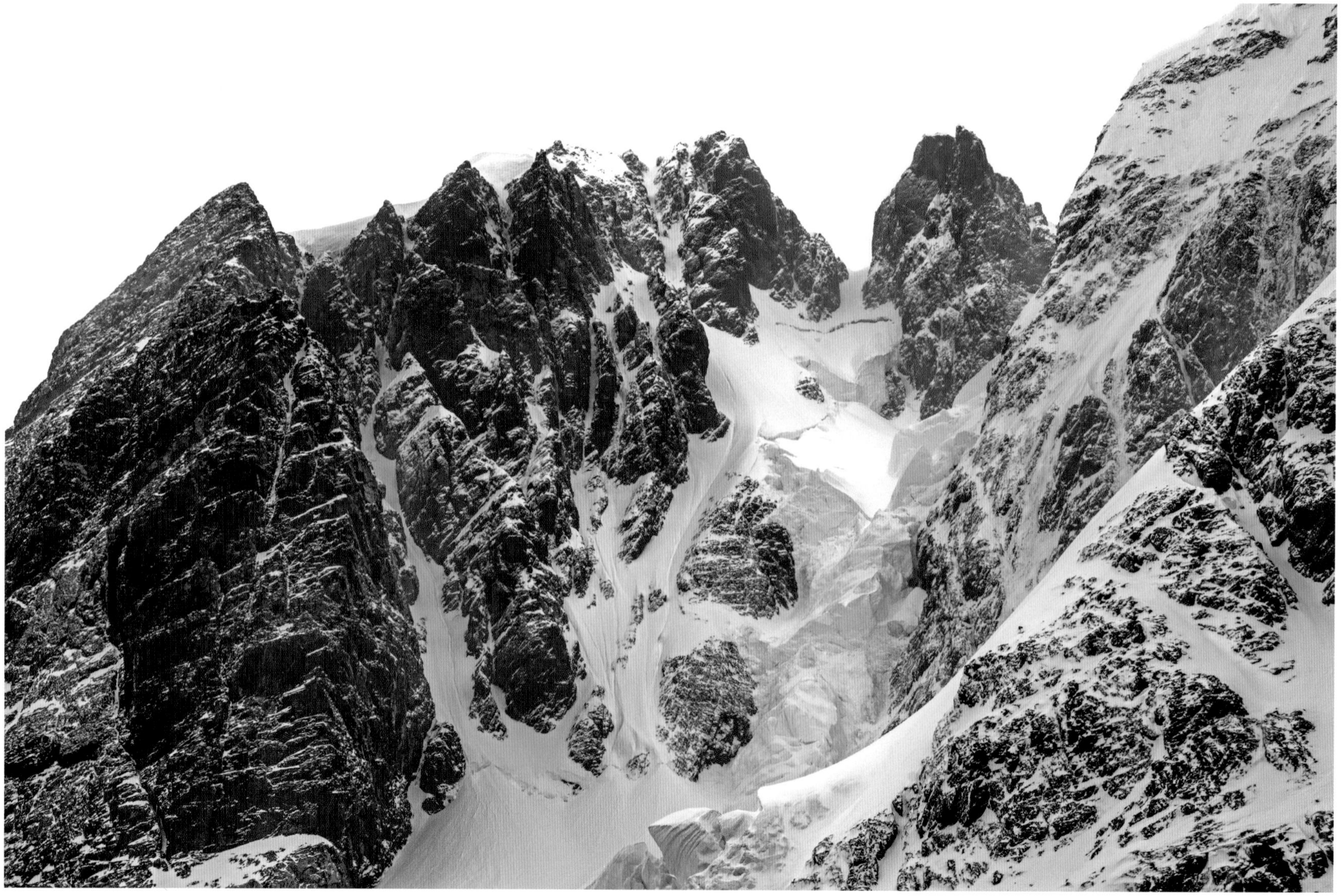

Dark rock

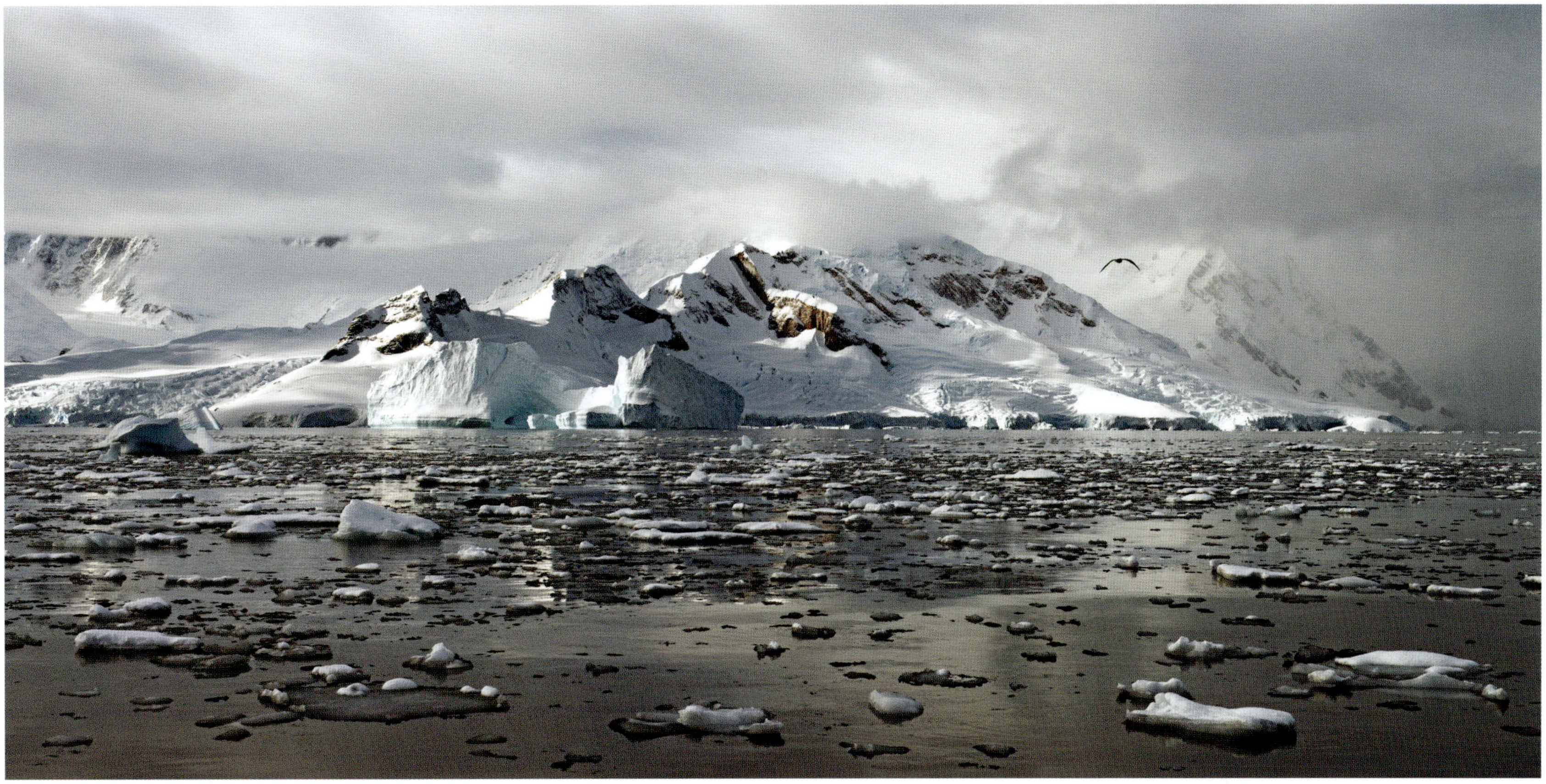

Homeward flight

I HAVE LANDED here three times before hoping for the right light and the weather conditions to go with it. Twice the sun dominated the landscape but I did not want a bright blue sky to compete with the deep pigments of the coloured scoria and black ash nor did I want to lose the more subtle colours that such a landscape could offer.

Two previous visits failed in a landing due to the inclement winds making even a stable Zodiac landing too hazardous. The light was right but not the weather conditions. So just being there was not enough. Things had to come together. In this case five visits and some luck and persistence improved my odds.

The overcast sky and the subdued brightness on this day made what was in previous visits just interesting, into an image that was now intriguing. It has a very painterly effect. Bold strokes of red ash stream down the mountain and delicate greys find their way through the snowbanks. On top, in the middle of the picture, a vast snowfield sends up its own light into a similar sky and in the foreground between two shallow streams coloured ash punctuates the ground.

Examining the picture your eye follows a subtle "S" curve movement; from the top of the red mountain moving to the right and then left to the meltwater stream flowing hard to the right. A pleasing effect in many pictures.

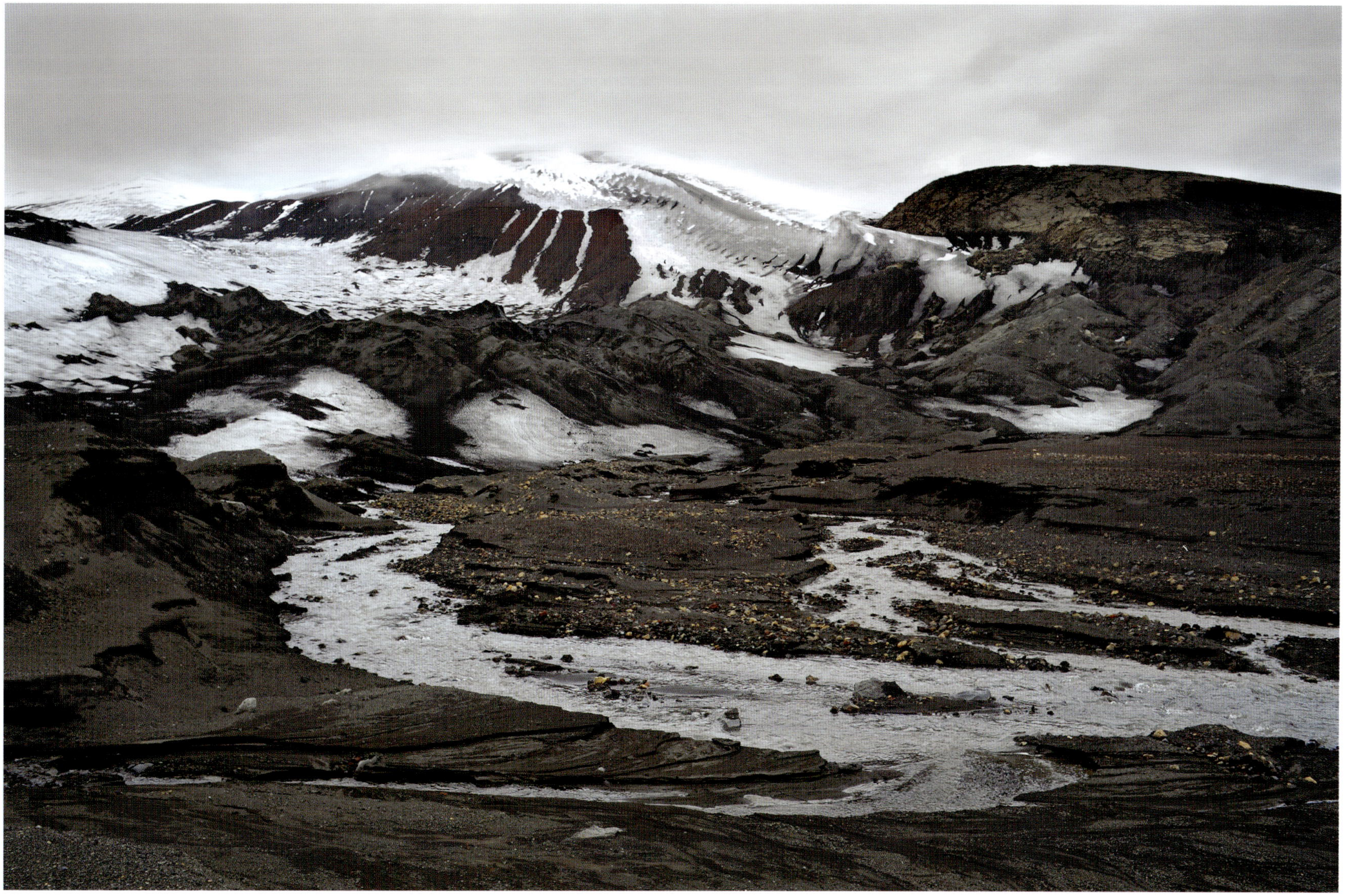

Volcanic rim and glacial meltwater Deception Island

ONCE ON BOARD ship you get a different perspective of the landscape. Being farther away and higher up reveals a different image that can just as well be a close up of an inky oil spill or a potentially destructive slurry of pyroclastic material on the slopes of a volcano. Steely grey water adds darkness to the already primeval elements and I am reminded of how unpredictable and destructive this place could be. Ships entering here into this flooded caldera have reported water so hot that the paint was peeling off their hull.

Last major eruption was in 1970.

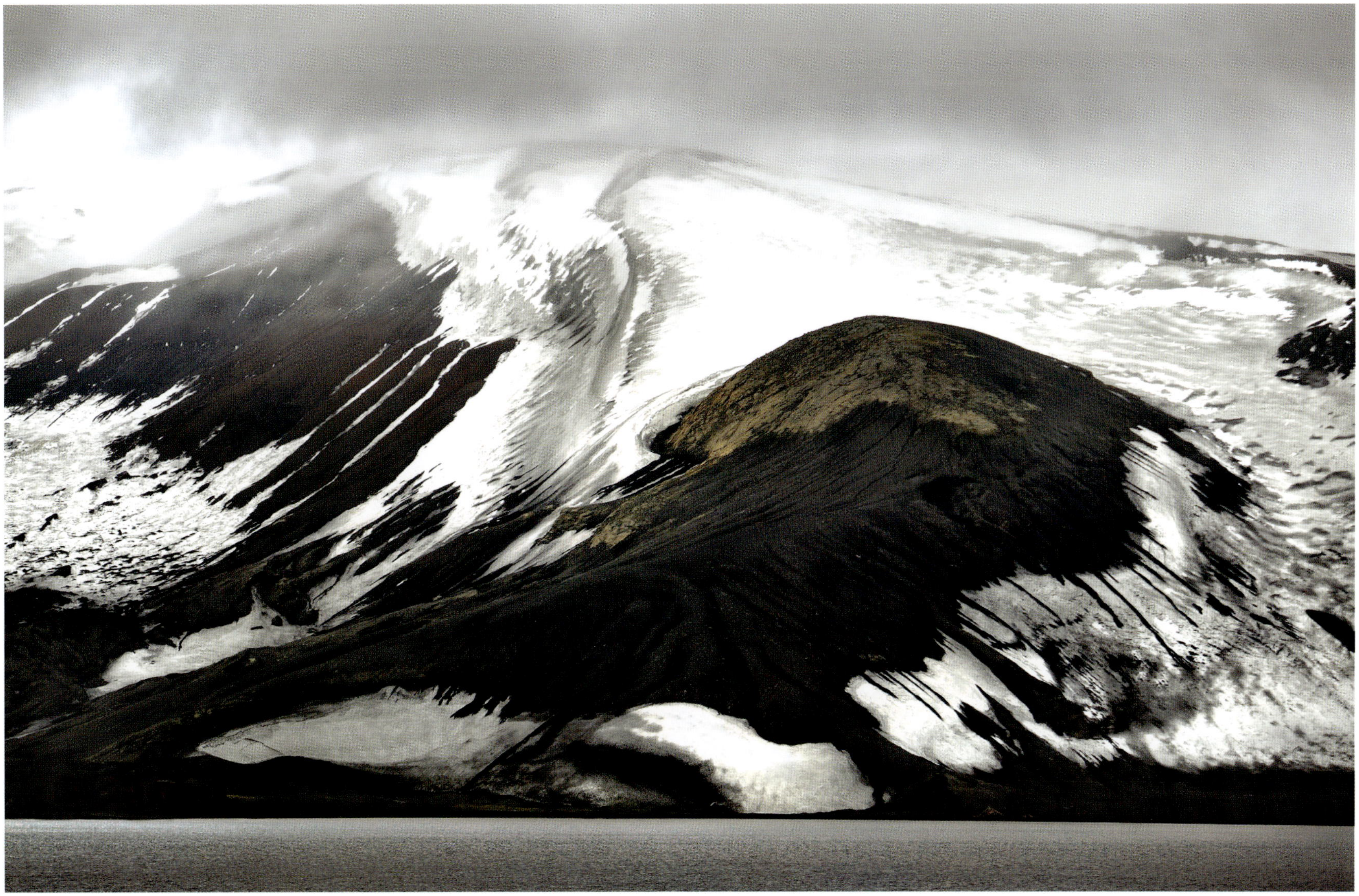

Cinder spill and lahar

THERE ARE NOT too many places in the world where you could sail a ship into an active caldera. Here at Deception Island in Antarctica you can. Part of the South Shetland Islands chain and separated by the Bransfield Strait from the Antarctic peninsula, Deception Island is a large volcanic crater that rose above the sea floor and then partially collapsed. A weaker section of the volcanic rim fell into the sea leaving an opening approximately 390 metres wide. Through this gap, called Neptune's Bellows, the ocean poured in flooding the ash filled crater forming a 6 x 8 kilometre harbour.

Captain Osiroff of the ship *Ushuaia* explained that for now the entrance is navigable but extreme caution must be observed because in the middle of the entrance part of the sea floor rises to 7 feet below the surface. Smaller cruise ships however can enter by hugging the craggy cliffs to the starboard. Two

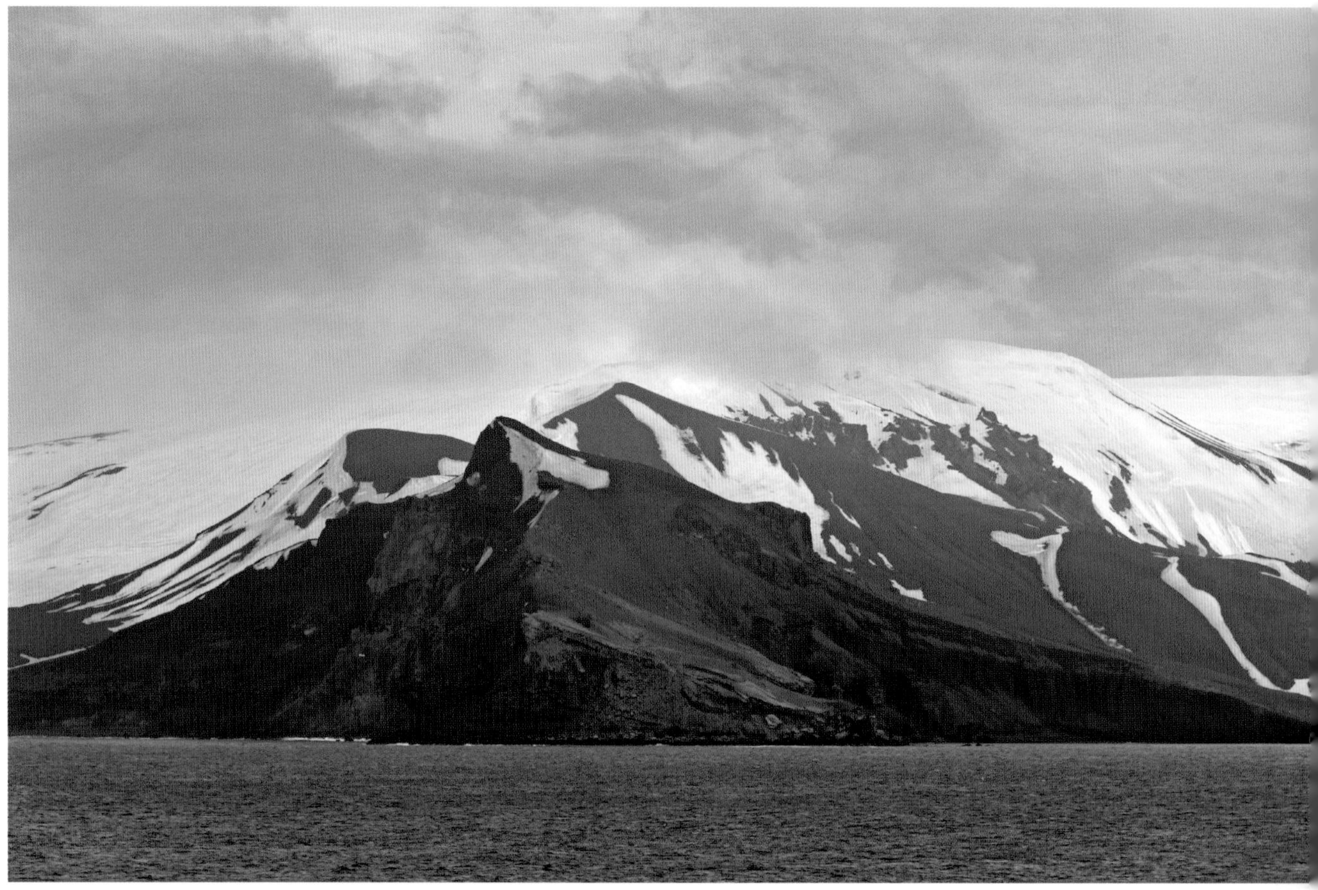

Entrance to Deception Island, Neptune's Bellows

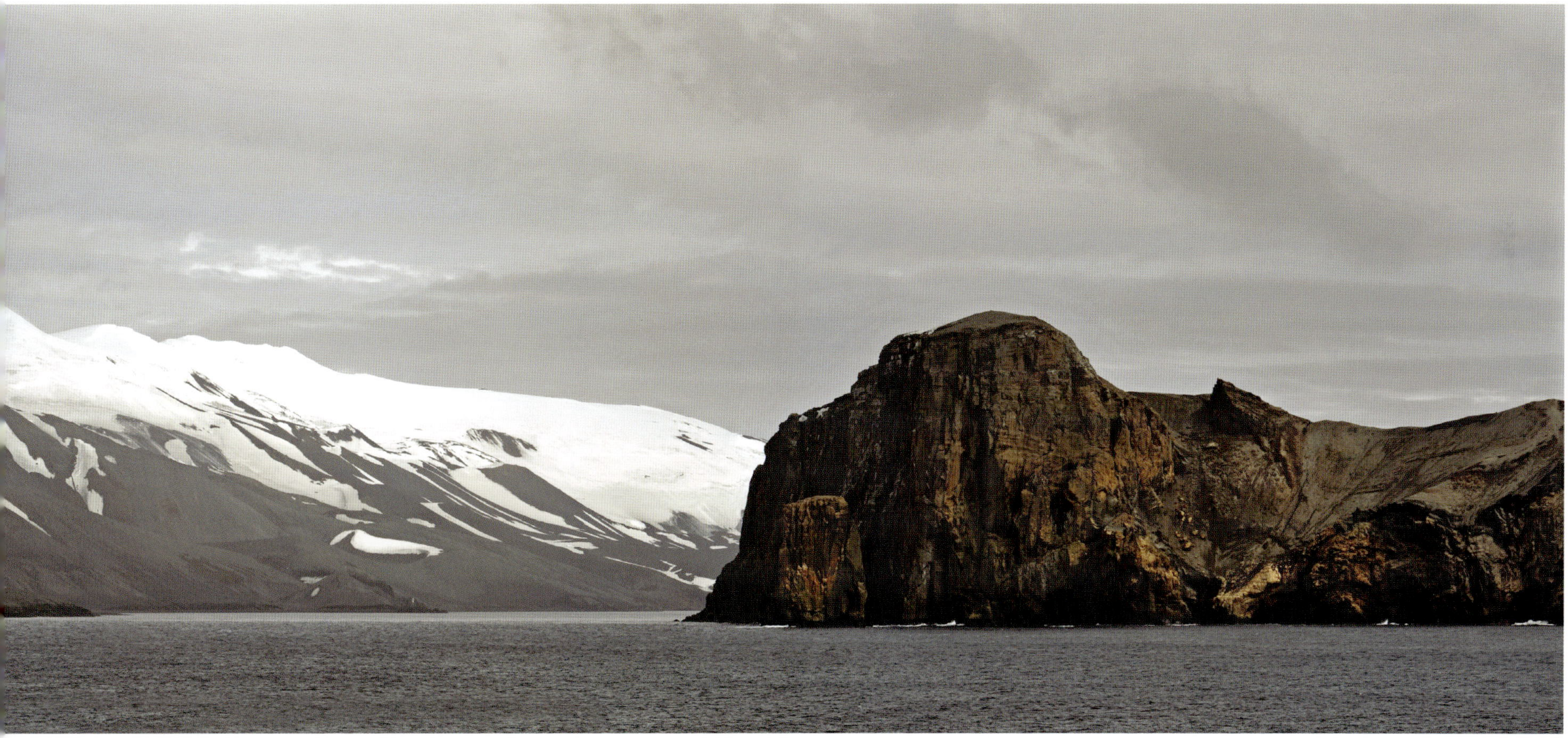

research stations from Spain and Argentina keep a summer vigil monitoring the volcano's activity. A good shake could easily change the caldera's configuration and trap a visiting ship. At the end of the season the research stations cease operations. All is locked and secured and things revert to the way it was centuries ago.

I experienced the proximity and reality of this active volcano years ago when I waded into water a foot deep. It was warm and pleasant unlike what Antarctic waters should be. I stuck my hand deeper into the gravelly beach and pushed down till it reached my palm. I immediately pulled back. It was like touching a hot stove.

Outer shore Deception Island

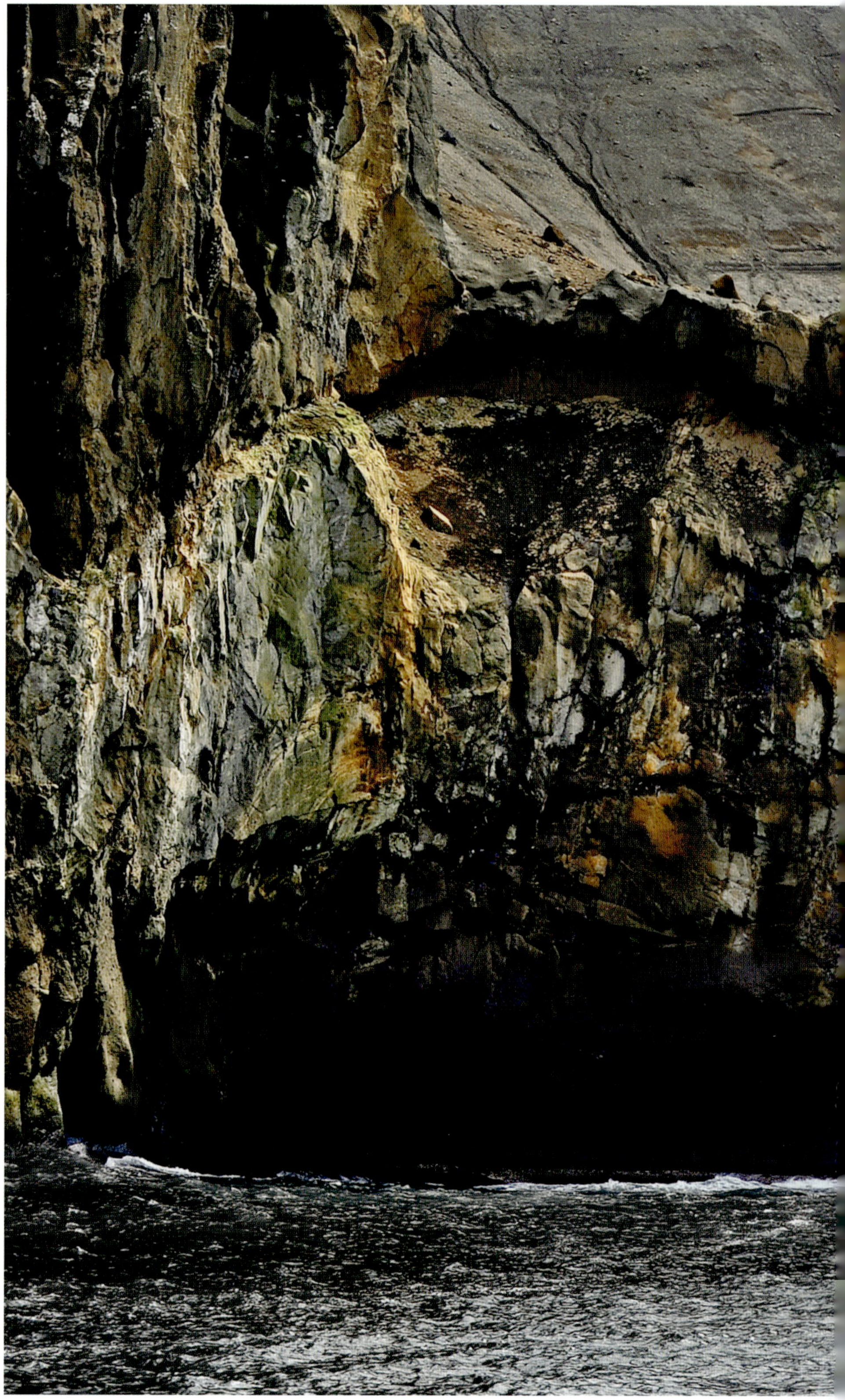

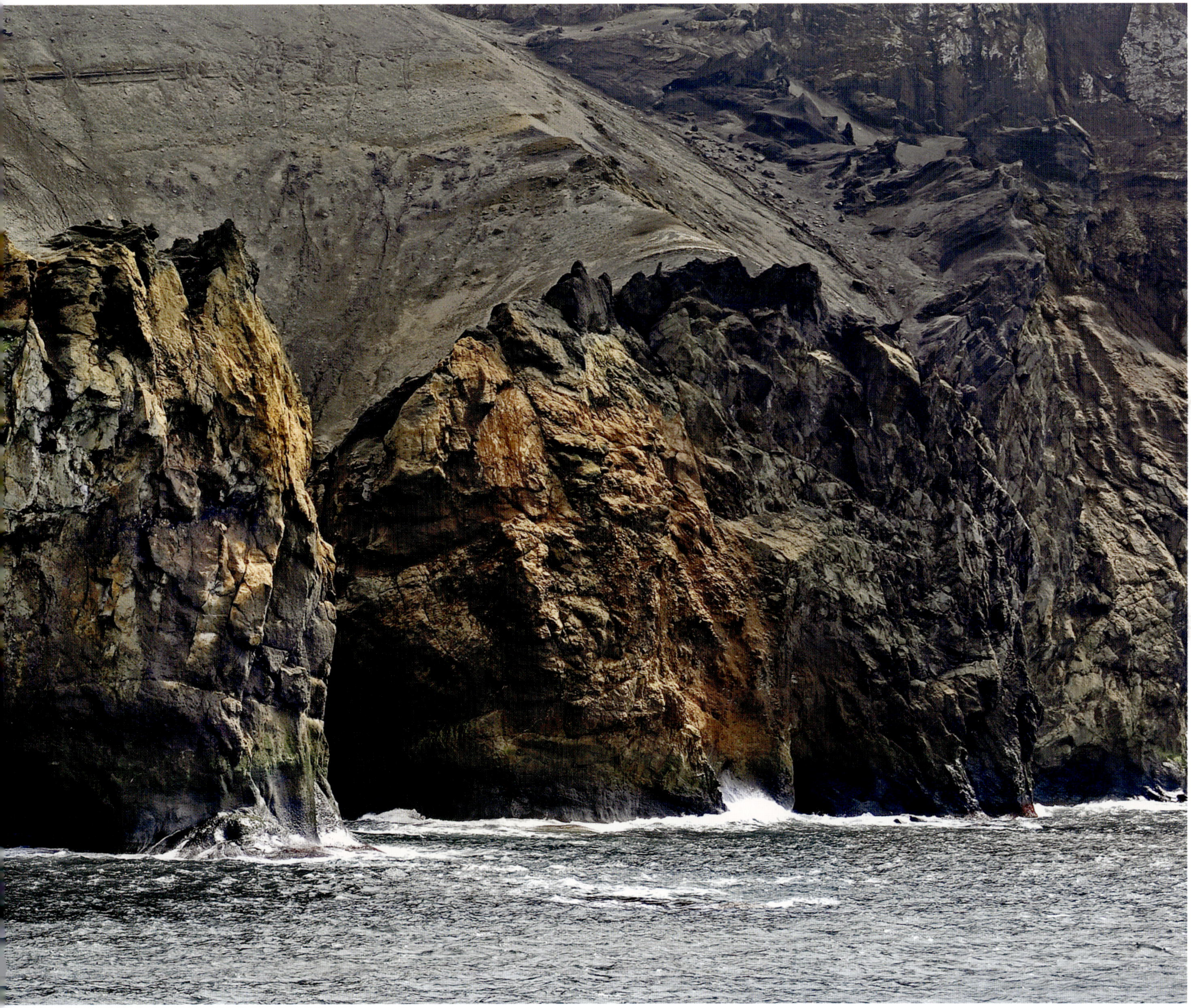

BEING ALONE IN a new environment can energize and inspire creativity. Undisturbed on this Antarctic beach the photographic possibilities seemed endless. Every shape and nuance of colour was motivating. Wandering the coal black beach the coarse ash would crunch beneath my feet and lead me around ejected lava bombs the size of trucks. It was hard to get over the fact that I would not meet someone else. After sharing a ship's cabin for a week it again felt good to be alone.

Soon the shore narrowed and the rising tide forced me back to a wider part of the beach where I began. I expected to continue photographing in the opposite direction when I felt the wind rise quickly. I looked out to the horizon to see Grant Redvers bouncing over the waves in the Zodiac anxious to pick me up. Reports of strengthening winds forced the decision. Disappointed to be whisked away so soon I hung on tightly to my camera equipment as we tacked our way back through the spray and choppy sea.

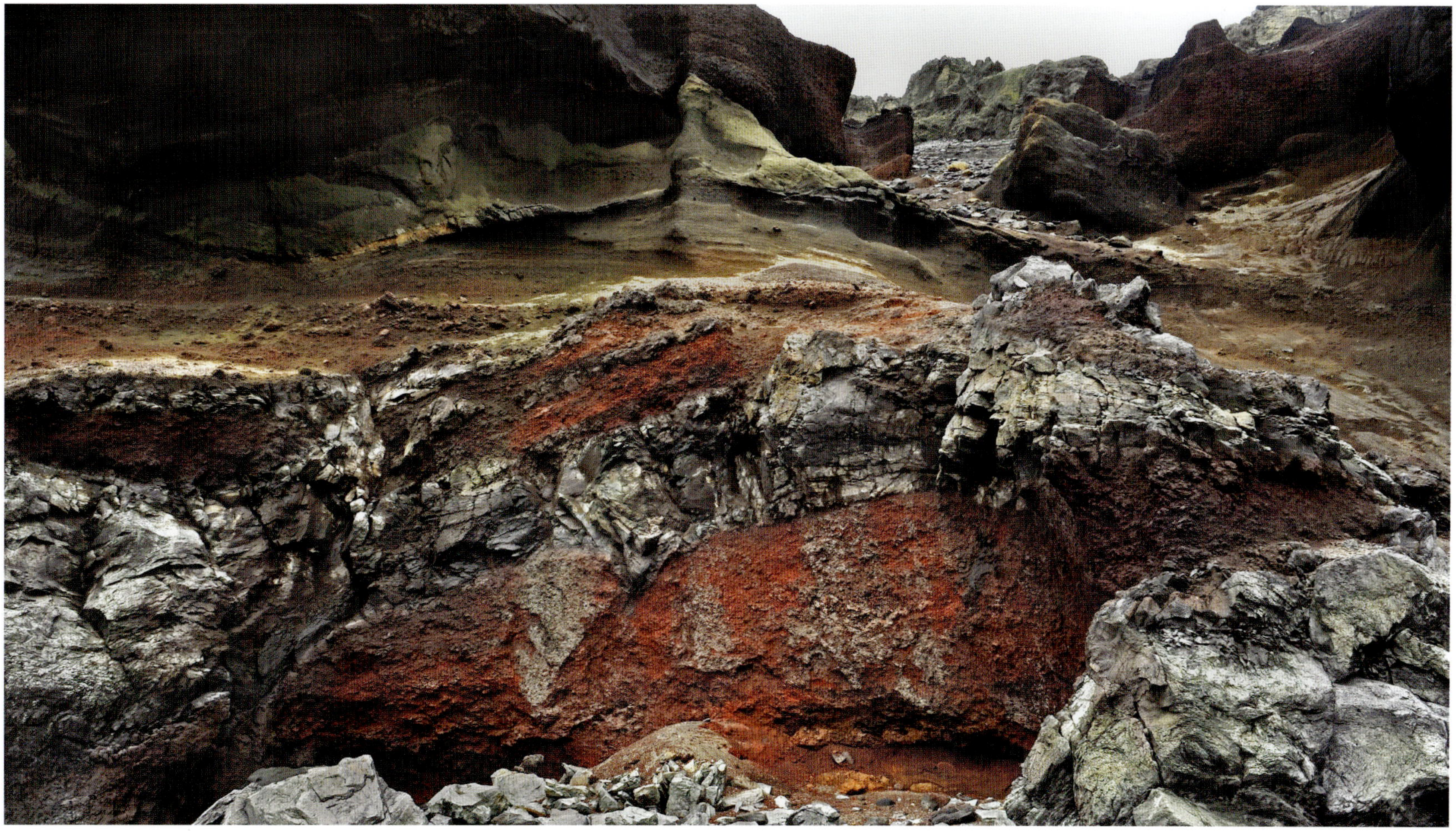

Igneous rock

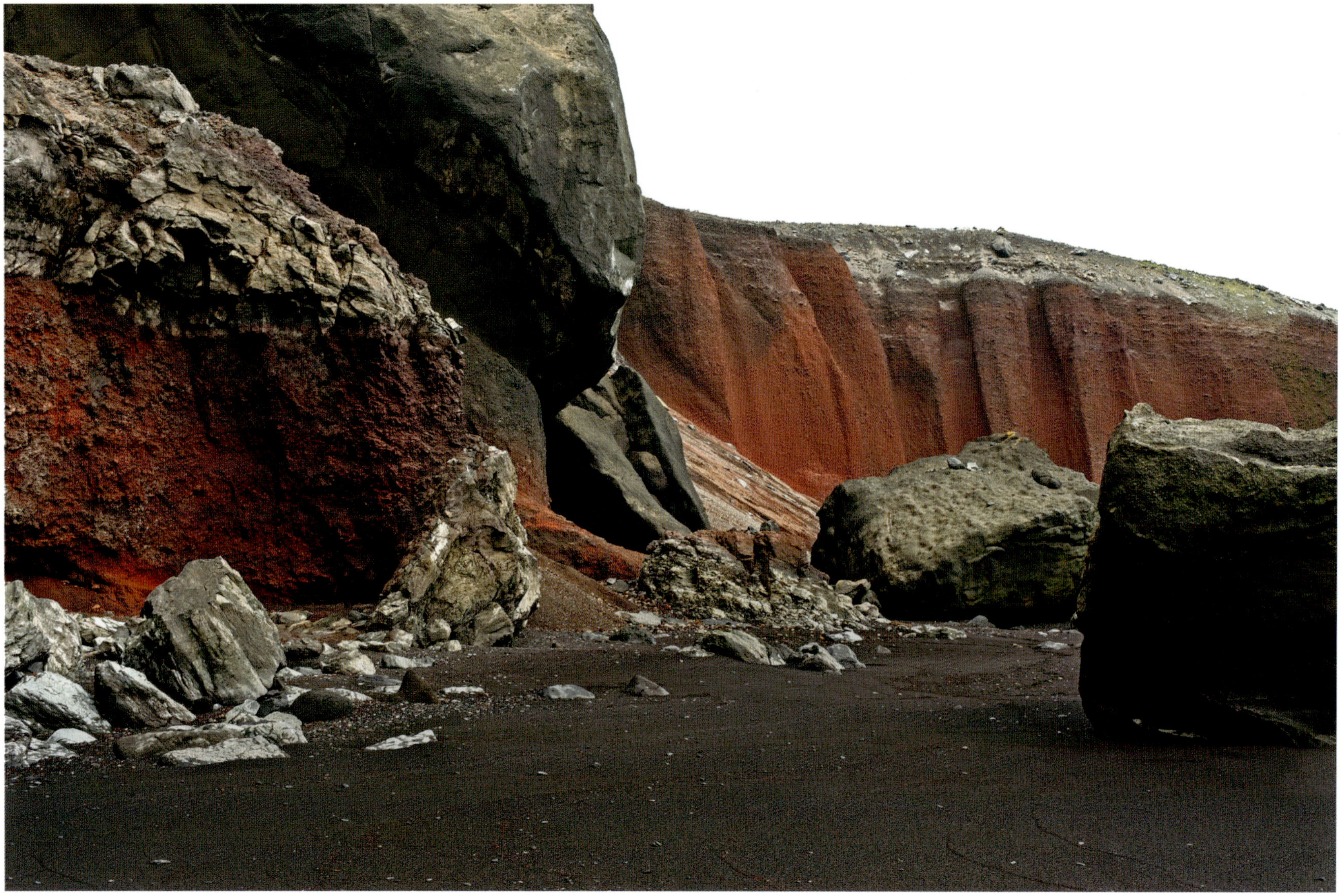

Black beach

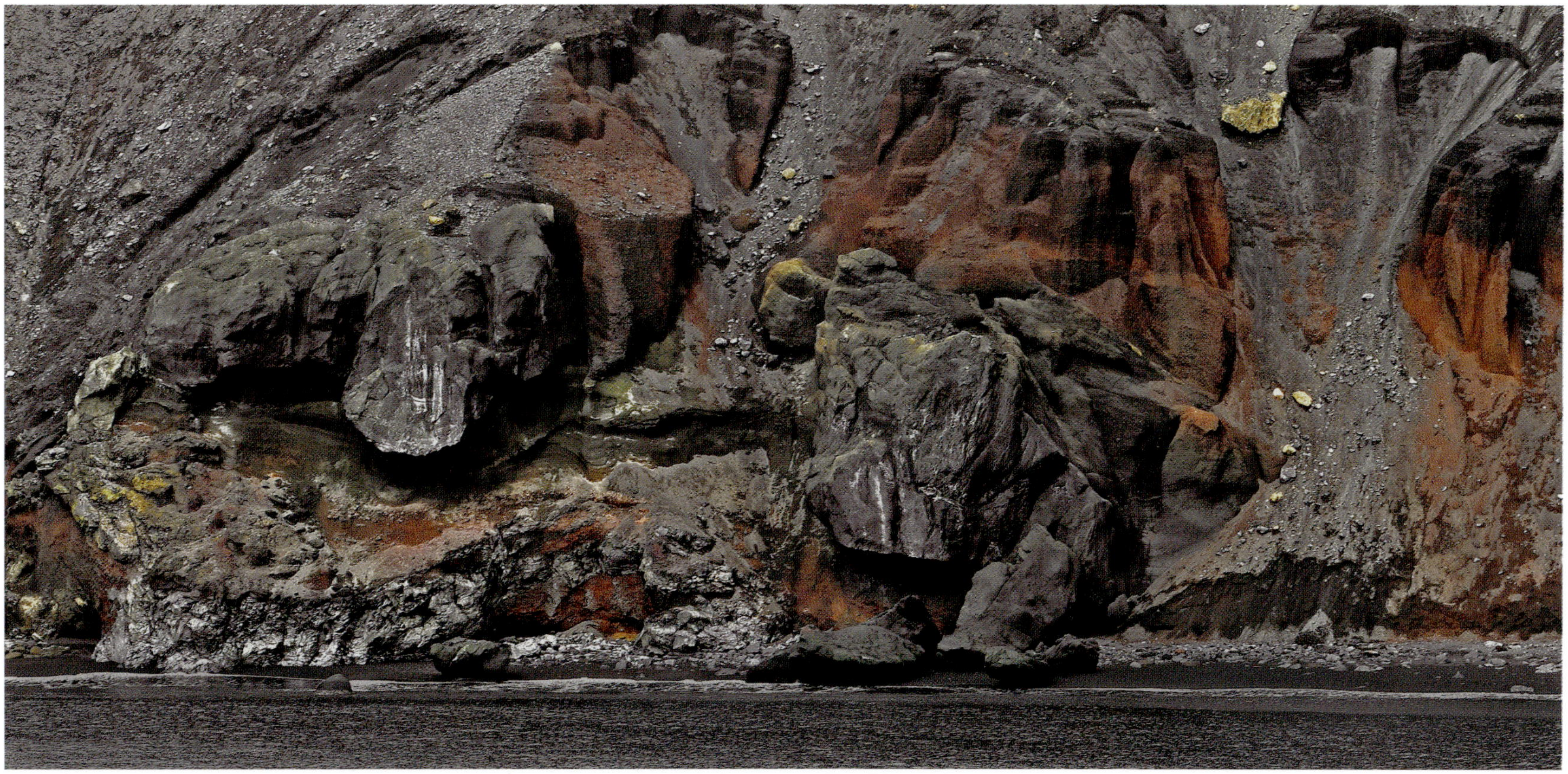

Ash and cinder conglomerate

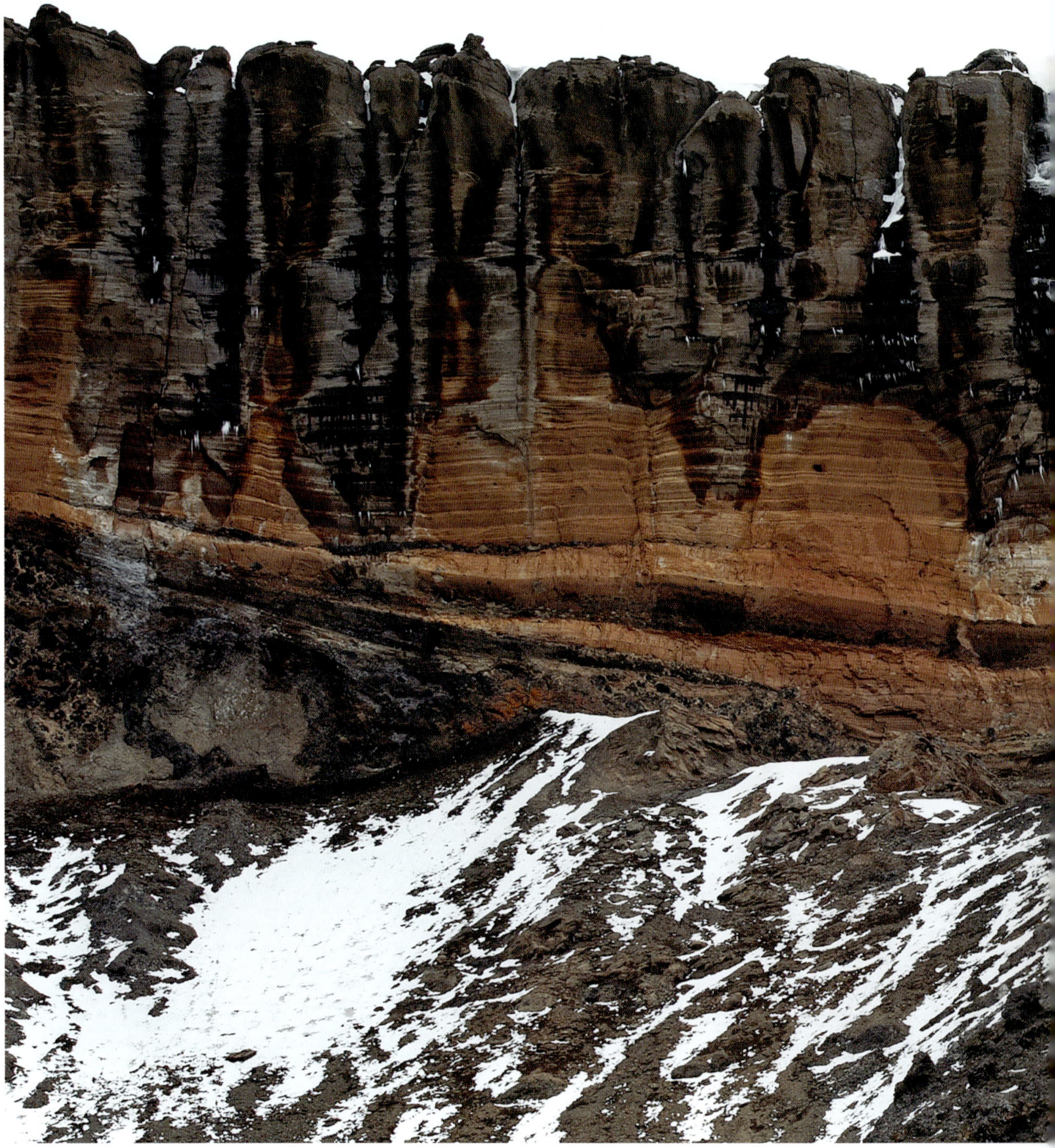

Stratified rock Brown Bluff

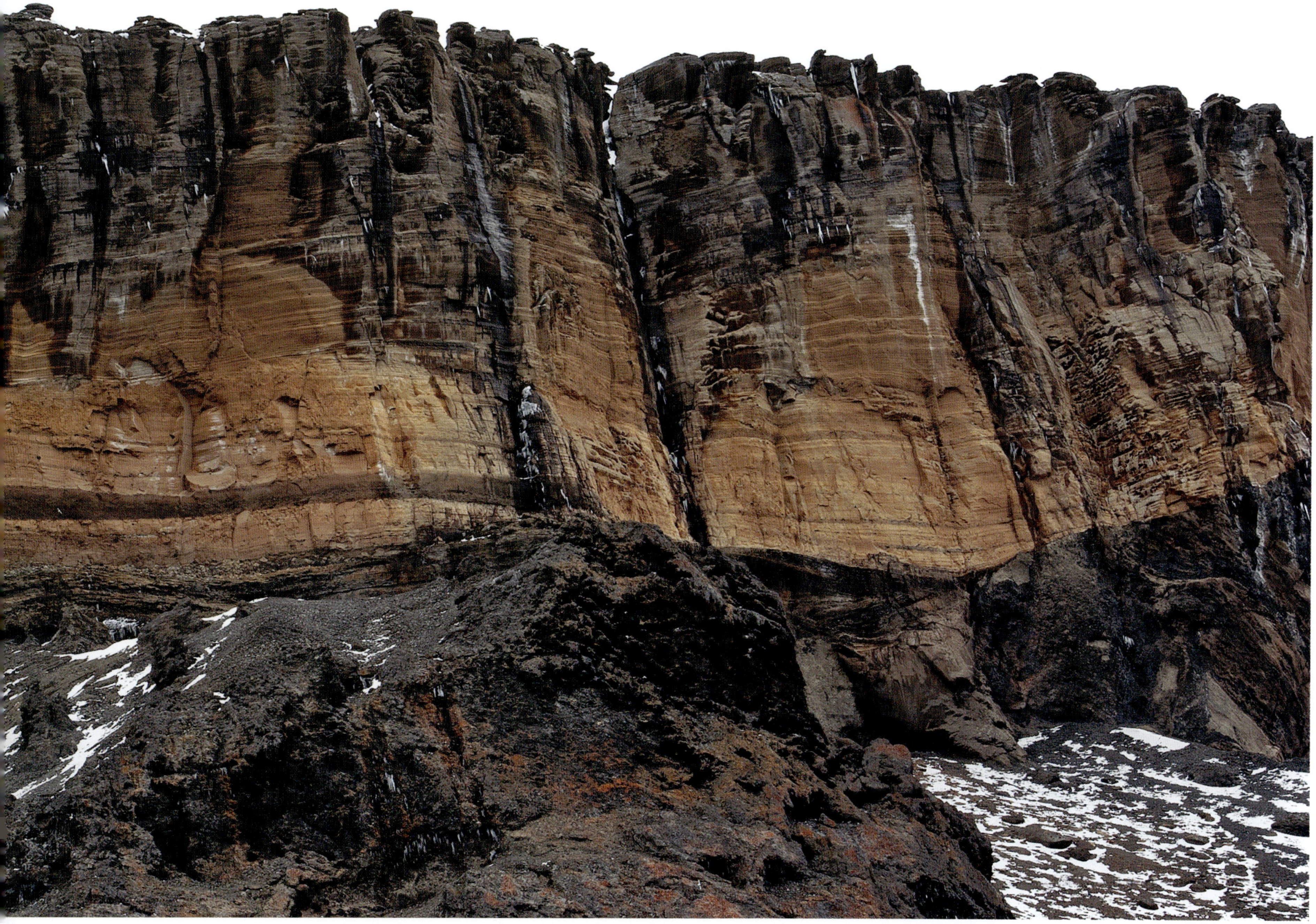

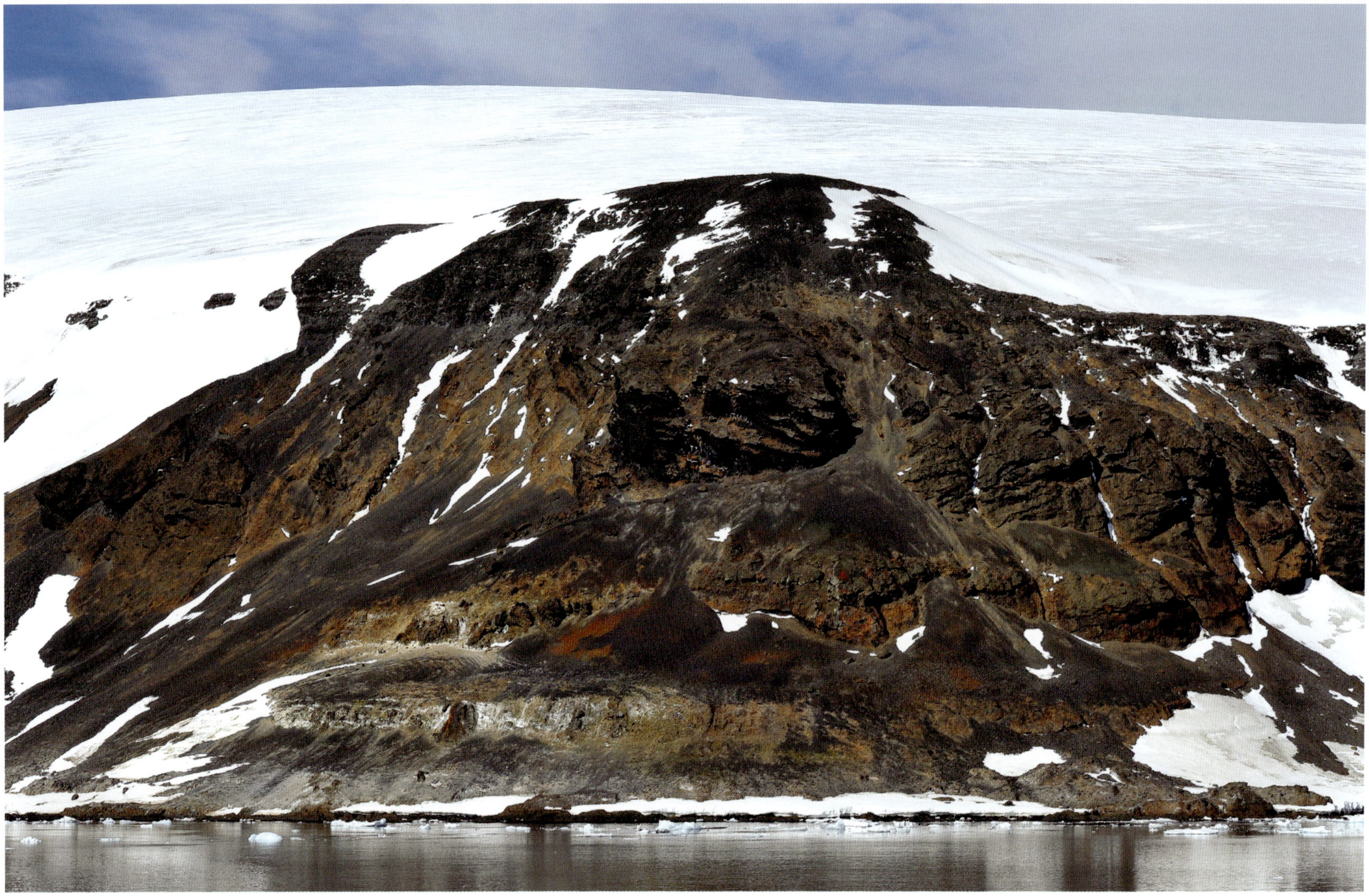

Exposed mountainside

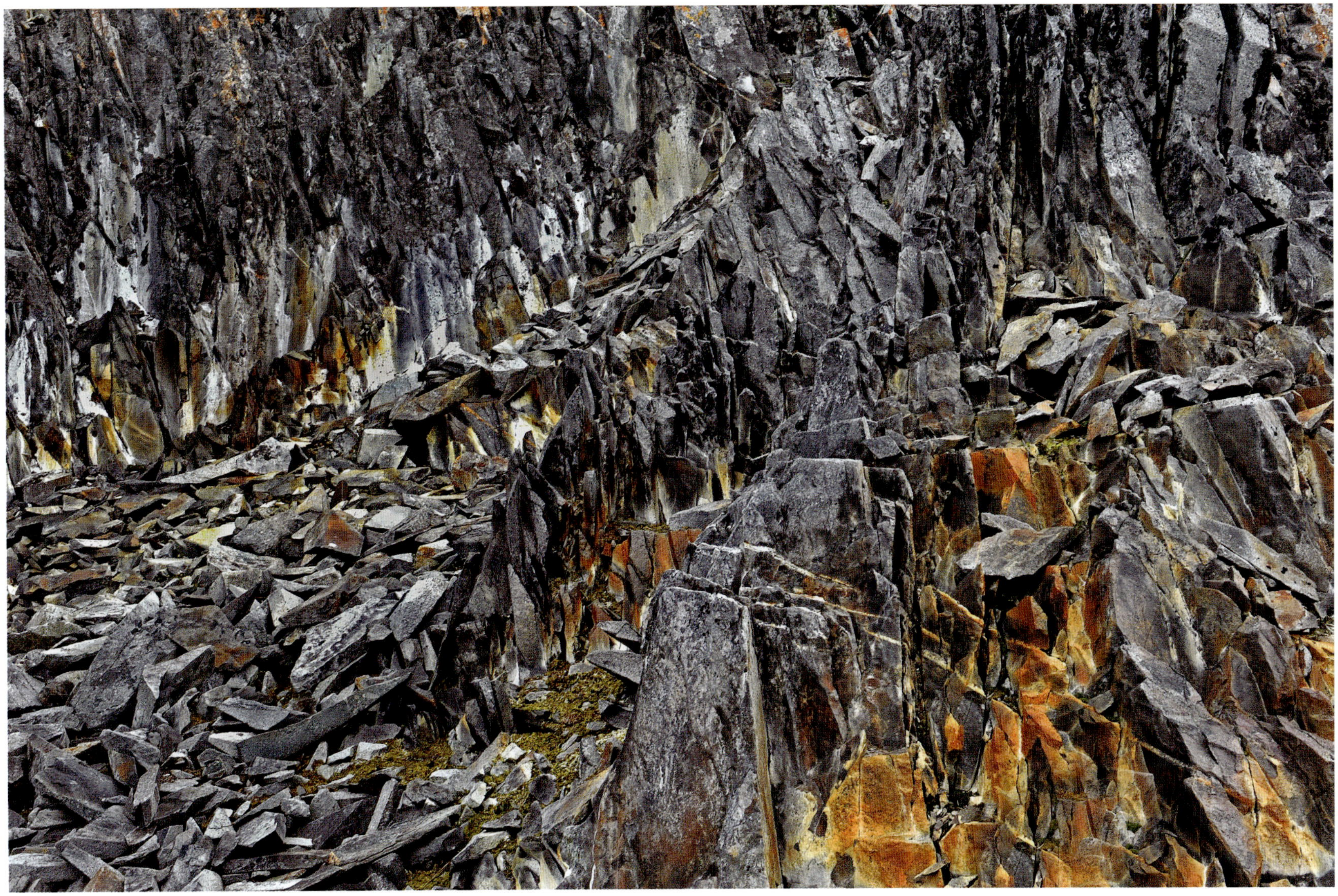

Frost shatter

ANTARCTICA HAS NEARLY two hundred years of human involvement. Over this short period the variation of contrasting events that occurred here have been wide-ranging. From yesterday's reckless exploitation and misuse of wildlife to today's international agreement on an Antarctic treaty, our presence now has a softer footprint and our attitude is more one of protection and concern. With a heightened awareness of the fragility of the environment and the potential impact on the health of our planet our decisions have become more measured and cautious. Global awareness and education have become the reigning factors. Responsible tourism and valuable research are now the prime activities in Antarctica.

The best example of this turn-about in attitude is found here on a wide beach on Deception Island. First used as a haul-out for rendering whales into fats and oils, the same facilities were later converted into a research station. Due to volcanic activity this also was abandoned. It has now become the focal point of many adventure tours. Here in Whalers Bay surrounded by evidence of the not so distant past, holding tanks and buildings that once housed whalers and scientists alike, are slowly corroding and crumbling into the black volcanic ash. Among the bleached remnants of scattered boards and rusty barrels bright crosses remind us of the younger lives that ended here. Nothing else remains.

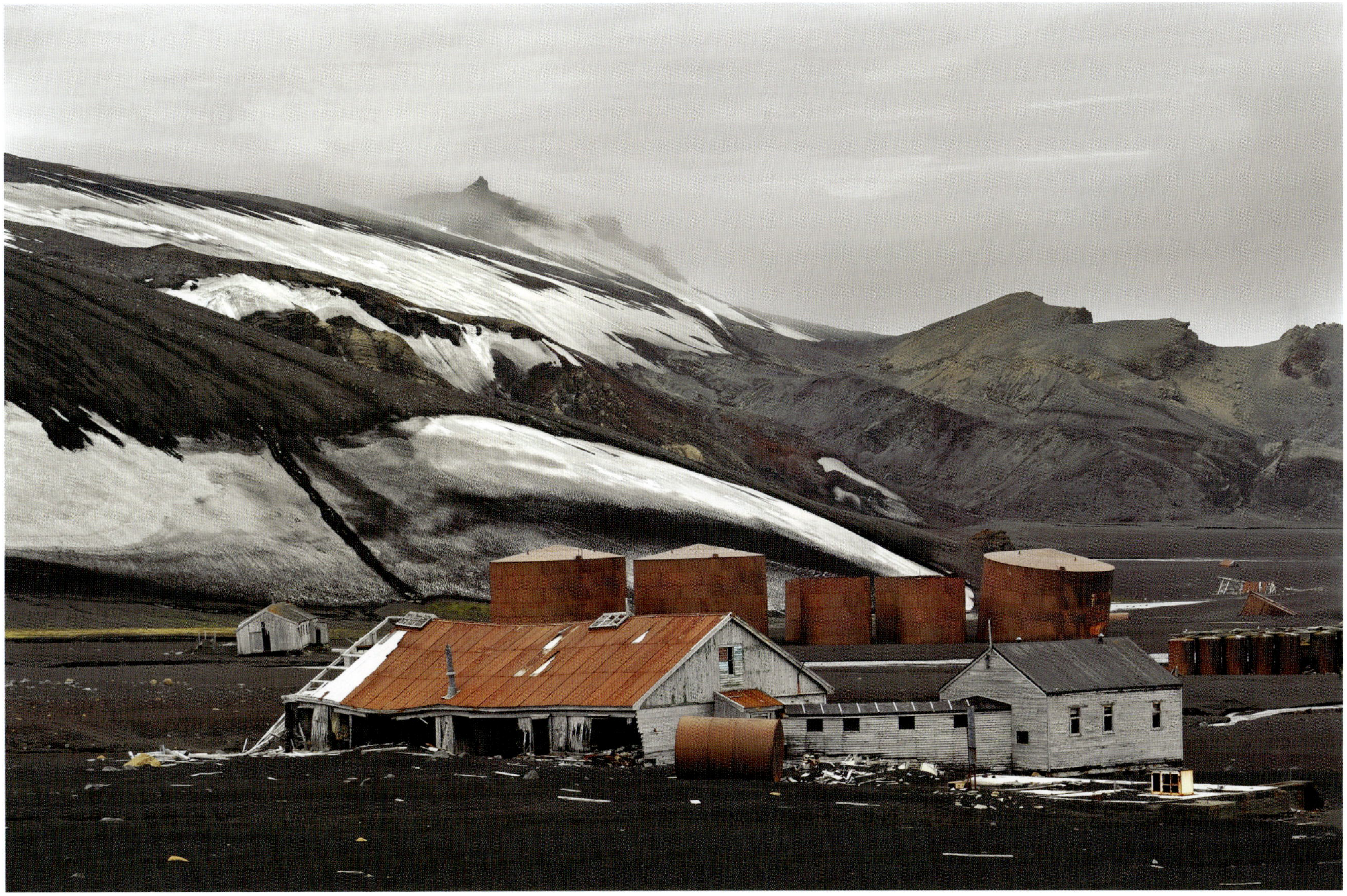

Abandoned British station Whalers Bay

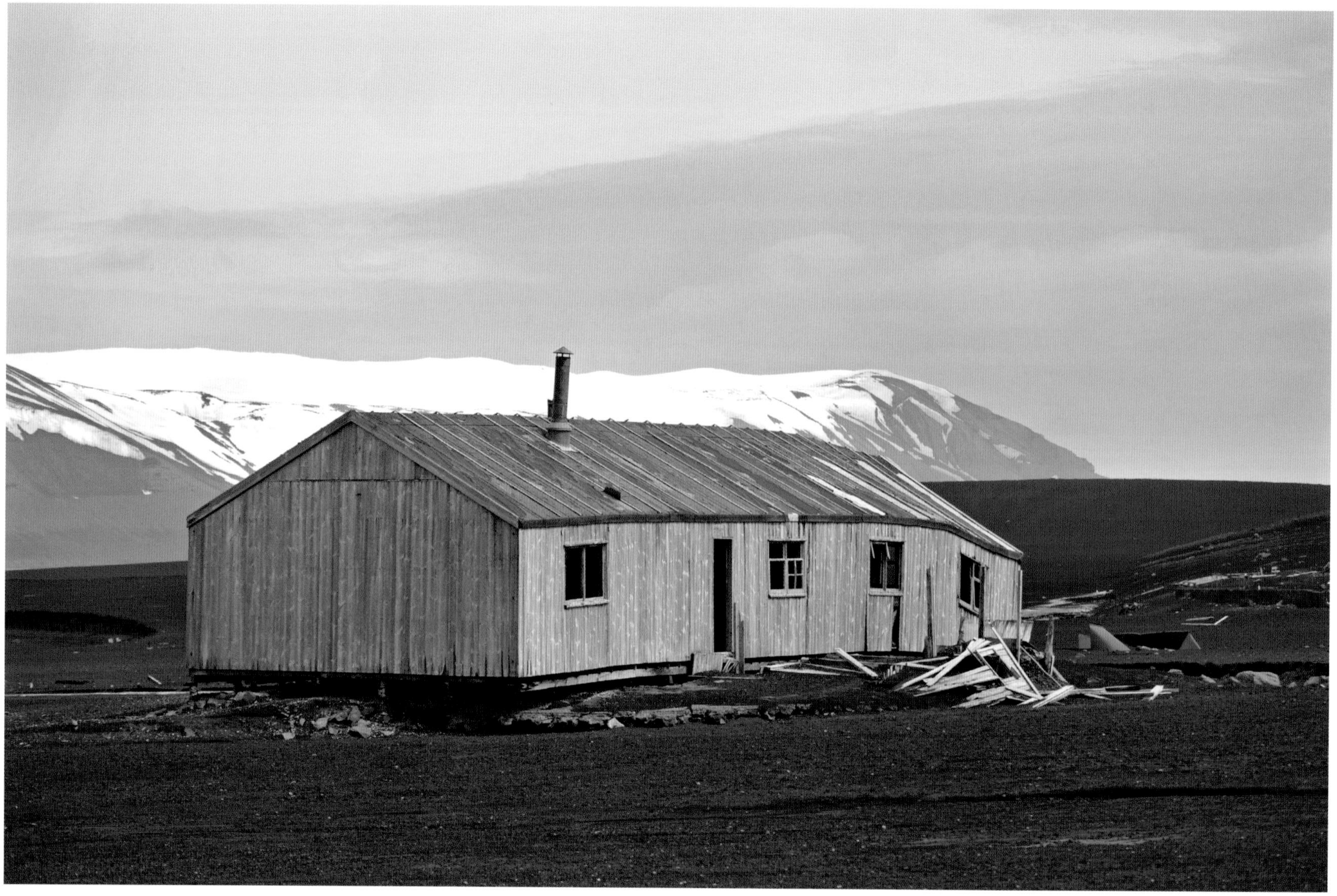

Abandoned building British Antarctic Survey

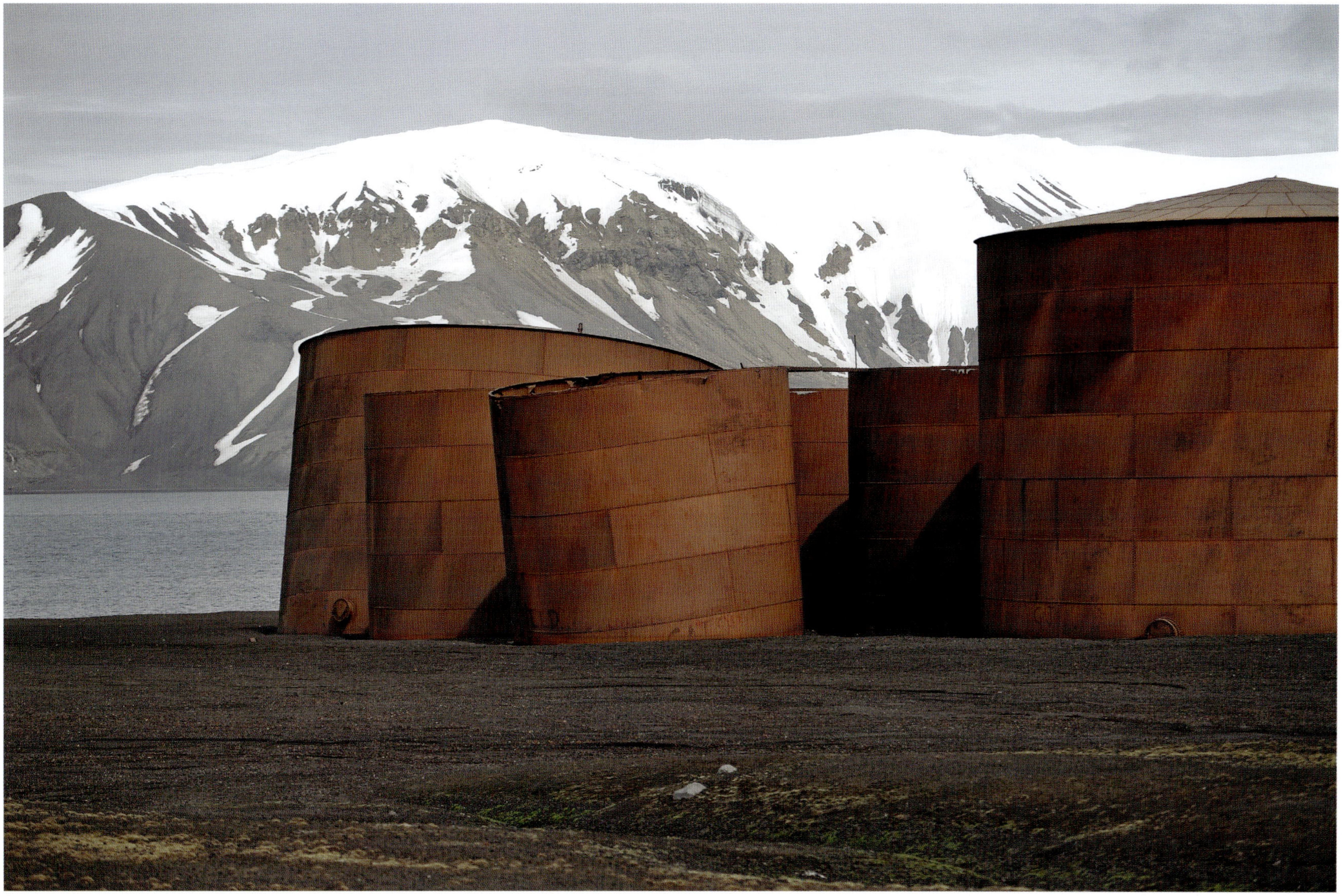

Decaying storage tanks

LATE FALL BRINGS snow to the islands of the Antarctic peninsula. This bright ash mound rising 300 metres from the ocean stands in stark contrast to the black ash hills of an austral summer. With the forecast becoming now ever more unpredictable itinerary plans change often and the weather sets the pace and direction of travel. What plans seem obvious at one hour are discarded the next.

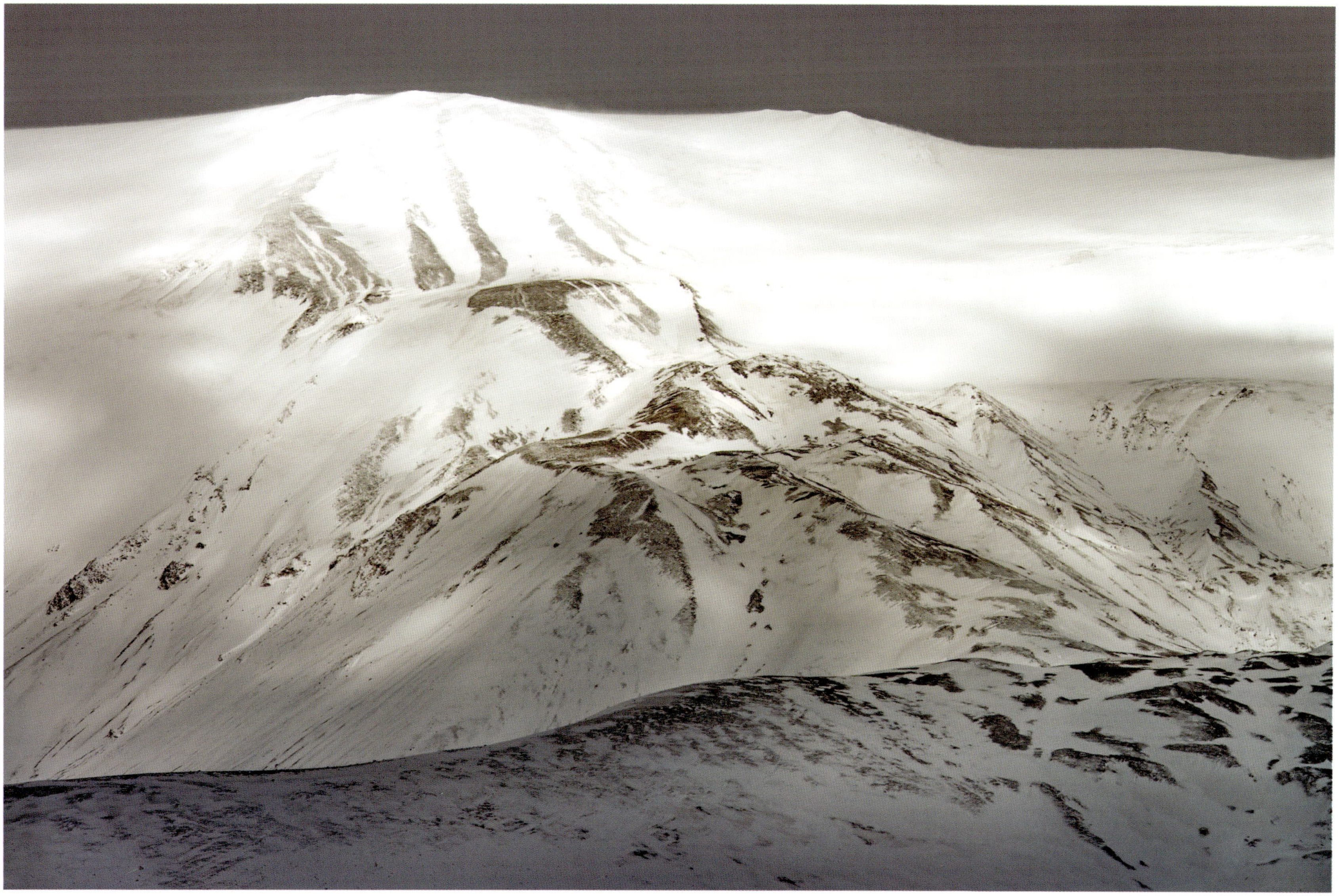

Volcanic rim

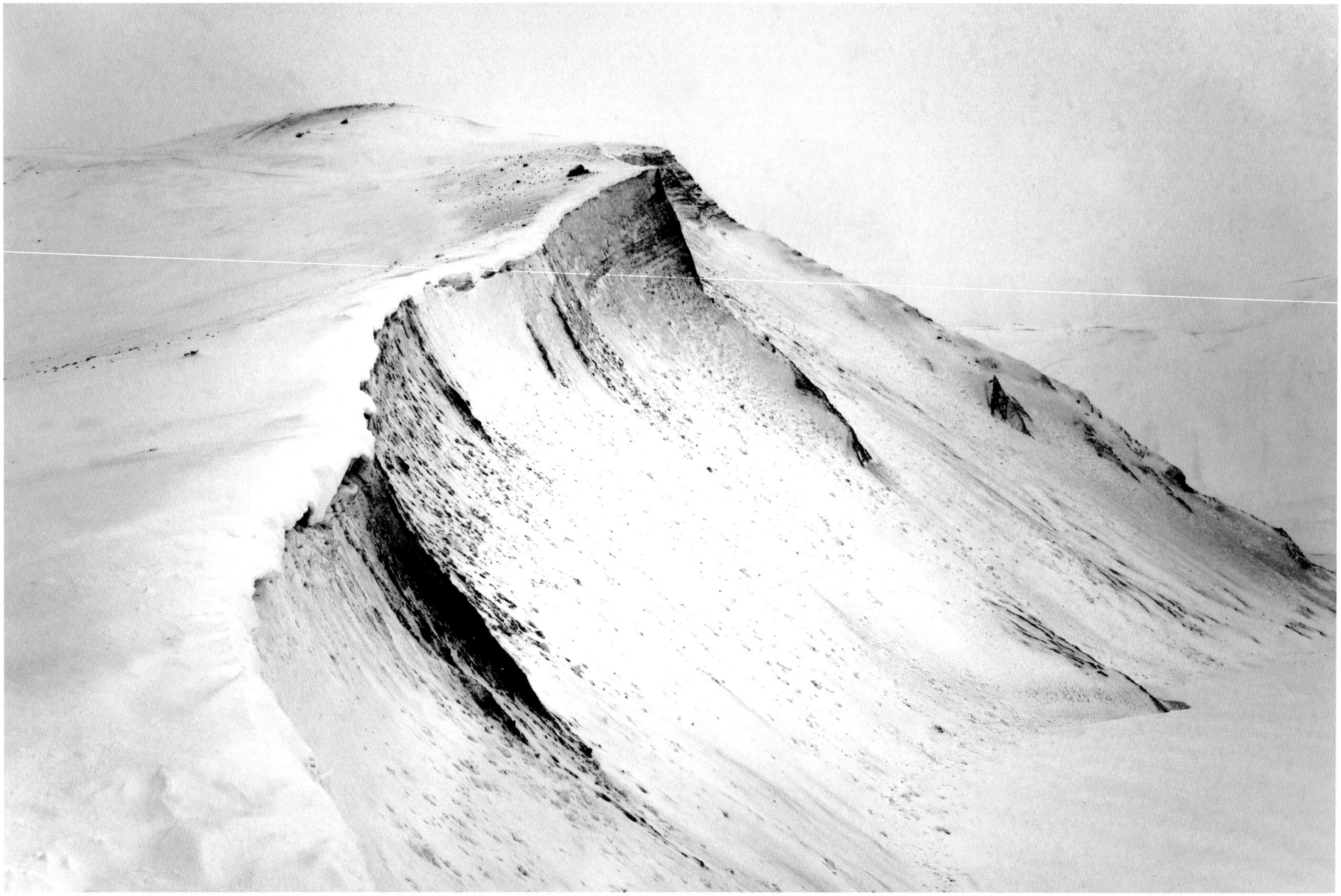

Crater's edge

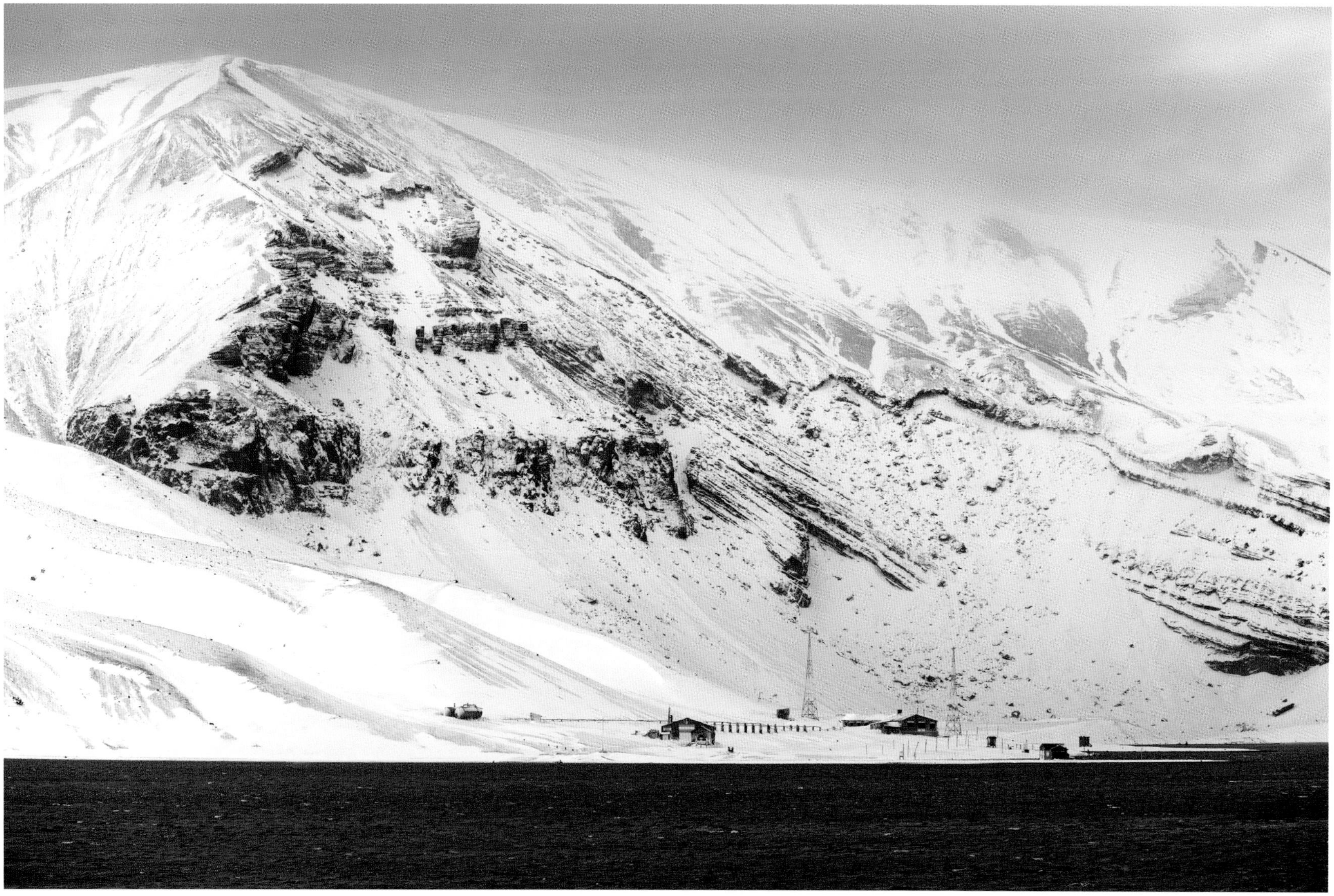

Argentinian research station closed for the winter

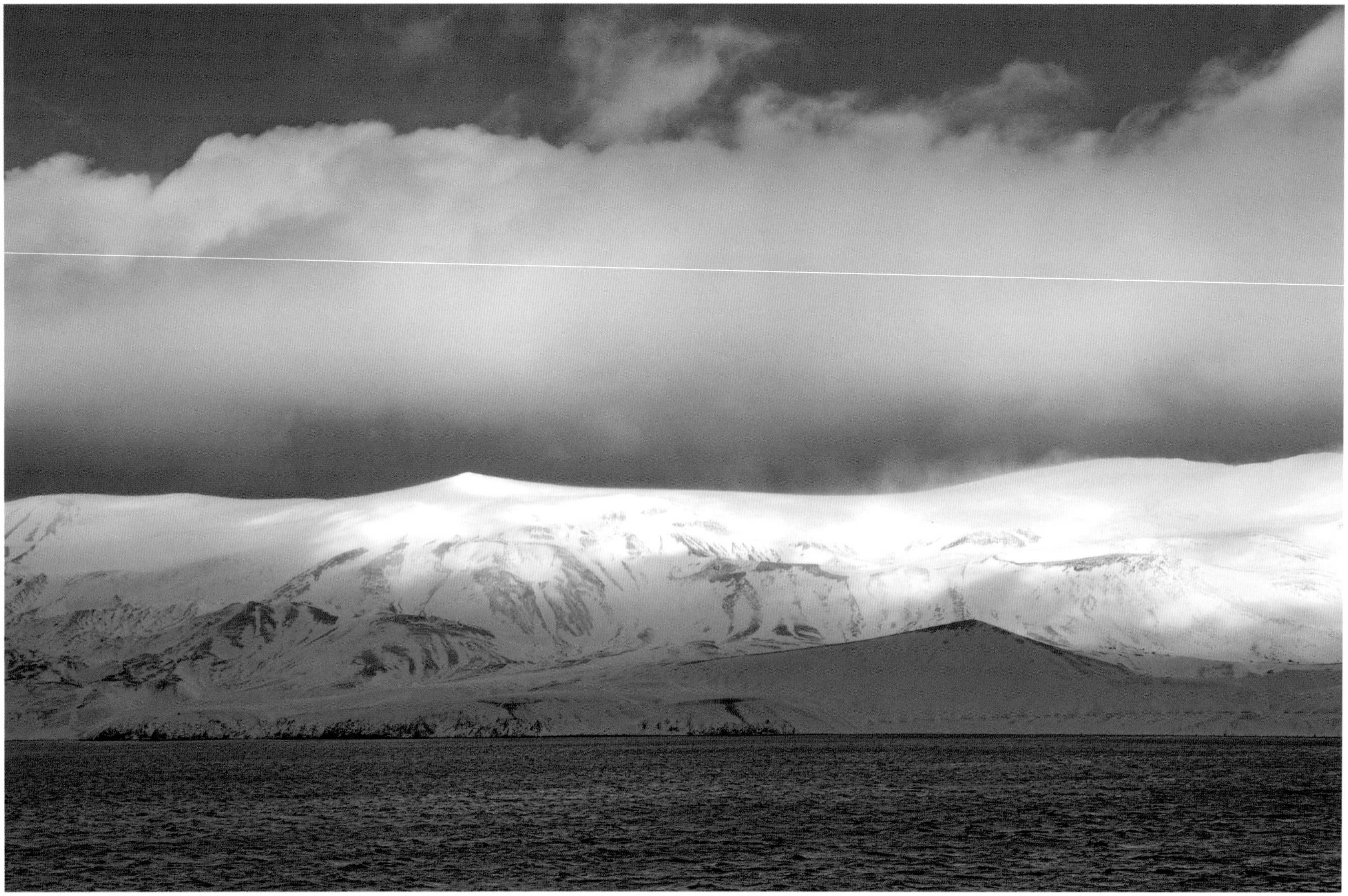

Cloud bank over Port Foster Deception Island 2014

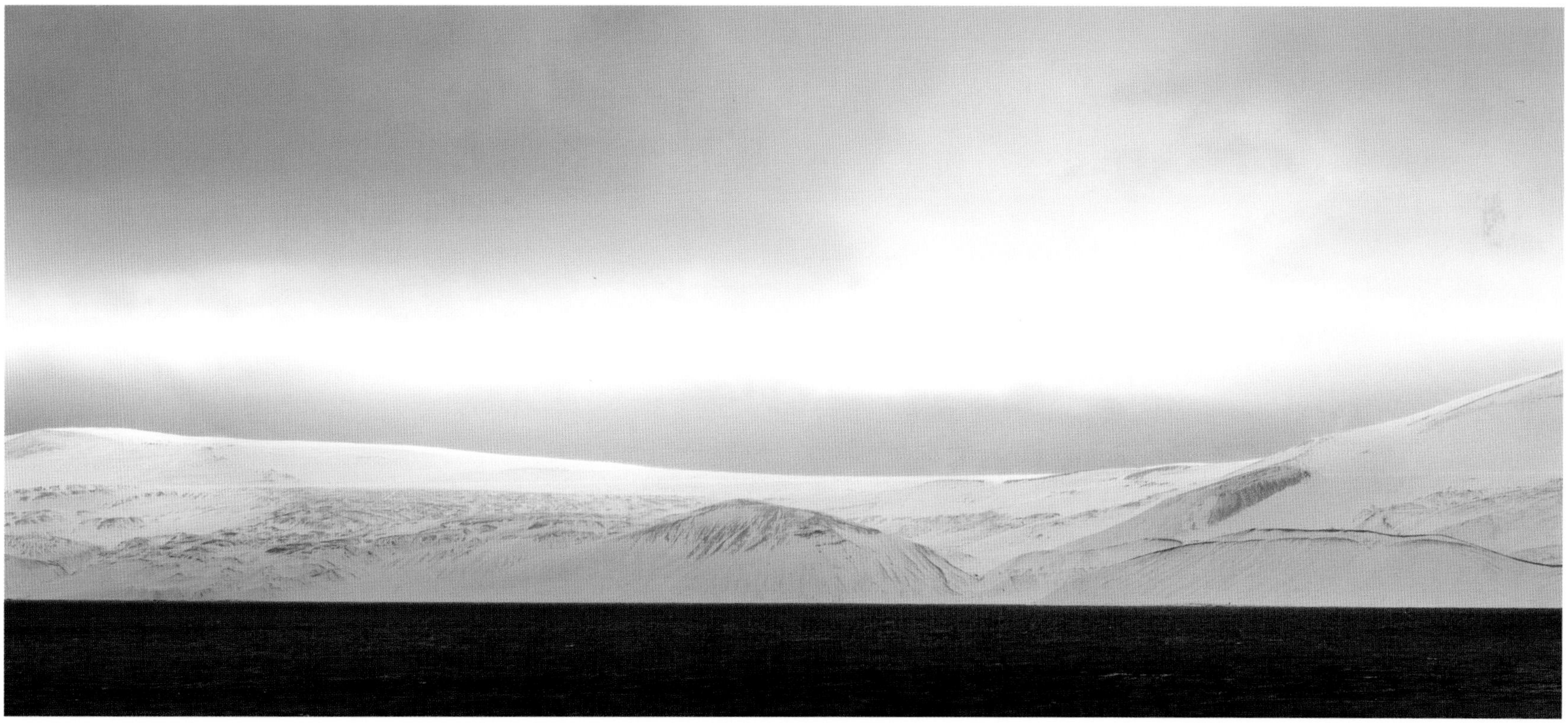

Inside the caldera Port Foster Deception Island 2014

THIS MOUNTAIN PEAK was enveloped in cloud and fog. All was an even grey as we sailed between calving glaciers and smoky icebergs. The morning had passed without colour and the sky and sea looked mundane without texture. Eventually the darkened sky lost its opacity and began to diffuse and scatter a stronger penetrating light. The ship moved slowly and cautiously wound its way through the icy maze.

I could see shadowy mountain peaks appear and disappear above me. When the light broke through it was blinding white. Momentarily the moving ship positioned us in the peak's shadow partially blocking the sun. Immediately the fog filled with a spectrum of diffuse colour spreading through the haze. The sky shone blue and the ice gleamed yellow. Two quick images and the vivid colours were gone. I had witnessed this effect before but was never ready enough to capture its effect. It reminded me of the illusive green flash that given the right conditions one could see at sunrise over the ocean's horizon.

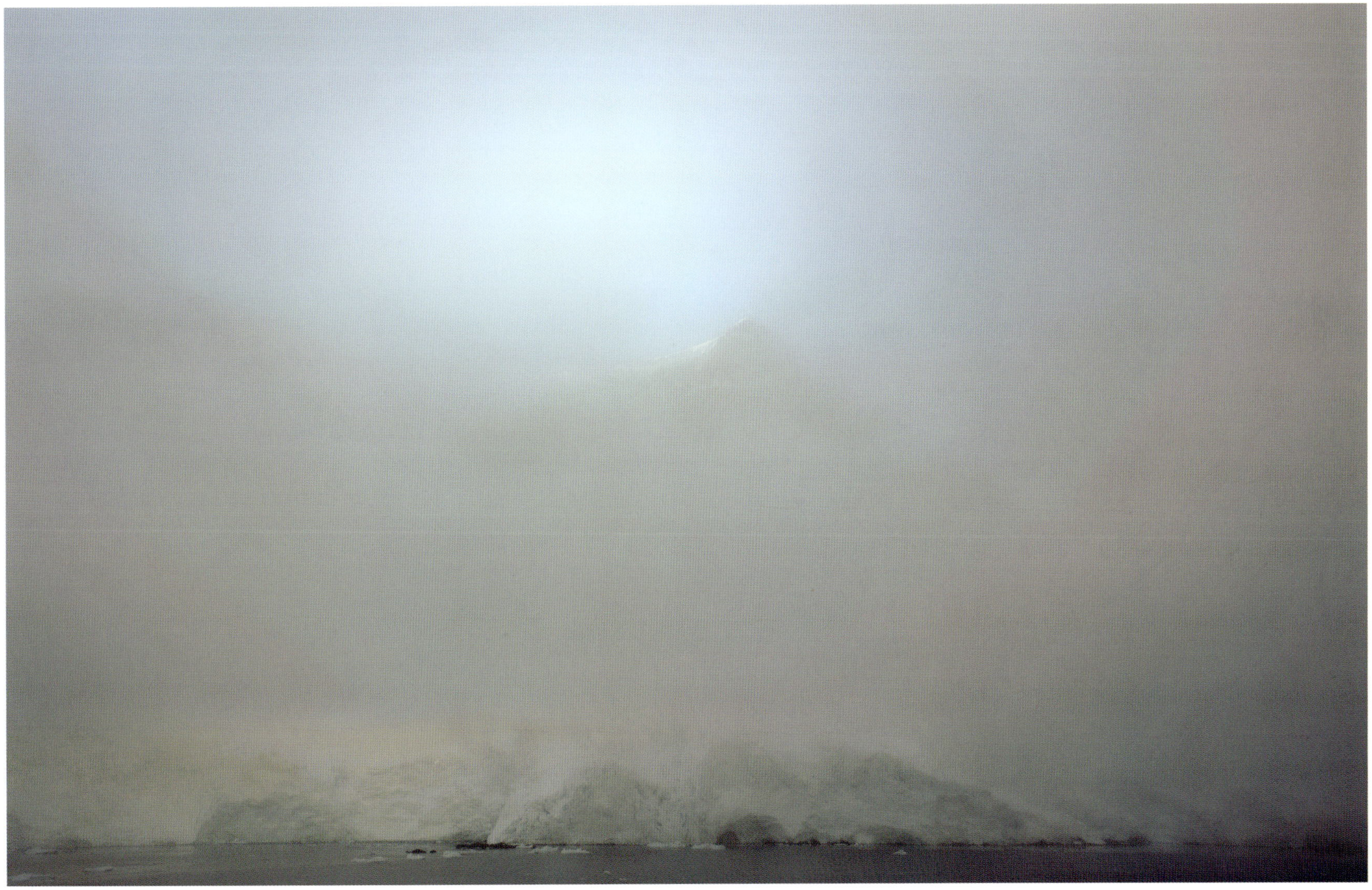

Flash of prismatic light

seeing your first albatross.

In 1917 Robert Cushman Murphy recognized the albatross's special status and the privilege, especially at that time, of witnessing their wide graceful flight. In his log aboard the ship *Daisy* he wrote, "I now belong to a higher class of mortals for I have seen the Albatross"

Nowadays it might be easier to reach those southern latitudes and be privileged to witness the soaring of the albatross but this privilege can be short lived. Too many sea birds are accidentally caught on the thousands of fishing hooks that line the ocean. It is estimated that one albatross is killed every five minutes. There is a global concern and methods are being tested to avoid the casualties but their effectiveness is still in doubt.

With their population on the decline it again might very well be a rare privilege to see the albatross.

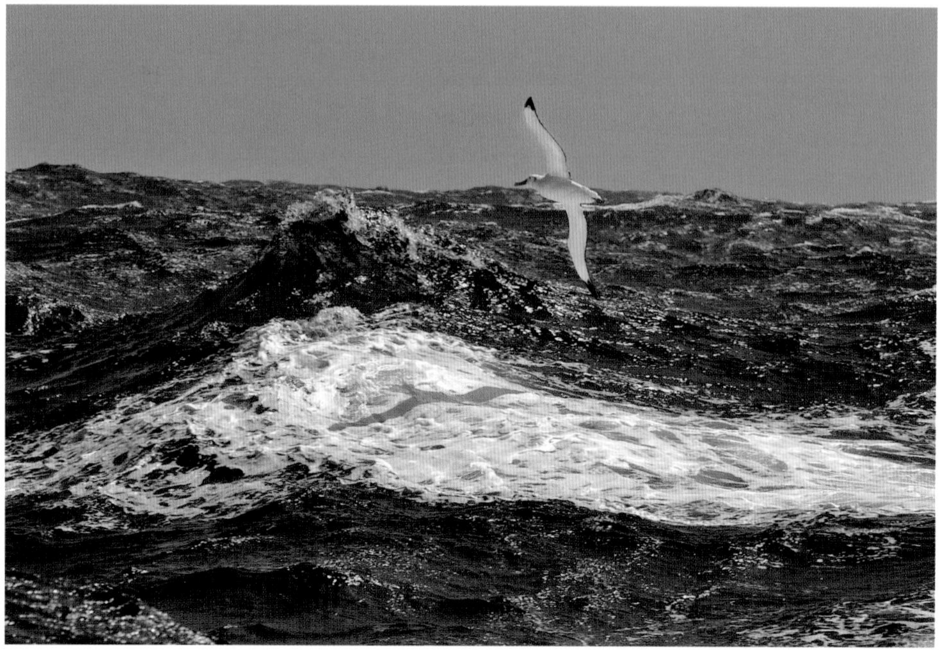

Shadow on the water

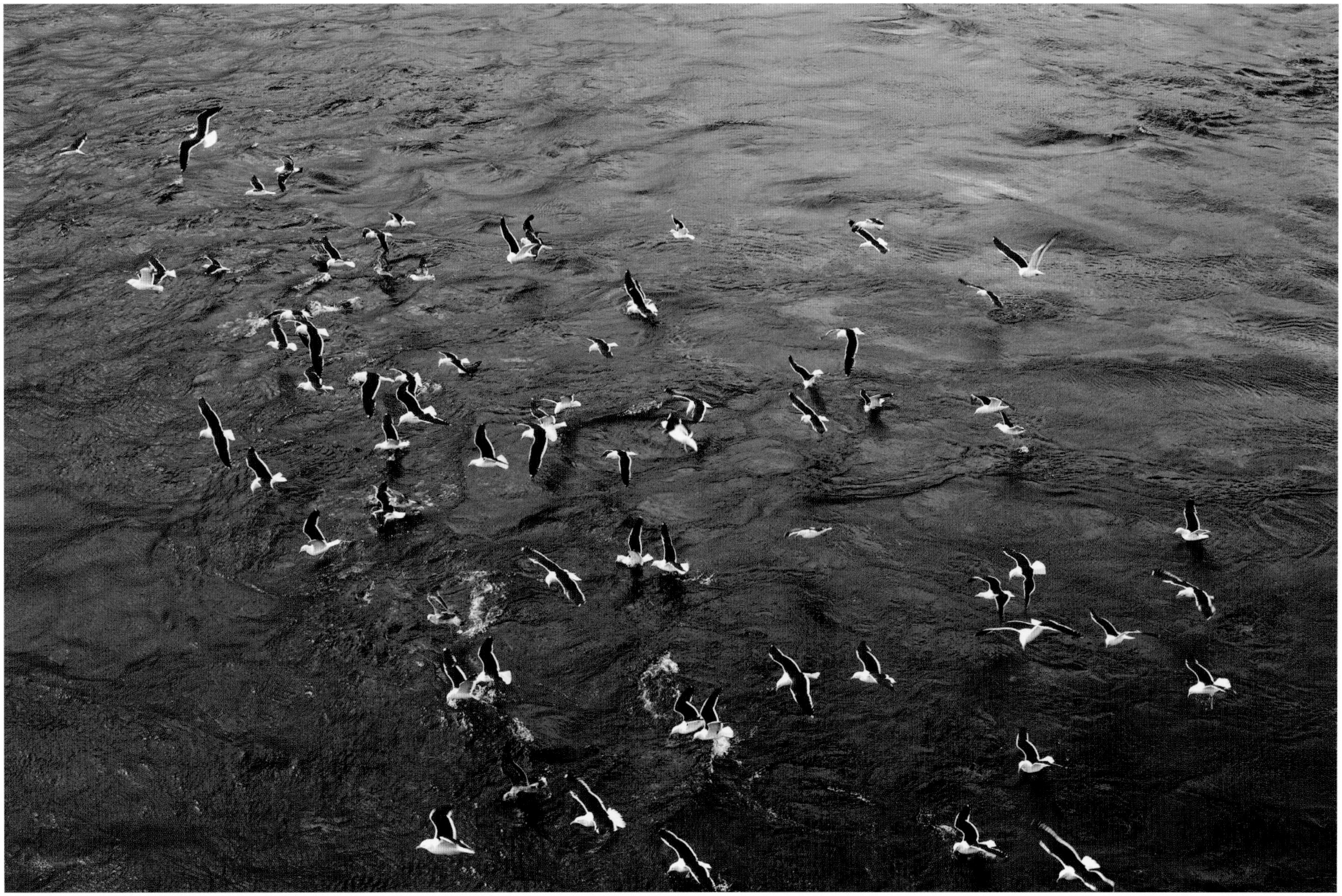

Kelp gulls Drake Passage

"PINTADO" FROM THE Spanish "painted" are true seabirds. Gathering in large flocks to feed on the surface of the ocean they peck and jab at fish, krill and squid. They appear suddenly alongside the ship darting over the waves and circling the hull. Fast fliers with dappled feathers on their back they are entertaining to watch maneuvring over the waves, jockeying for position. Part of the petrels they are the most numerous birds in the world.

Sheathbills are Antarctica's only bird without webbed feet. They also are the ultimate scavengers. There is nothing organic they will not eat. Listing what can be on their menu would make your stomach turn. Quick and bold they will enter a penguin or albatross colony to pillage and plunder and with their speed and agility outmaneuver its residents. Spilled food, damaged eggs and smaller chicks are instantly snatched and darting bills are skilfully avoided.

Nests can be made up of bits of moss mixed in with the picked over bones of recent carcasses. Even though they live on filth and refuse they are fastidious cleaners meticulously scrubbing their faces and preening their snowy plumage. They are the peninsula's ultimate waste managers.

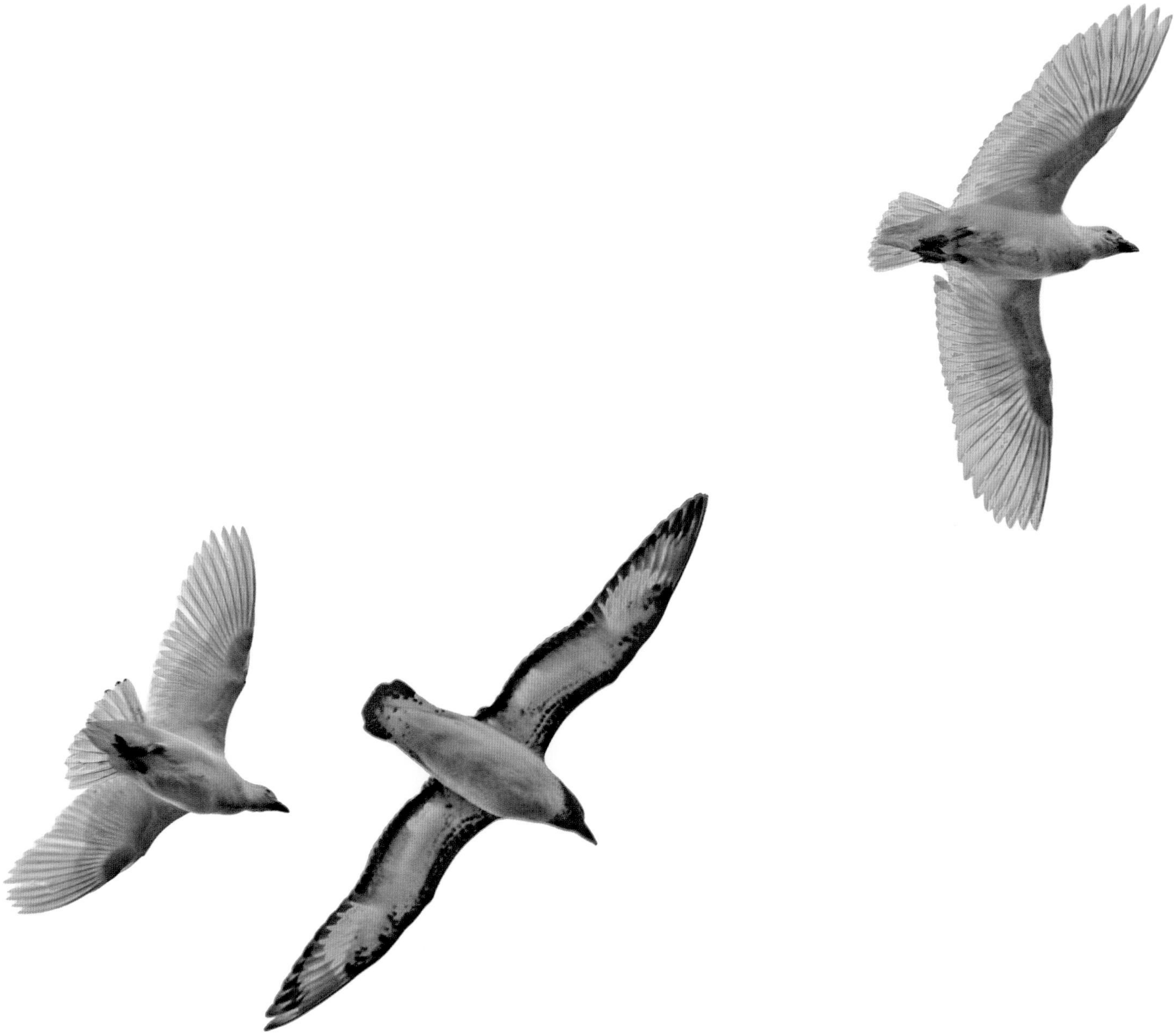

Pintado and sheathbills

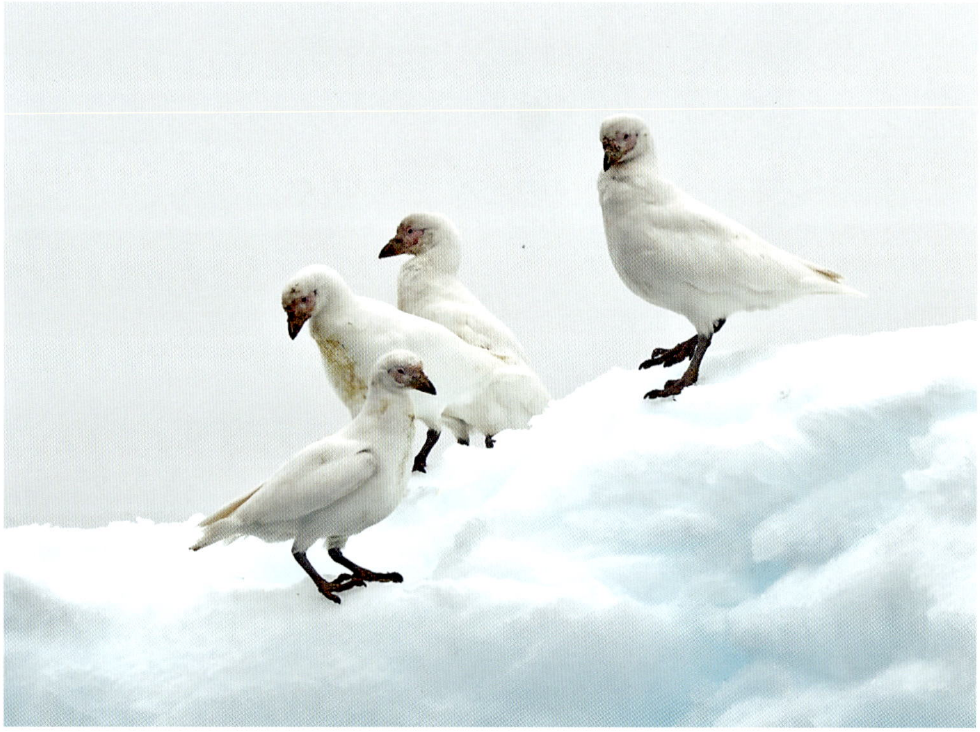

Sheathbills

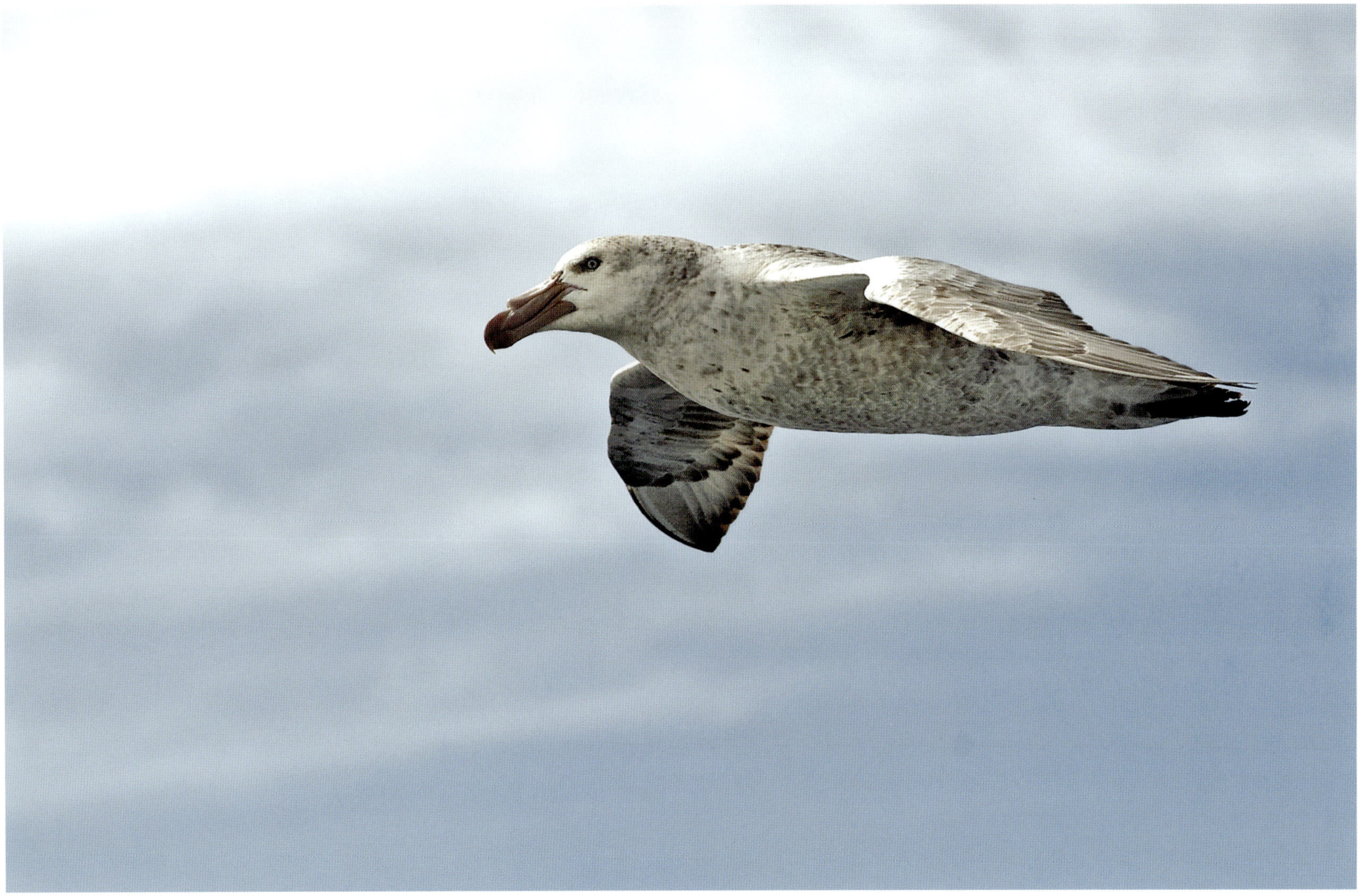

Giant petrel

WHEN THE EVENING is calm and the sun takes its long slow trajectory over the Antarctic peninsula I am always on deck to witness some of nature's most captivating displays. The low angled sun moving between the sharp mountain peaks spreads its twilight glow through the varying atmosphere with an always unpredictable light. Captured in this stillness and engrossing landscape it is easy to drift into thoughts of what happened here.

A little over a hundred years ago young men in their twenties and thirties with assigned responsibilities beyond their age enthusiastically took on a pursuit to risk their lives exploring this unknown. Their small sailing ships with unreliable machinery were their only lifelines. They could easily fit onto the bow of today's ocean liners yet with the bravery and optimism of youth they forged over perilous seas dealing with the unknown anxious for discovery and adventure. For some it would be their final journey.

I am close to Wiencke Island where a little over a hundred years ago such a fight-for-life drama took place.

In this stormy passage, unsure of the ship's ability to maneuvre through the strong winds, a sailor was on deck scrutinizing an approaching iceberg. The waves and wind had so rocked the little ship the cargo of coal spilled on deck. Loose coal slid and rolled across the planks plugging the scuppers that would drain the water that spilled on-board. The ship was severely over loaded. All sailors were ordered to stay below. However the young Auguste- Karl Wiencke felt it was his greater duty to

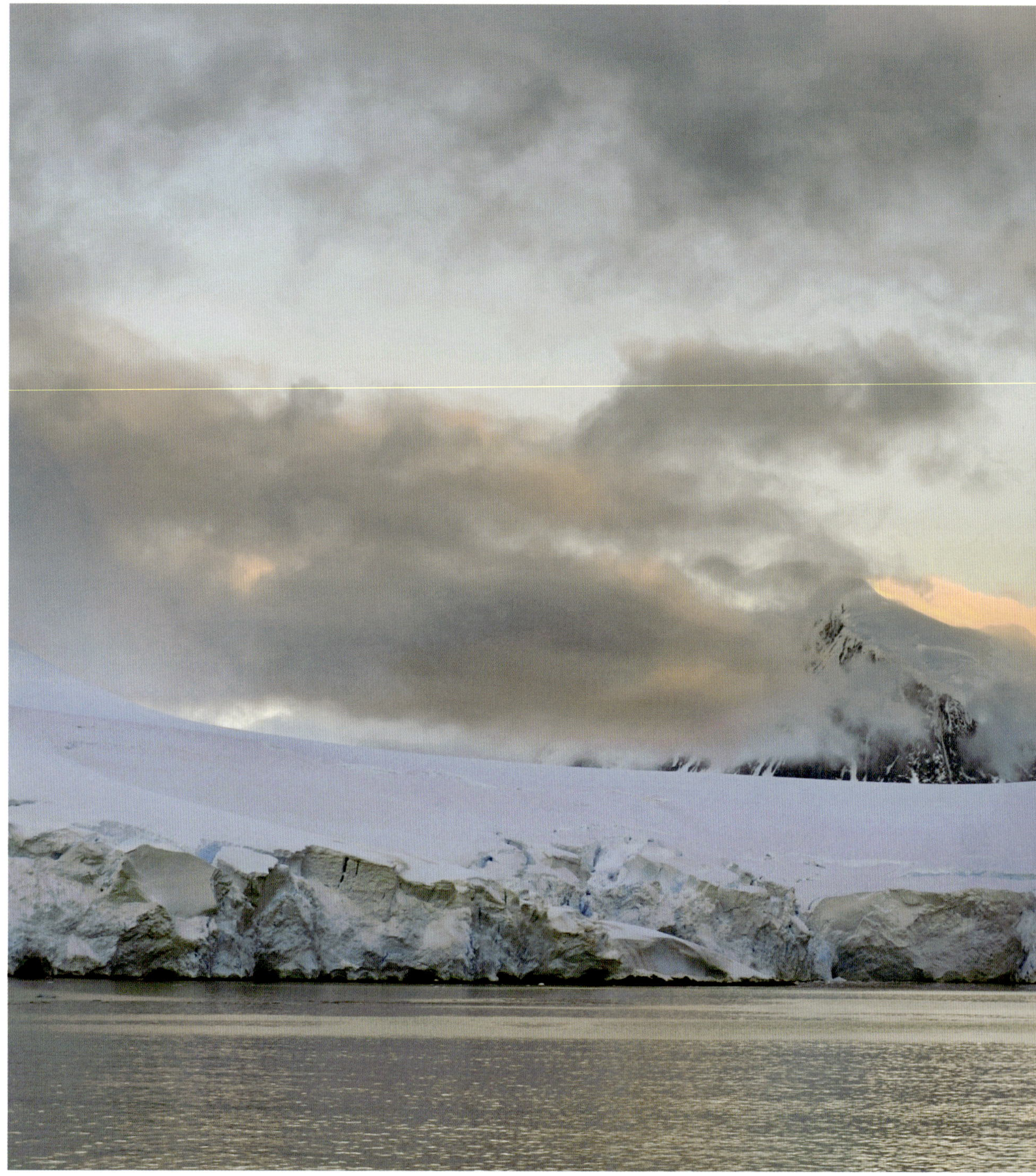

Gerlache Strait

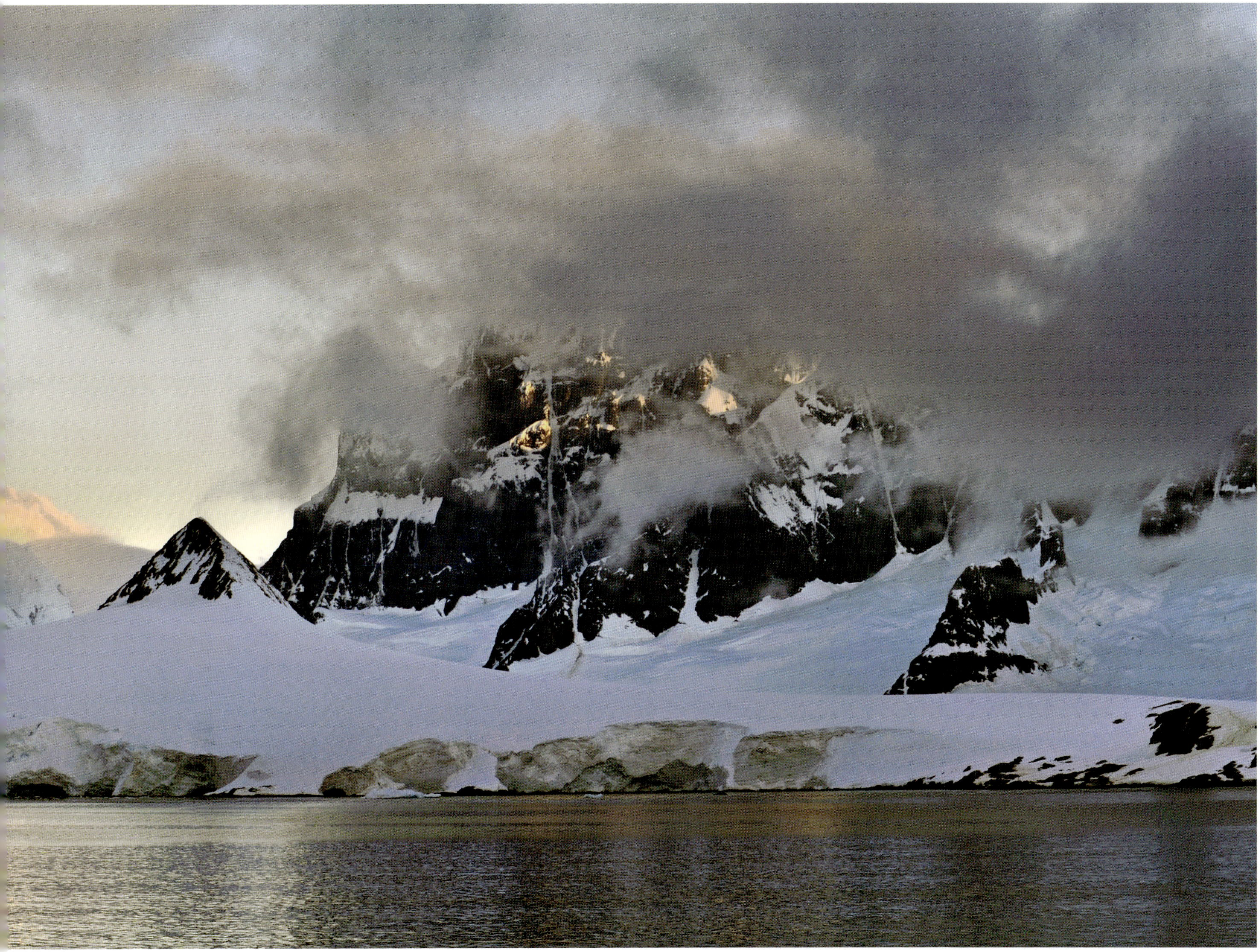

unplug the scuppers. Quickly he tied himself to a line and took action. His disobedience should have been spotted by the watchmen but he was preoccupied with the looming berg. The ship pitched violently when the watchman heard a frantic scream and turned to see Wiencke desperately struggling in the freezing water trying to pull his way back. The line came loose but another seaman tossed him the log line. Wiencke reached the small stick attached to the end providing an easy hold but the freezing water took away his strength. When he was nearly within reach and rescue he lost his grip and slid away from the moving ship. In a truly heroic deed another sailor with rope attached jumped into the bitter water. He was about to grab hold of Wiencke but a strong wave separated them and Wiencke, now paralyzed with cold, could do no more and sunk beneath the waves. He was but a twenty-year-old boy.

This beautiful island is named in his honour.

An interesting note here; the watchman preoccupied with the route the oncoming berg would take was Roald Amundsen who fifteen years later was first at the South pole and the man who threw out the log line was Dr. Frederick Cook of North Pole fame and controversy.

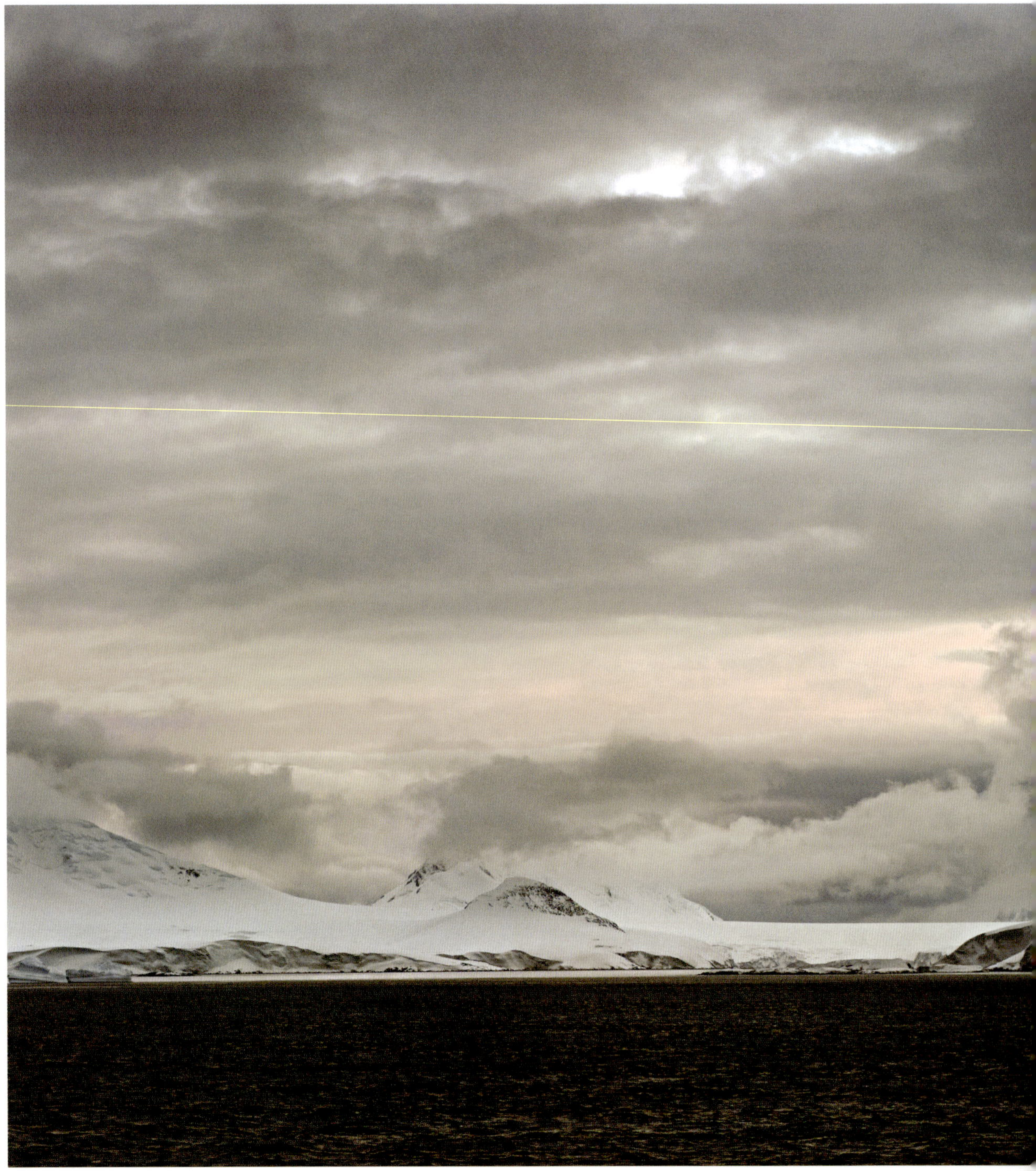

Wiencke Island

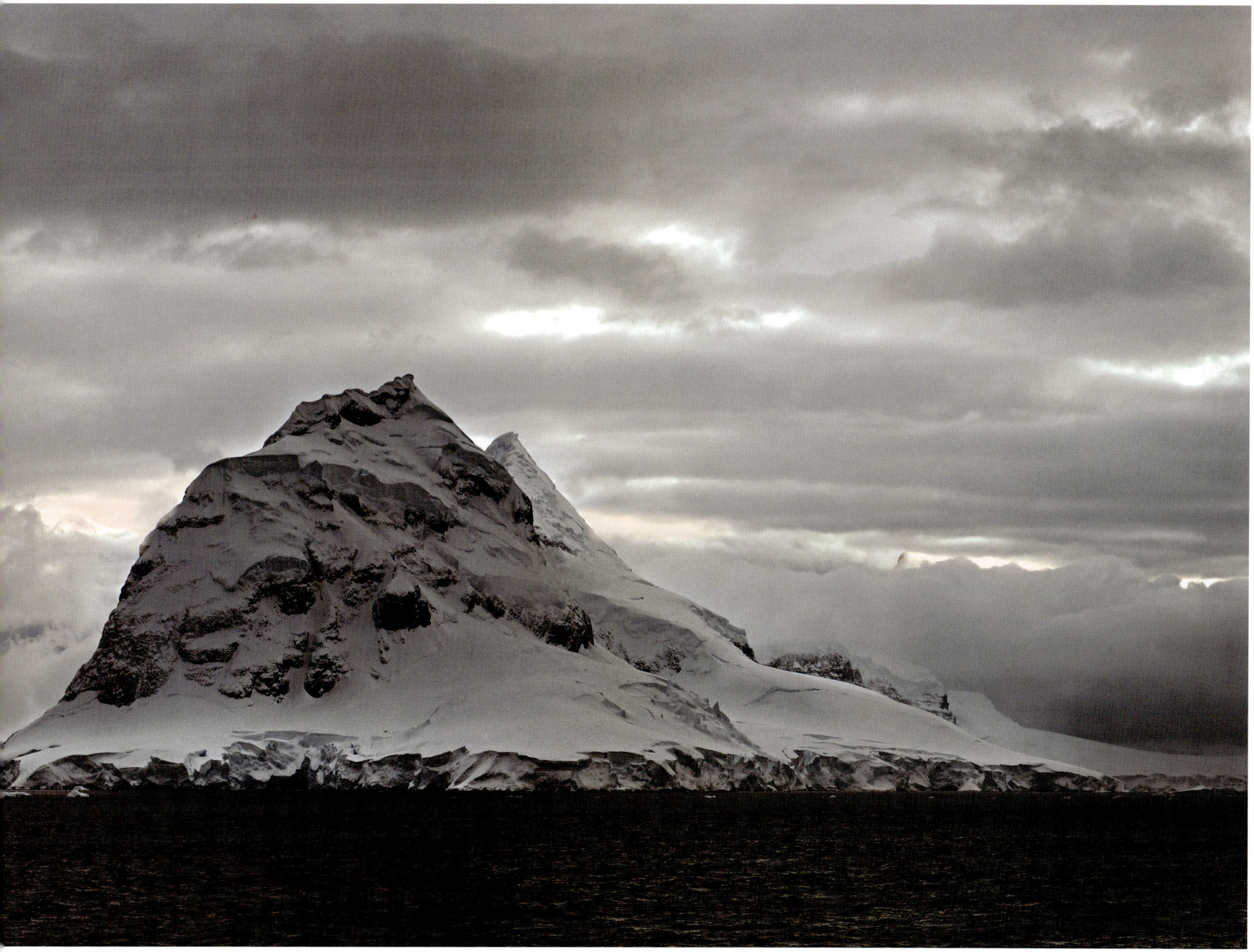

A DISTINCT RUSSET trail leads from the ocean to a penguin nesting site.

Locations that are snow free, dark and warmer are a rare commodity. Most of the shore is ice and snow but when I see exposed rock I usually scan it for signs of life. There are many conditions to consider when deciding on a home. Falling ice and rock must be avoided as well as the waves of a rougher sea rising to wash it all away. But above all the closer it is to a reliable food source the more desirable the location.

White guano or bird waste indicates a diet of fish while a pinkish red colour signals a diet of krill. Sometimes in the snow you see what seems like green guano. This is not guano but the regurgitated lining of their stomachs where the poisonous fluorine from the shell of the krill is absorbed and then expelled. Nature's clever way of avoiding trouble at the gate.

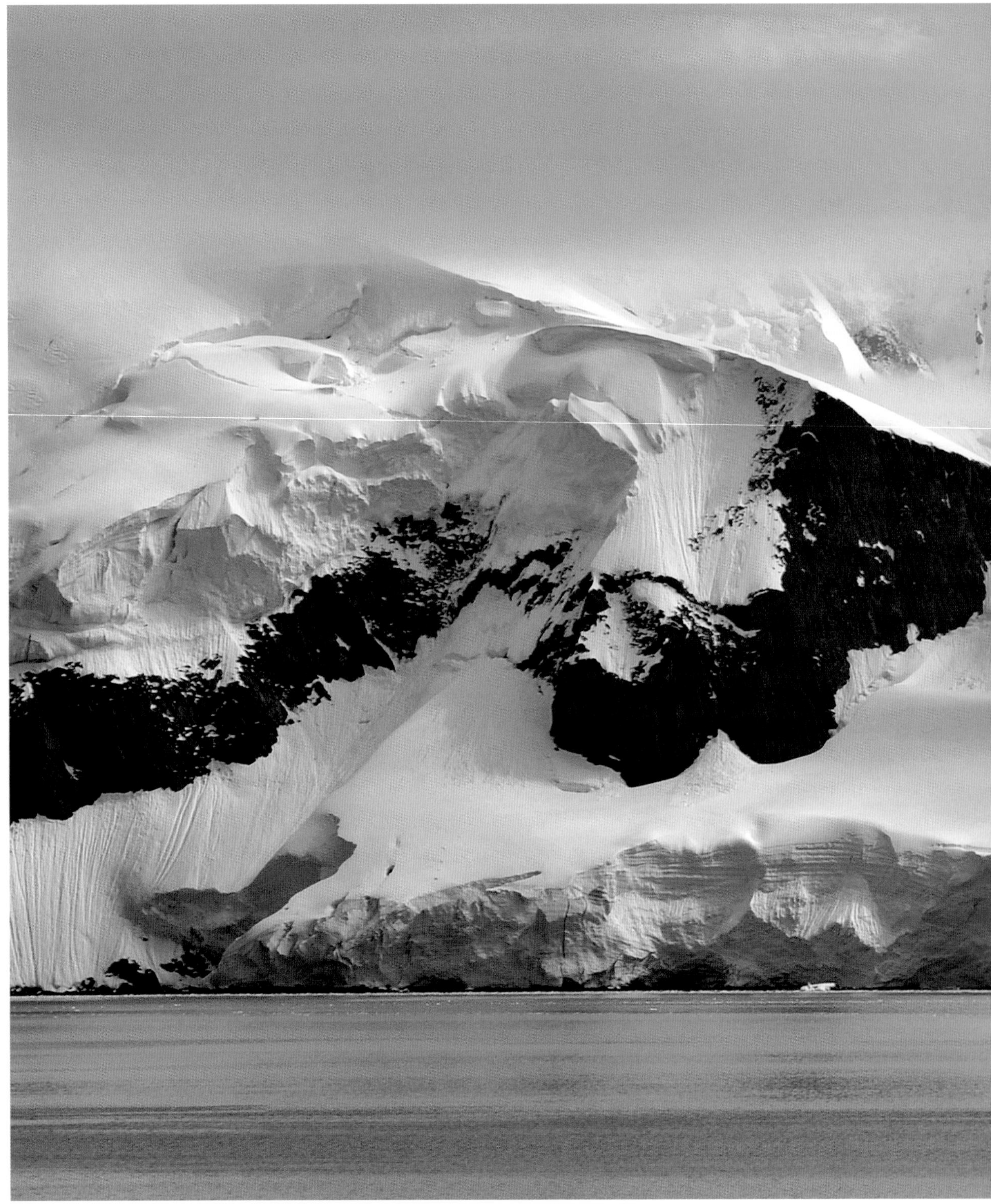

Penguin trail on an exposed shore

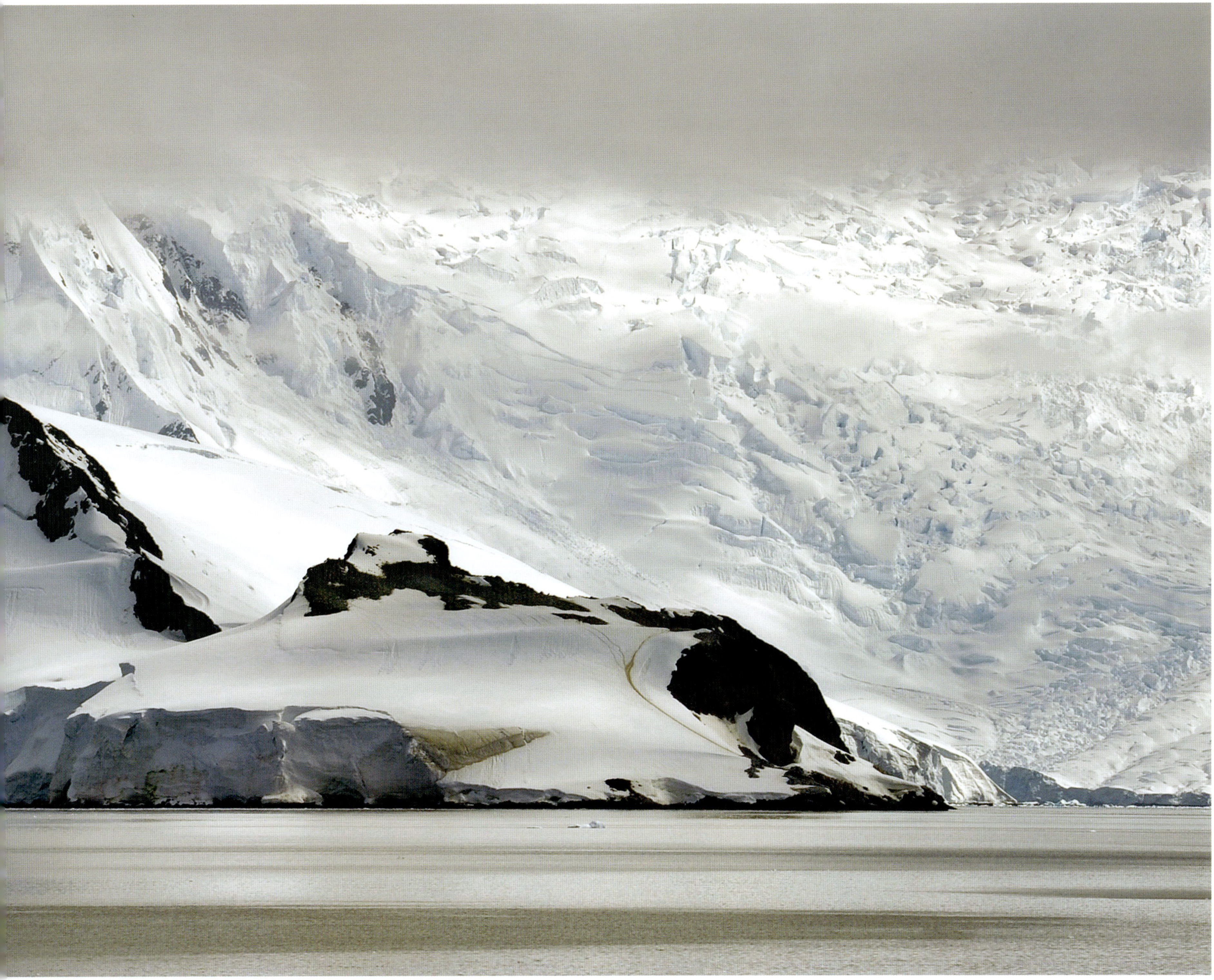

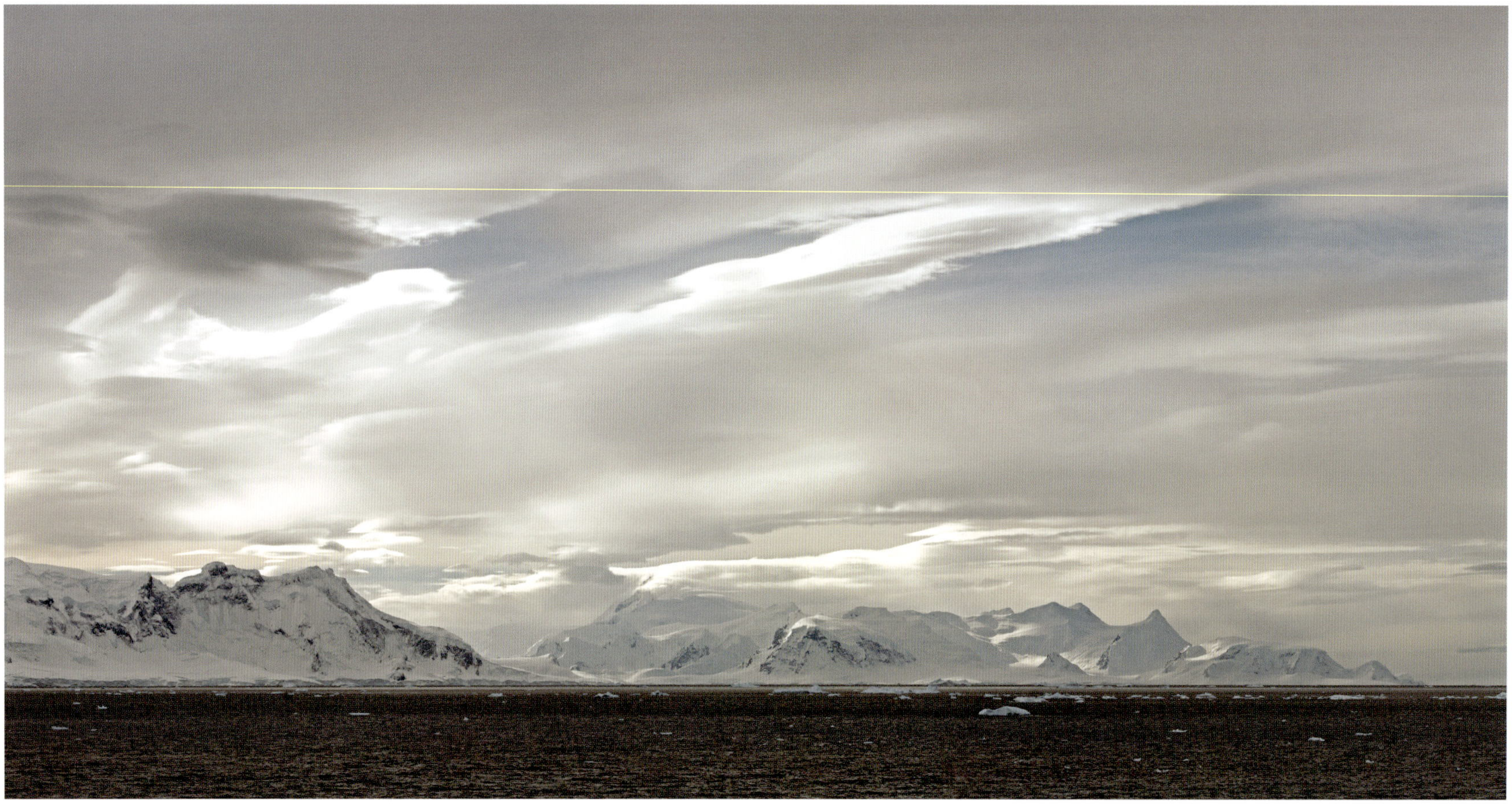

Alabaster sky

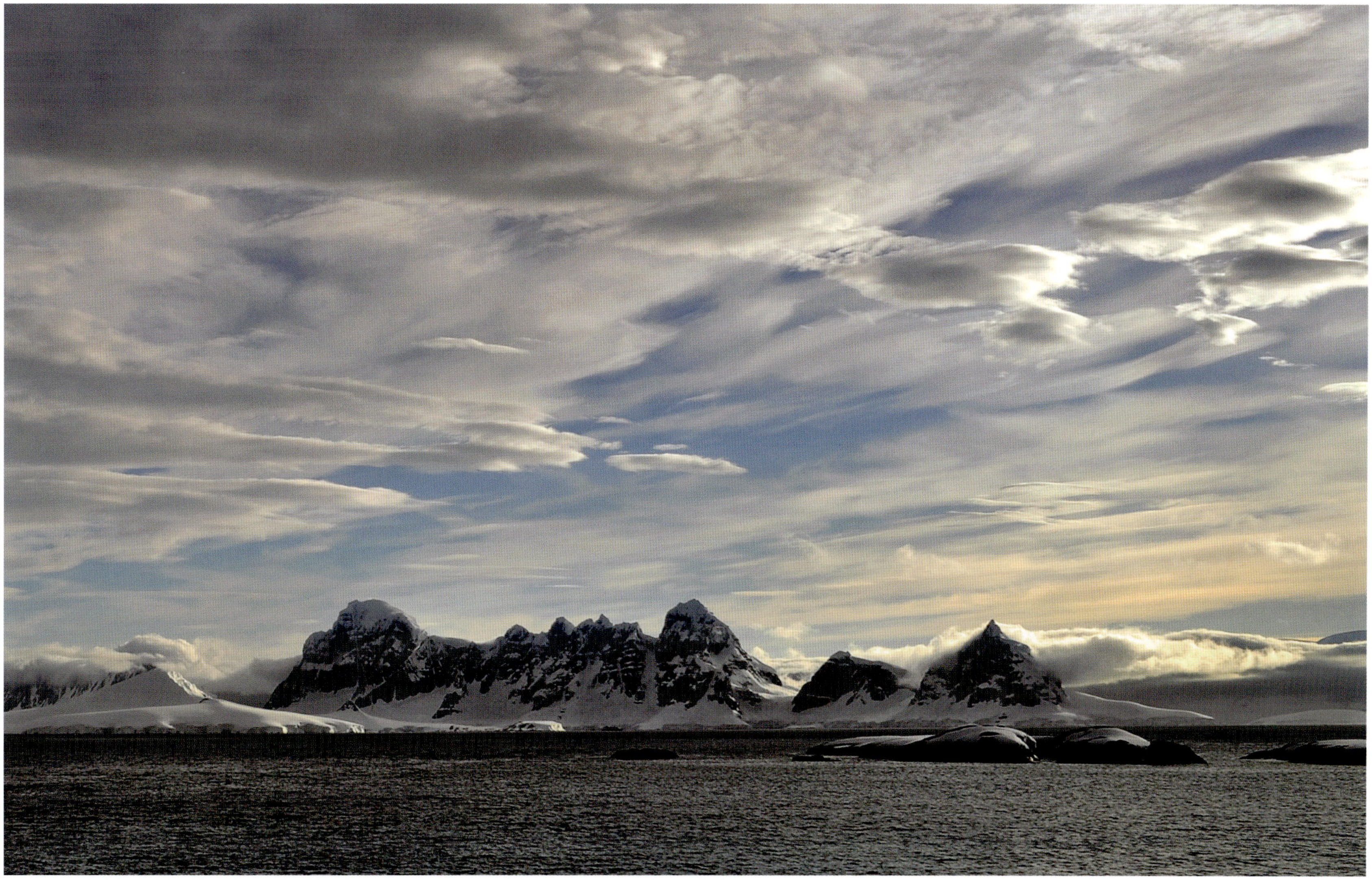

Bewildered sky

SOMEONE WHO HAD been to Antarctica many times advised me to remember these three useful words if I ever needed to describe to someone the exceptional scene that will periodically appear before me. I wasn't familiar with one word but I was glad to have written them down, just in case.

On a return from a voyage we all had the experience of being asked, "How was your trip? What was it like? What did you see?" Trying to describe and explain your experiences that revolve around the familiar and common may be easy but when challenged to convey the grandeur of an unfamiliar landscape enveloped by an awe inspiring light you know you will certainly fall short. You know from the very start that it will be nearly impossible and sometimes feel that it is not even worth trying.

Even a poet wielding the three useful words I was suggested, numinous, ineffable and ephemeral, would be no further ahead. Their meanings are as elusive as the image they are trying to portray. Using them you would be no closer in conveying the truth. Try manipulating the thousand words that are worth a picture and even with all their possible combinations they would in most cases not be able to communicate Antarctica's visual impact. Experience and direct involvement is the only strategy. Antarctica challenges you.

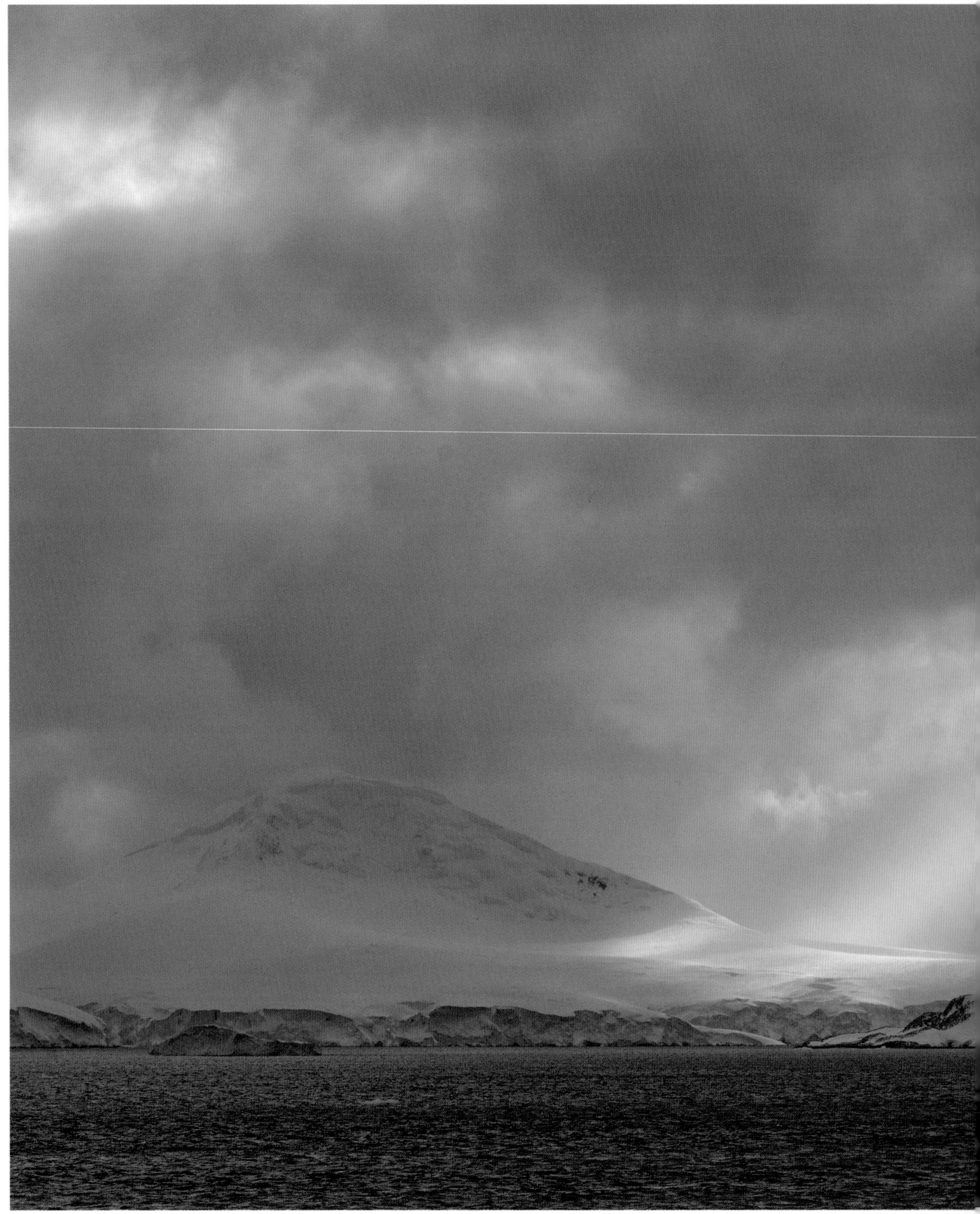

Numinous landscape

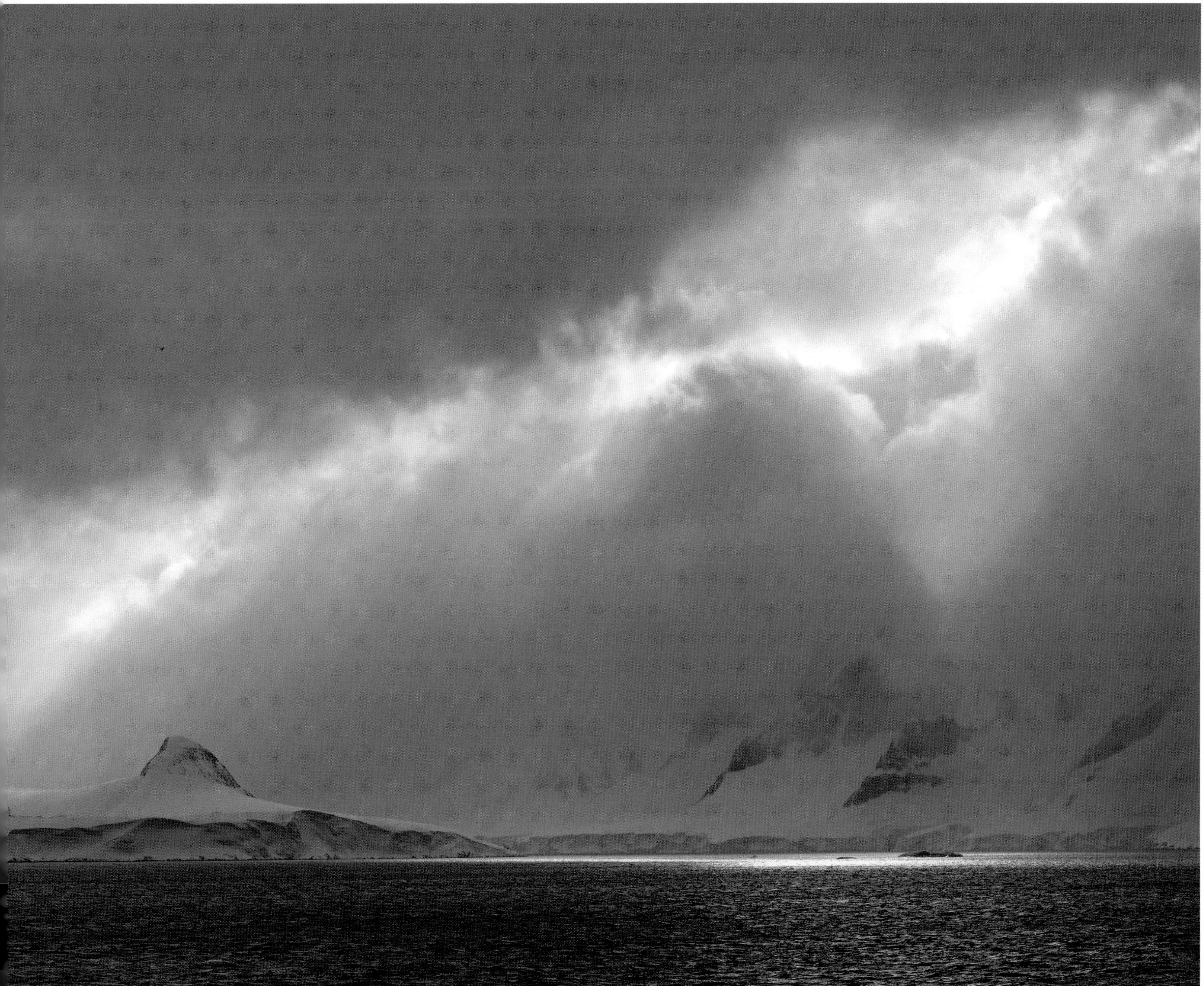

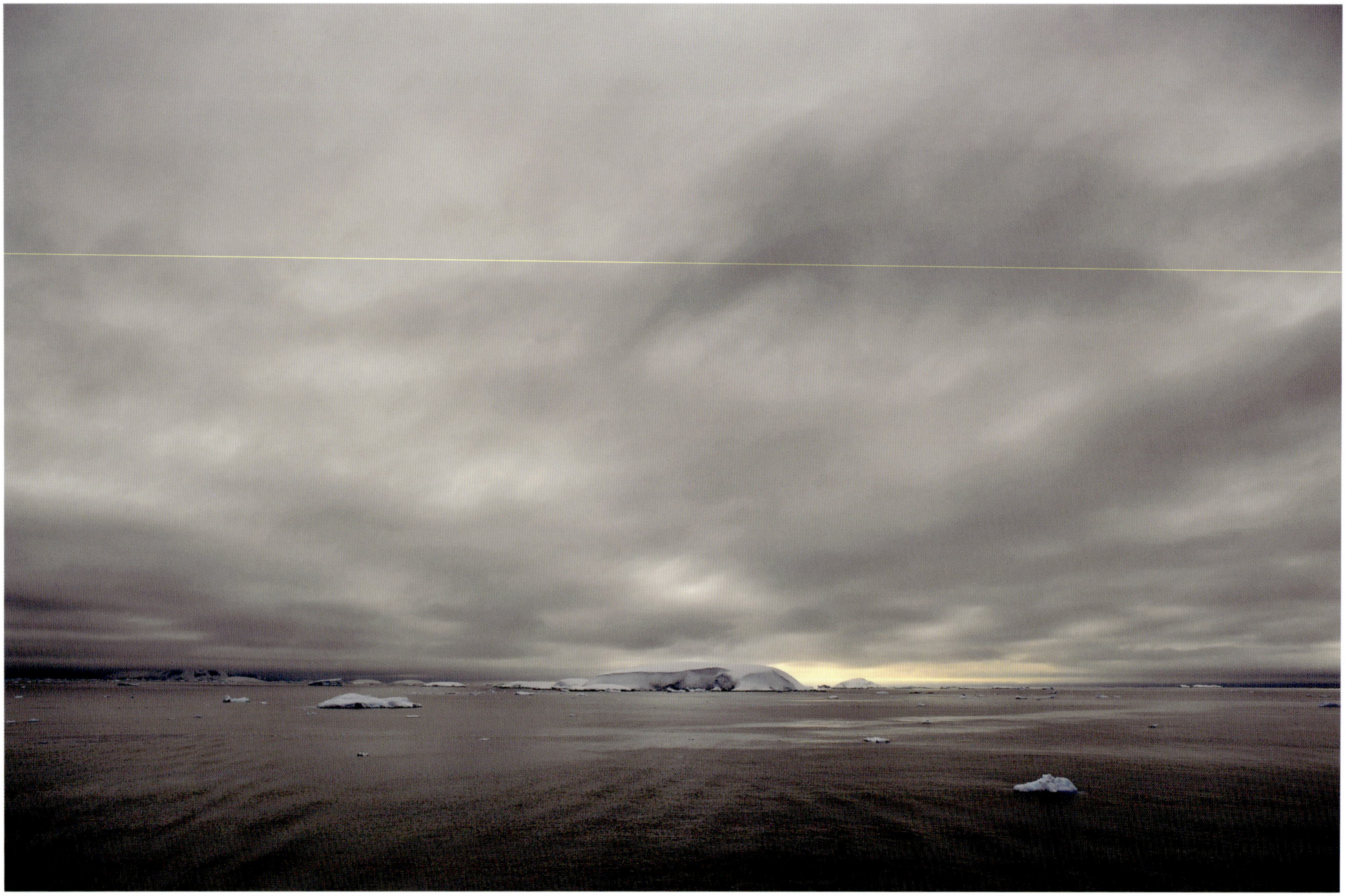

Grey sky evening light

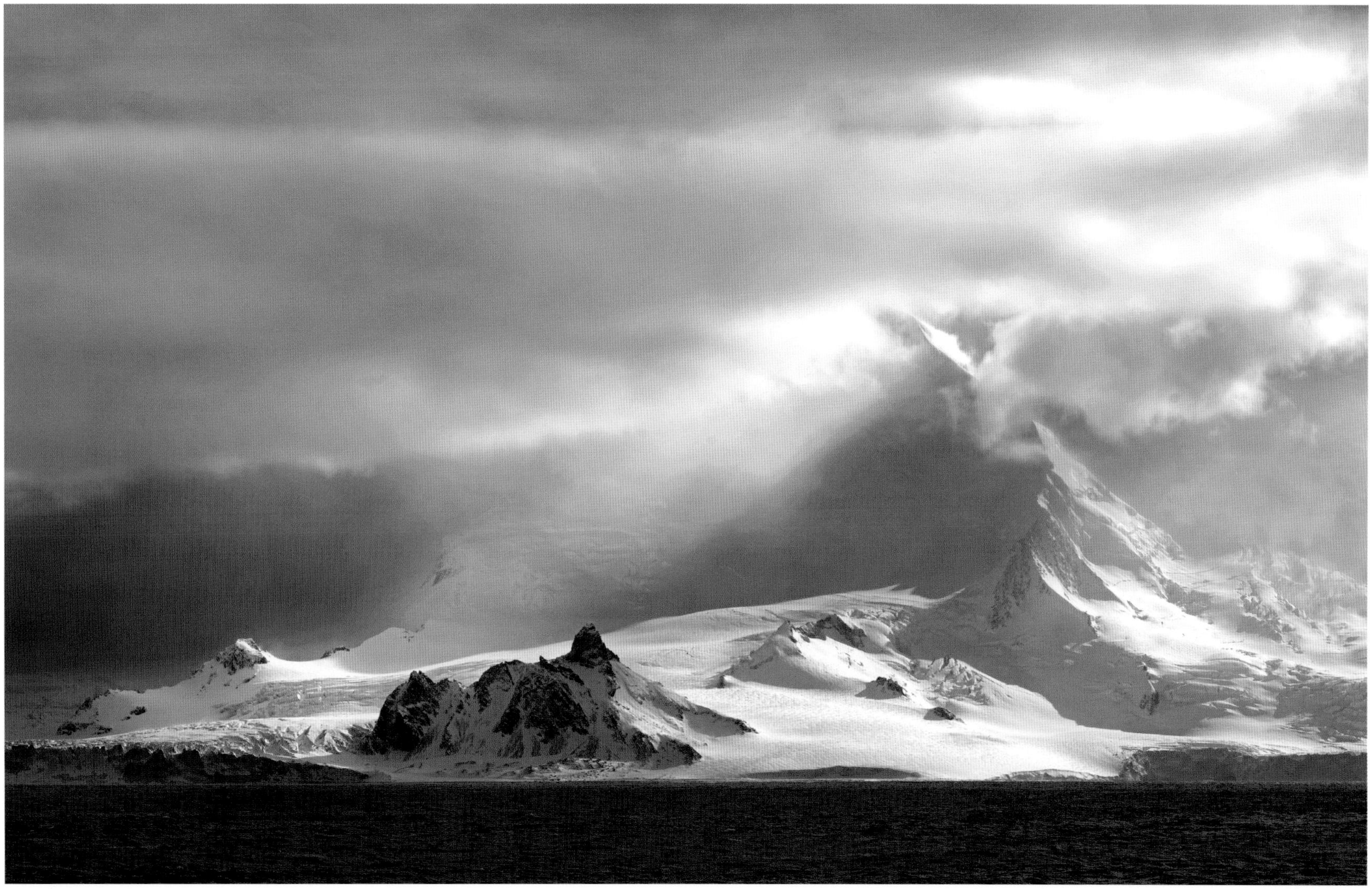

Break in the clouds

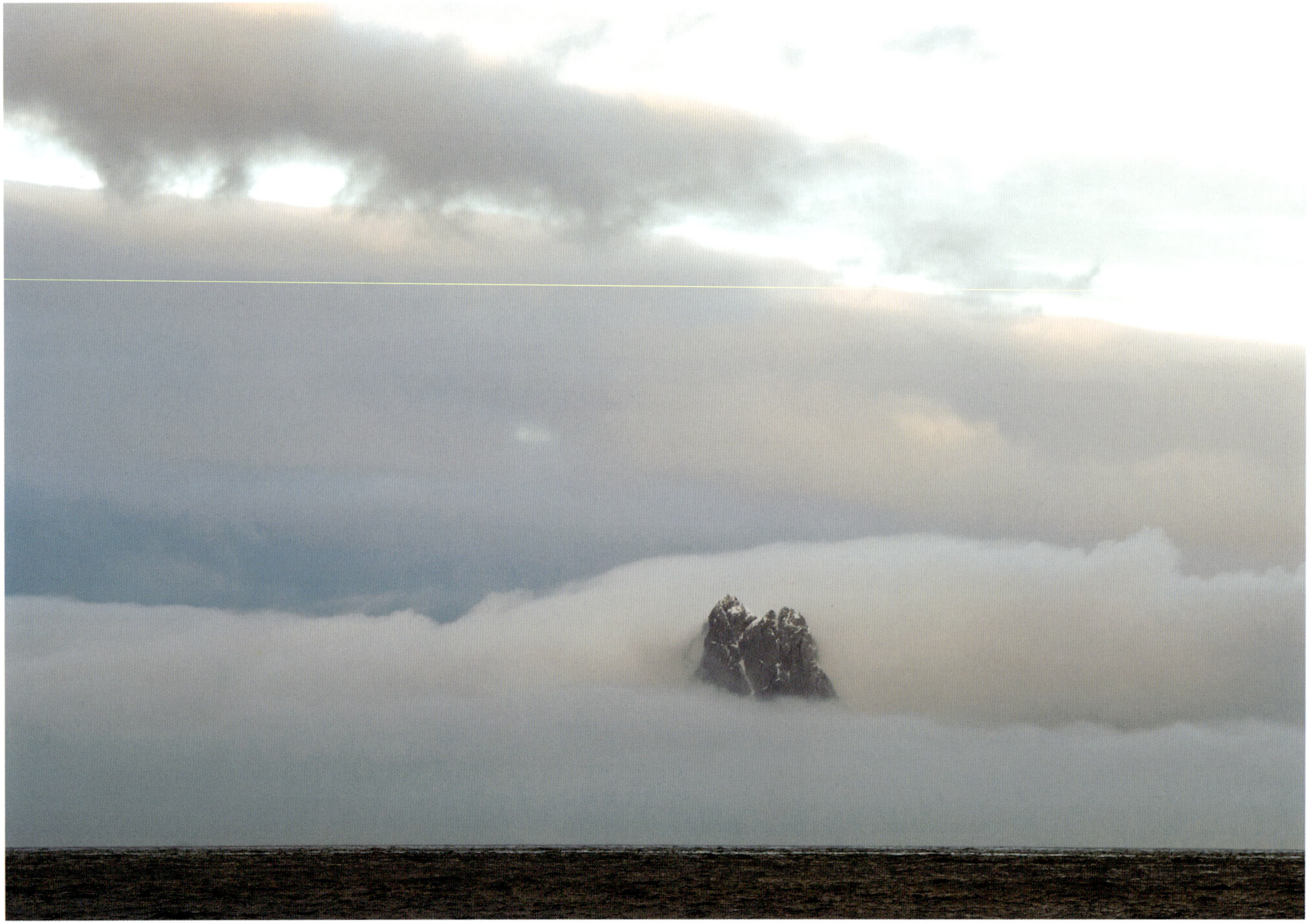

Hidden island

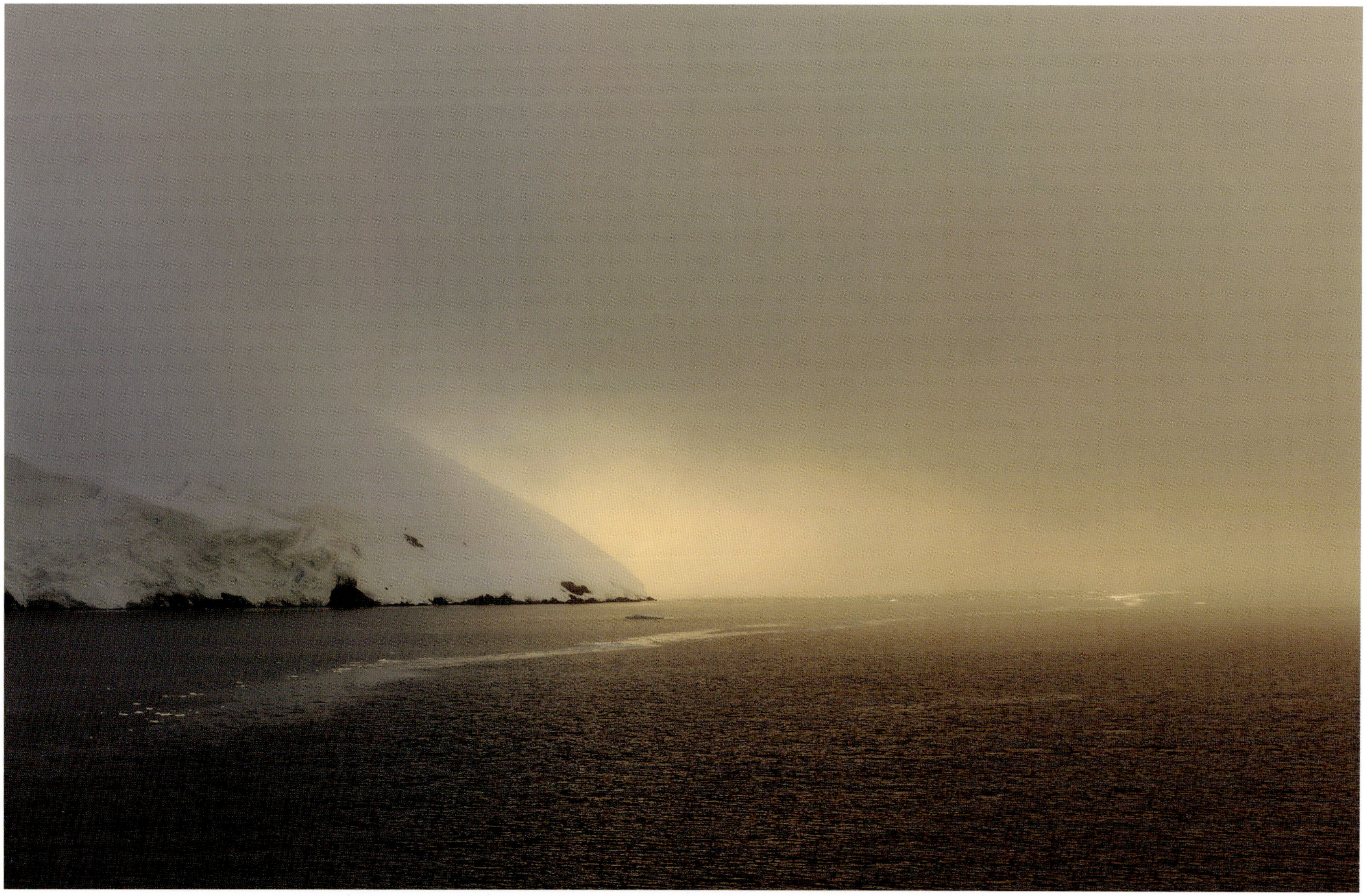

Rising mist

THIS IMAGE WAS taken from on top of a glacier on the Danco coast. Emile Danco, was a member of the impassioned expedition of 1898. Their ship, the *Belgica,* became trapped in sea ice and held fast for over a year. Above an unknown sea the men drifted aimlessly in the pervasive dark fighting all the challenges of an Antarctic winter. It was a harrowing ordeal physically and psychologically. Some went mad in the constant darkness. Danco suffered bodily and showed signs of shortness of breath for a long time but towards the end his heart could not endure the constant cold and in the twilight of an Antarctic winter was carried to a hole in the ice and delivered to the deep. It was the first death of the expedition and the absolute isolation severely affected the men.

Ineffable landscape

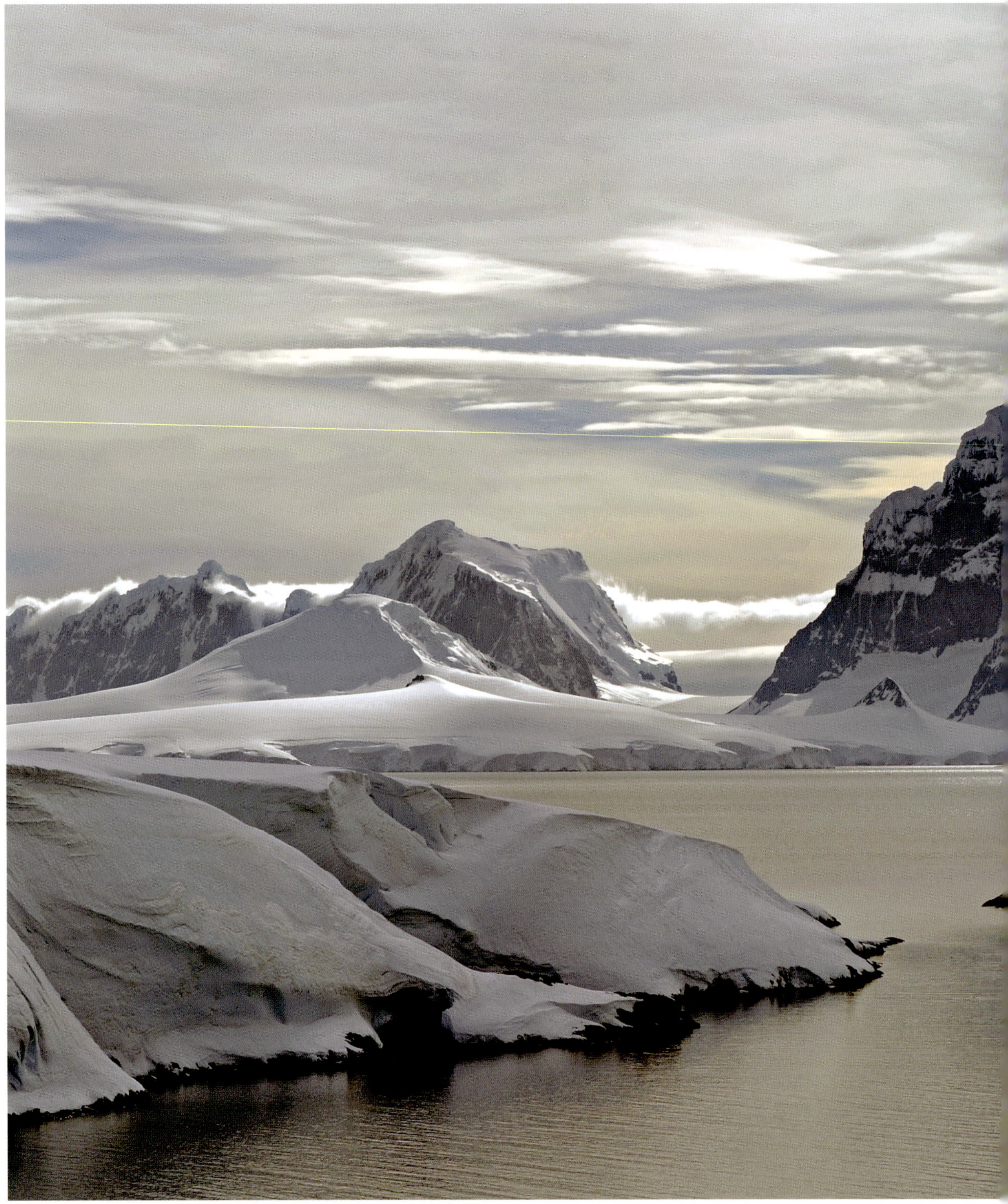

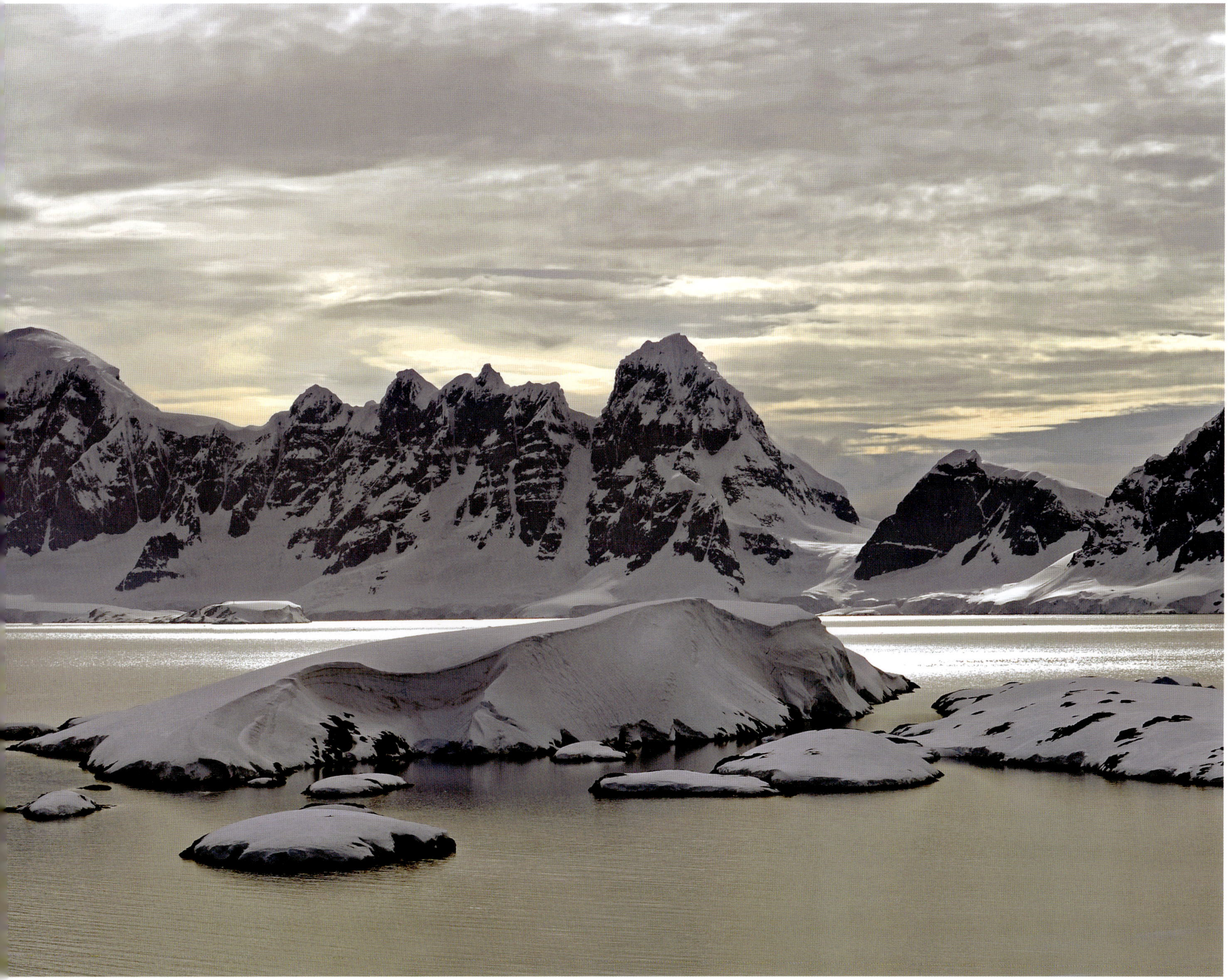

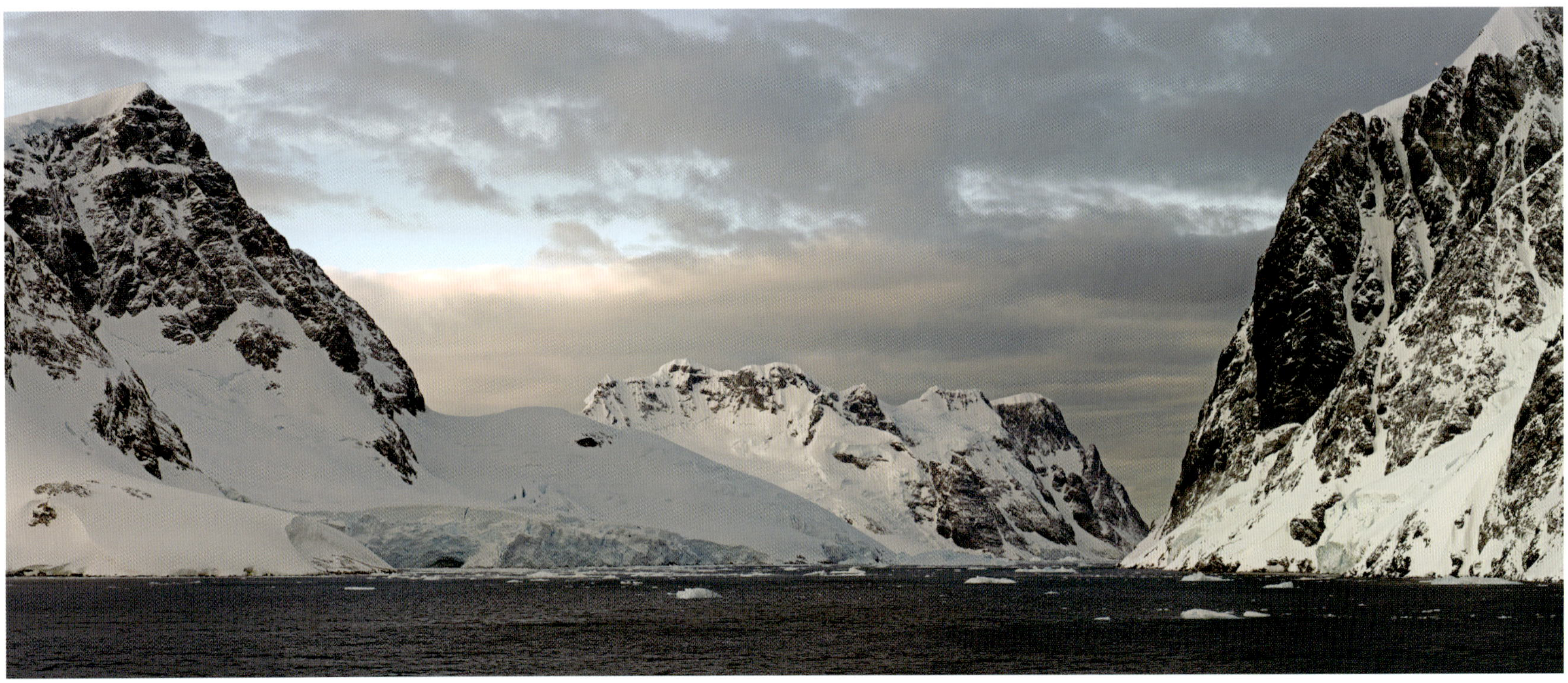

Entering the Lemaire Channel 2010

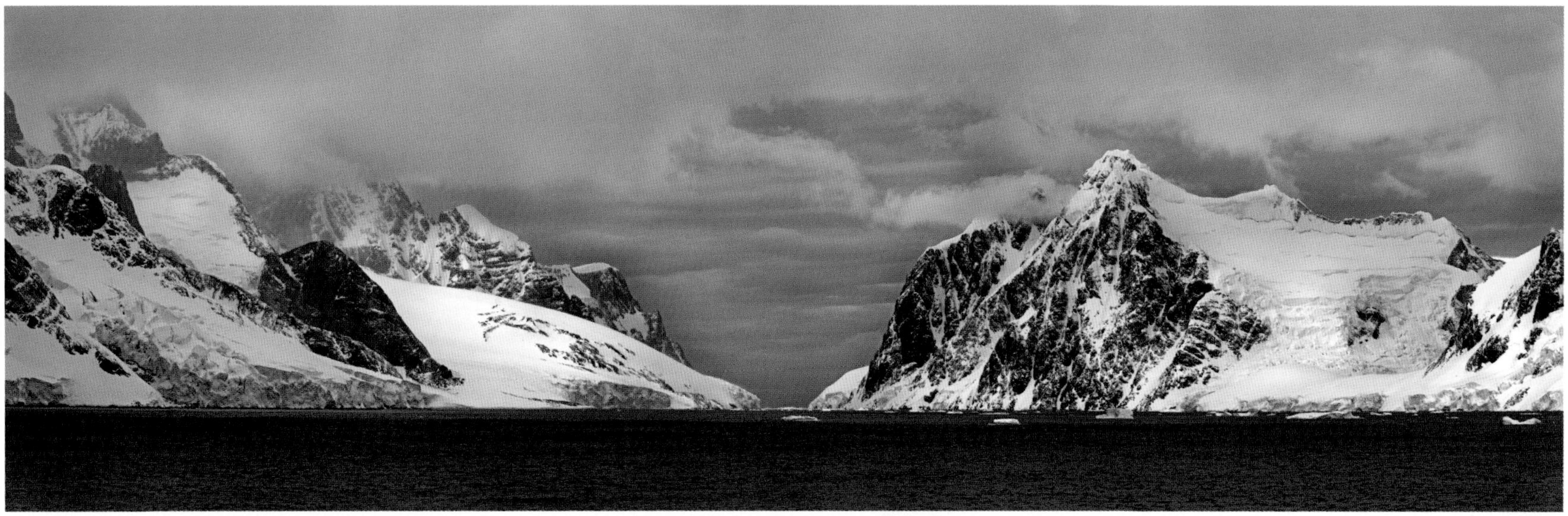

Entering the Lemaire Channel 2014

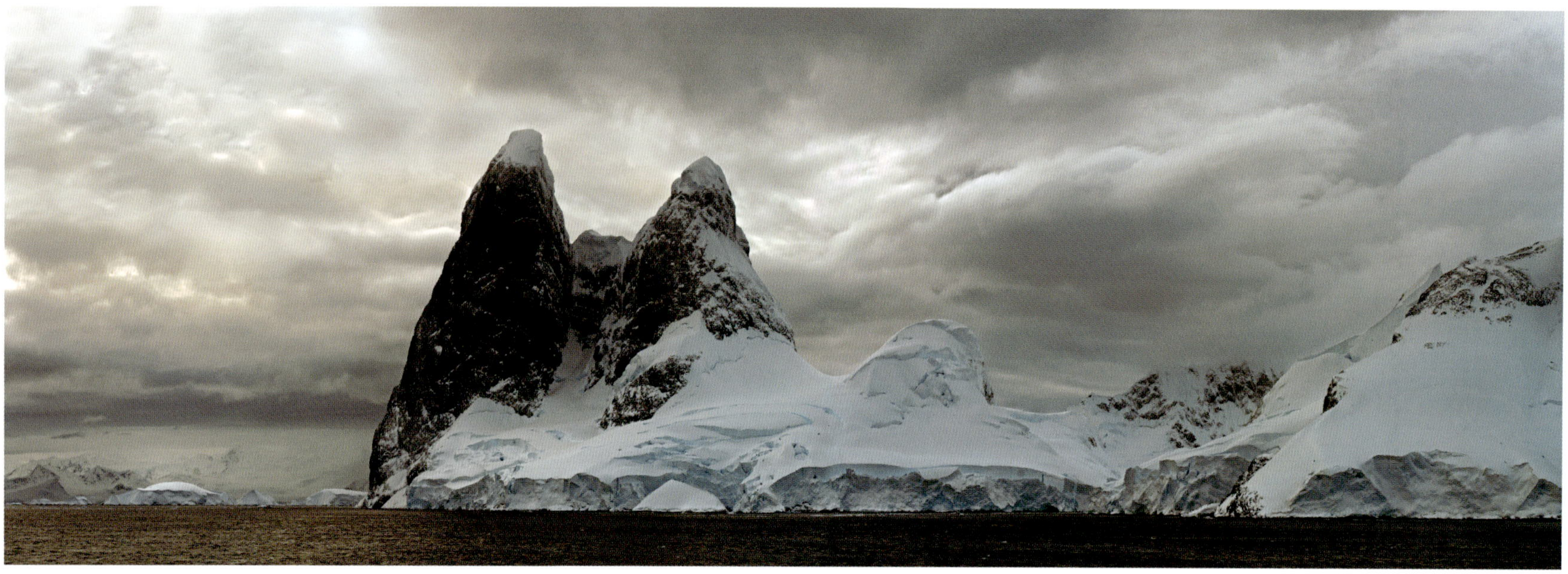

North entrance Lemaire Channel

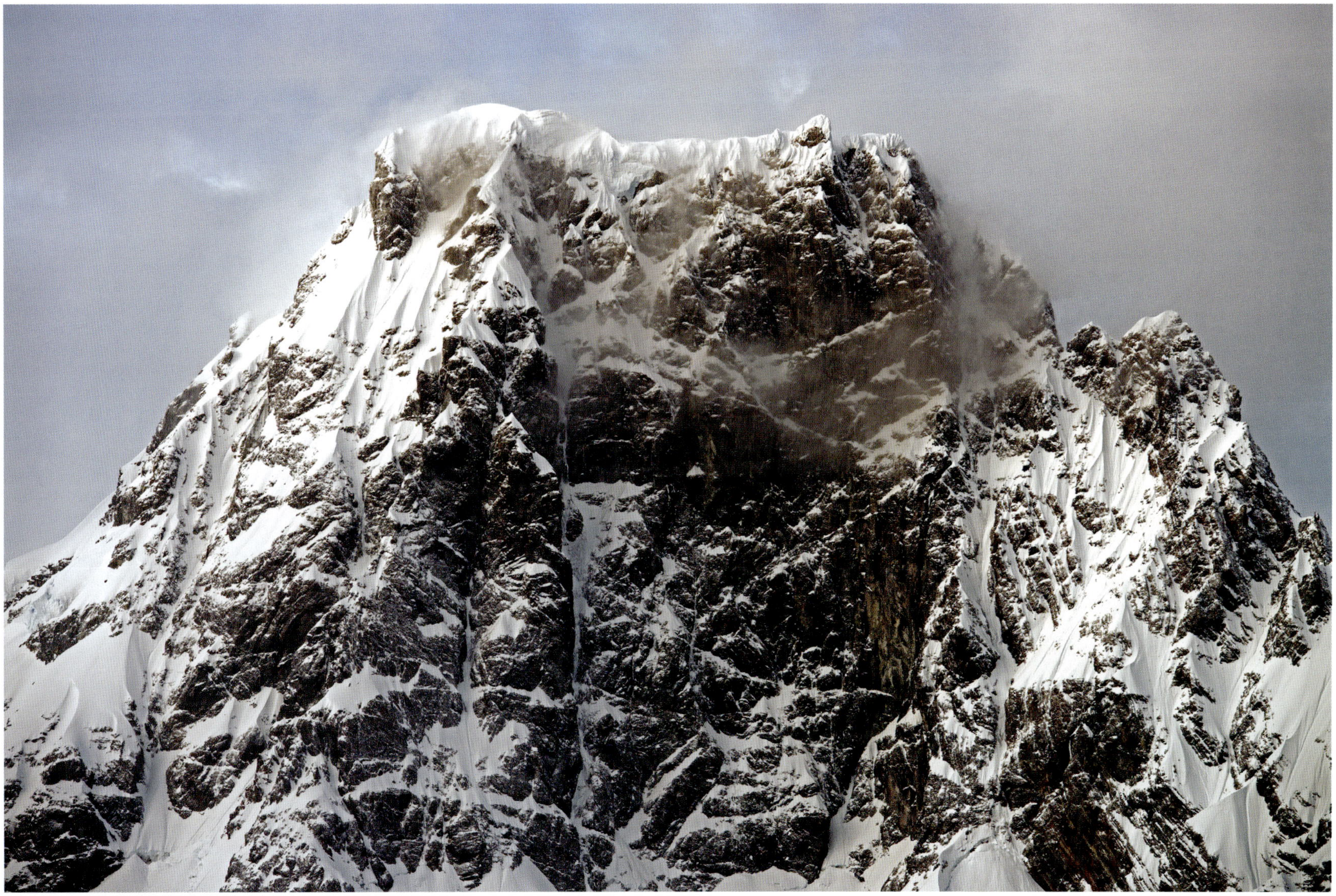

Wandel Peak Booth Island

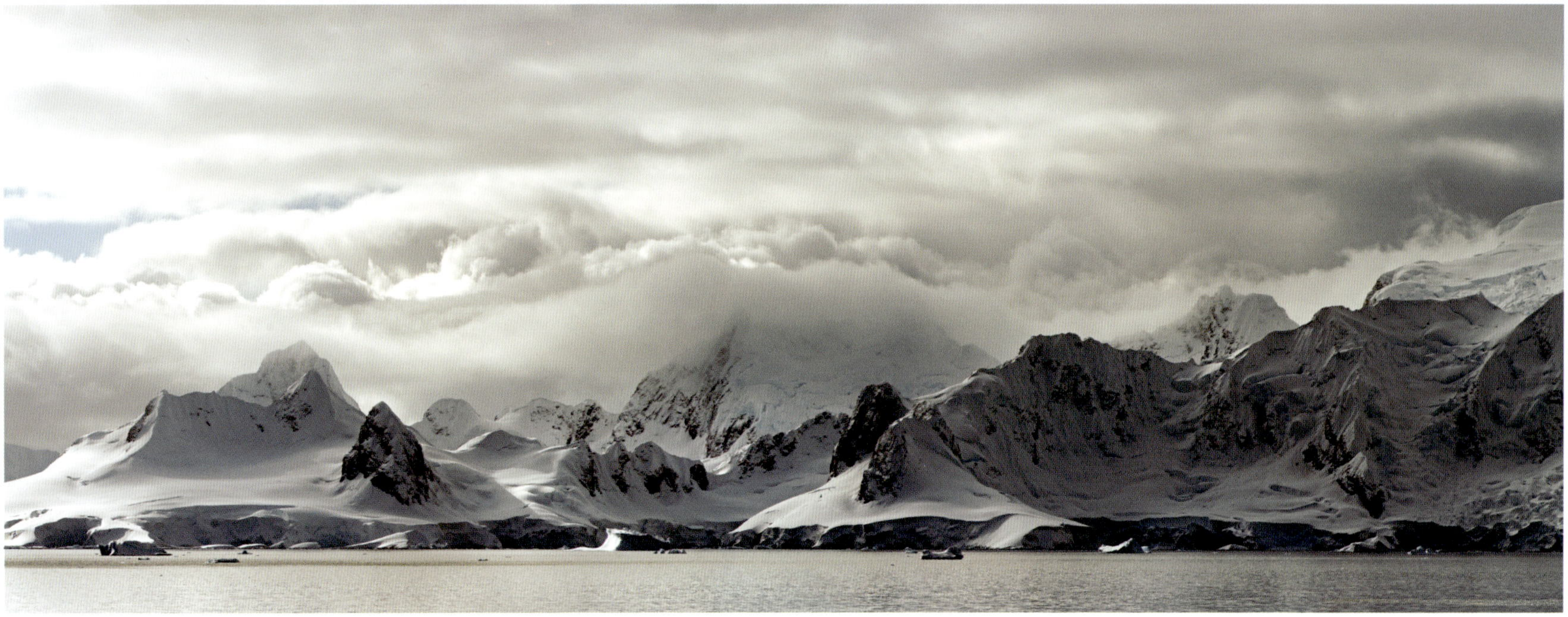

Storm clouds over the Danco coast

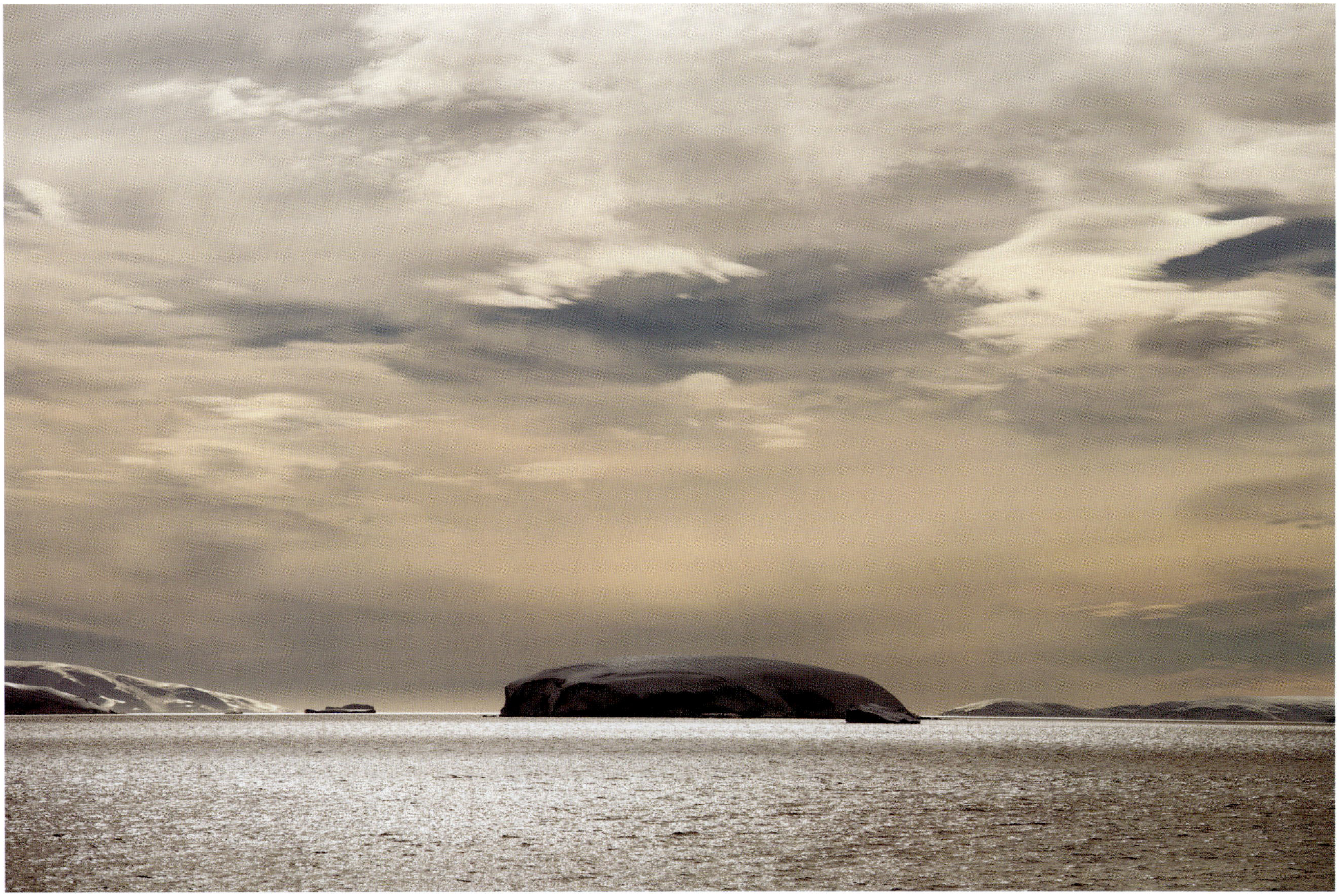

Ice covered island

AT CERTAIN TIMES the most tempestuous landscape in the world can reveal its serene and tranquil side. Displaying captivating colours and soft textures Antarctica will charm the travelling eye, luring it endlessly over unfamiliar landscapes. Enveloped in its mysterious features Antarctica not only can but will seduce and captivate. It can easily manipulate your mind from an uncomfortable position of physical cold into a realm of psychological warmth.

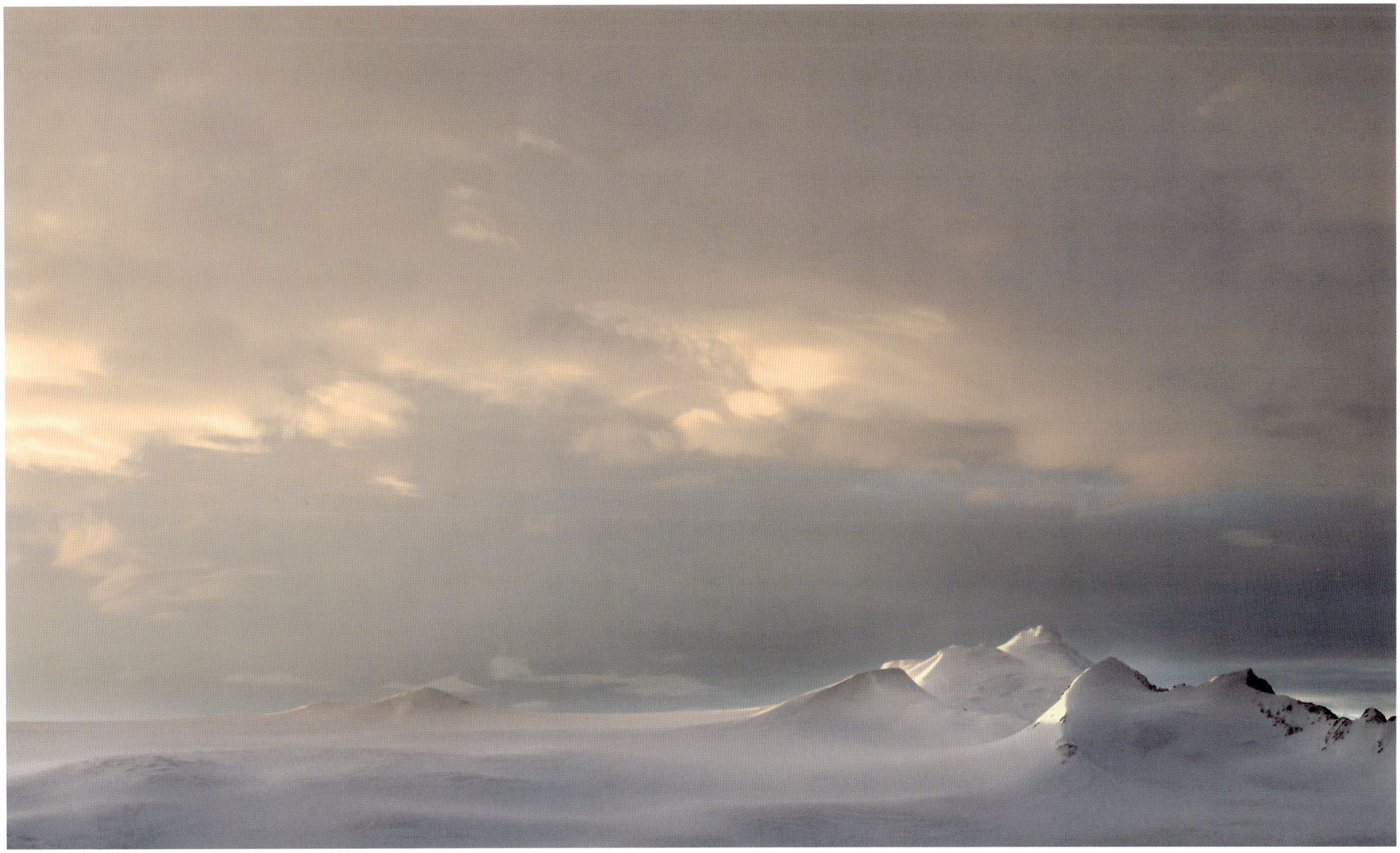

Antarctic twilight

THE WHOLE DAY could be lost to fog and mist with the sky remaining bland to the point of abstraction. All forms are lost to a dull, even light and depth perspective vanishes. There is no colour, no change, and all is monotone and timeless. All is still. Only the ship's wake and continuous vibration disturbs the scene. However, with the long summer days change is inevitable and when it happens the sky transforms into a flurry of activity and leaden clouds fill with unexpected colour and brilliance.

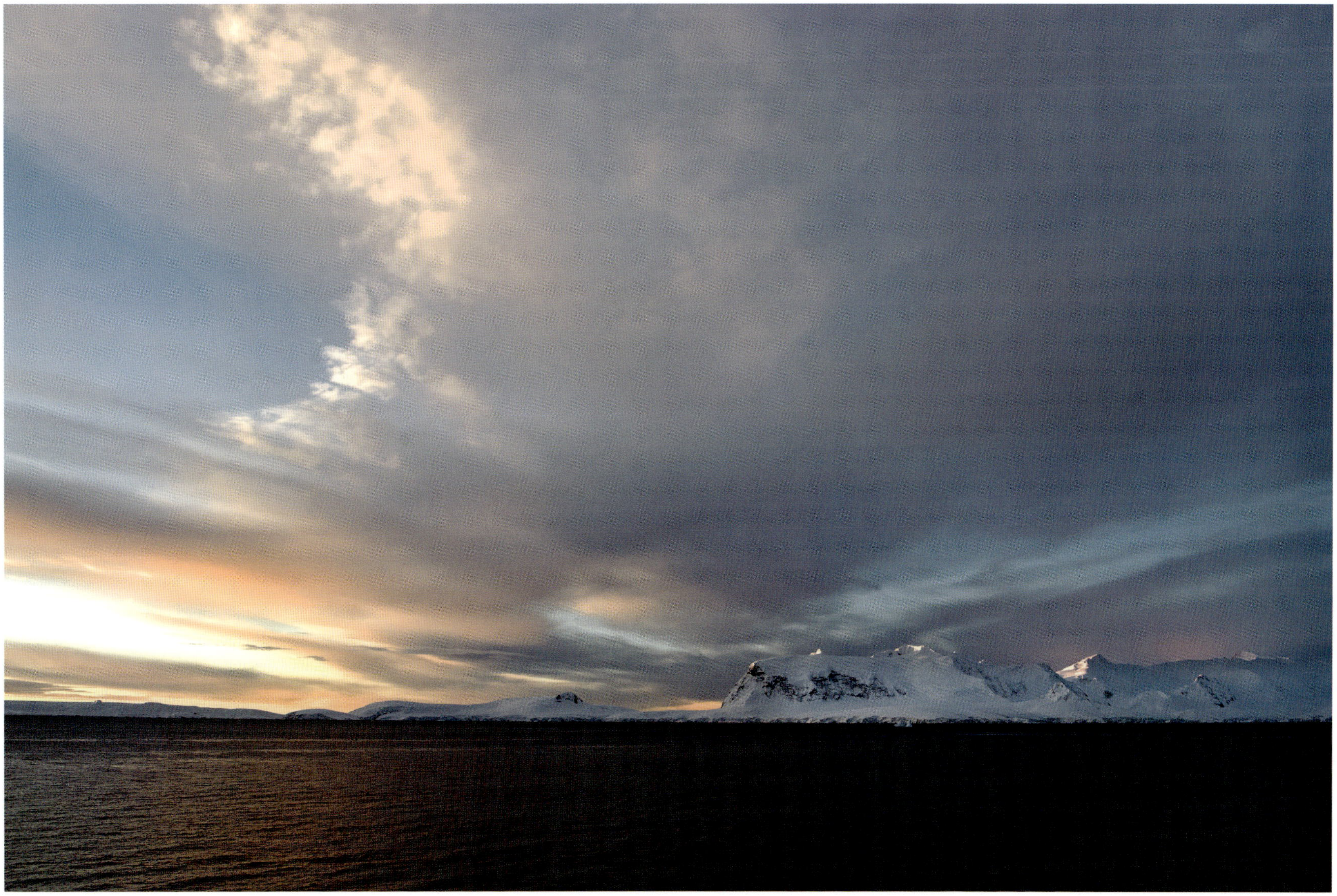

Evening

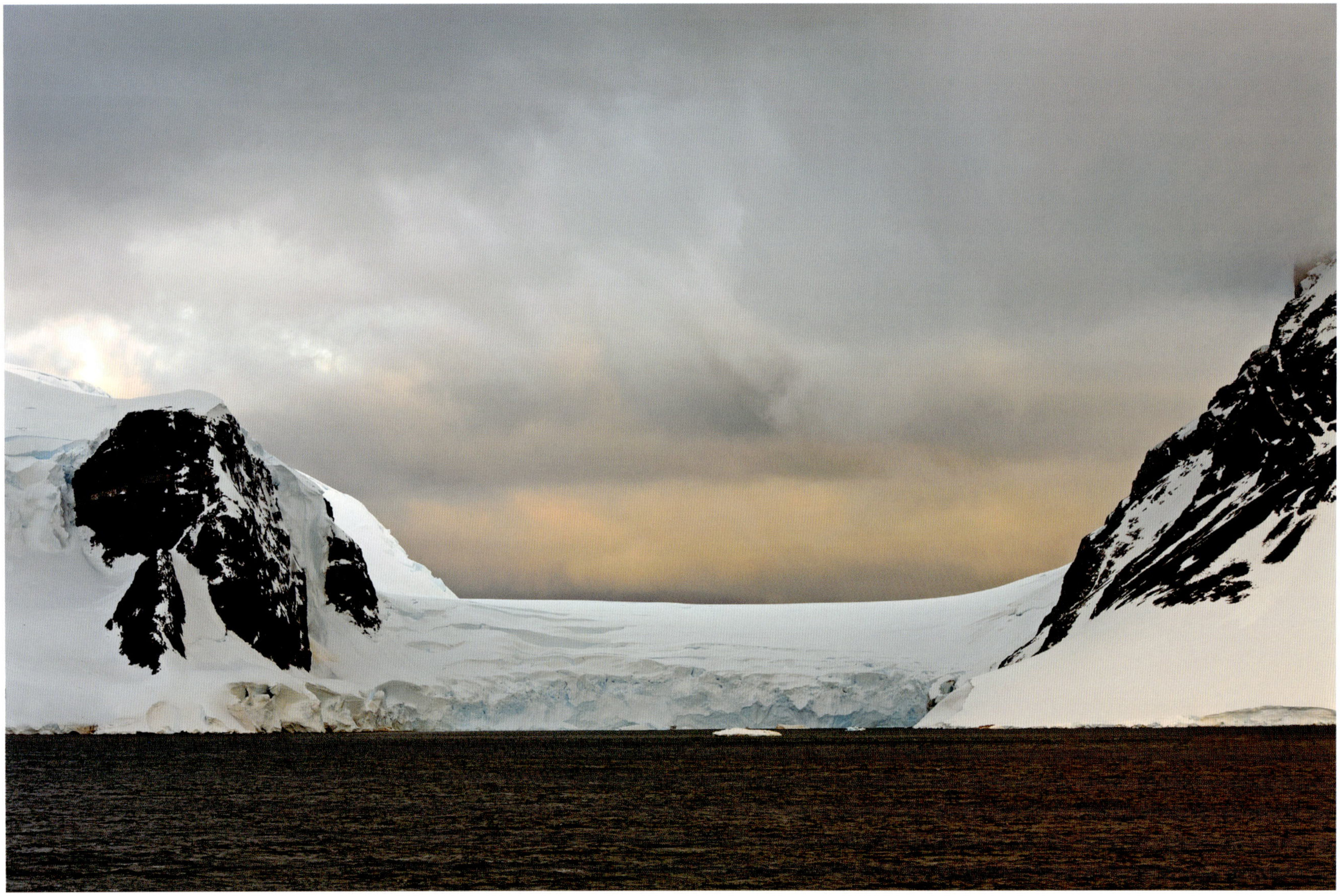

Glacier flow

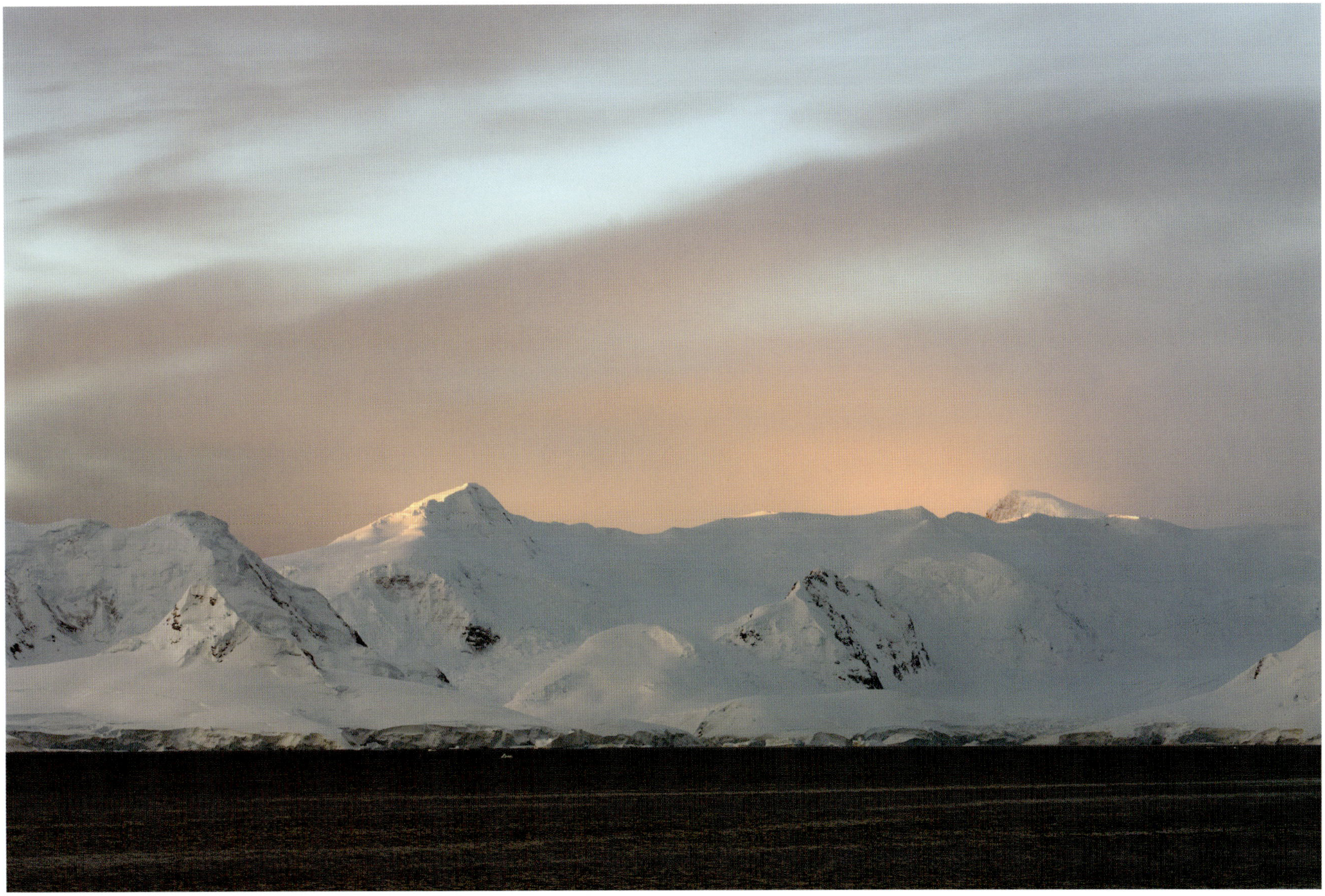

Antarctic sunset

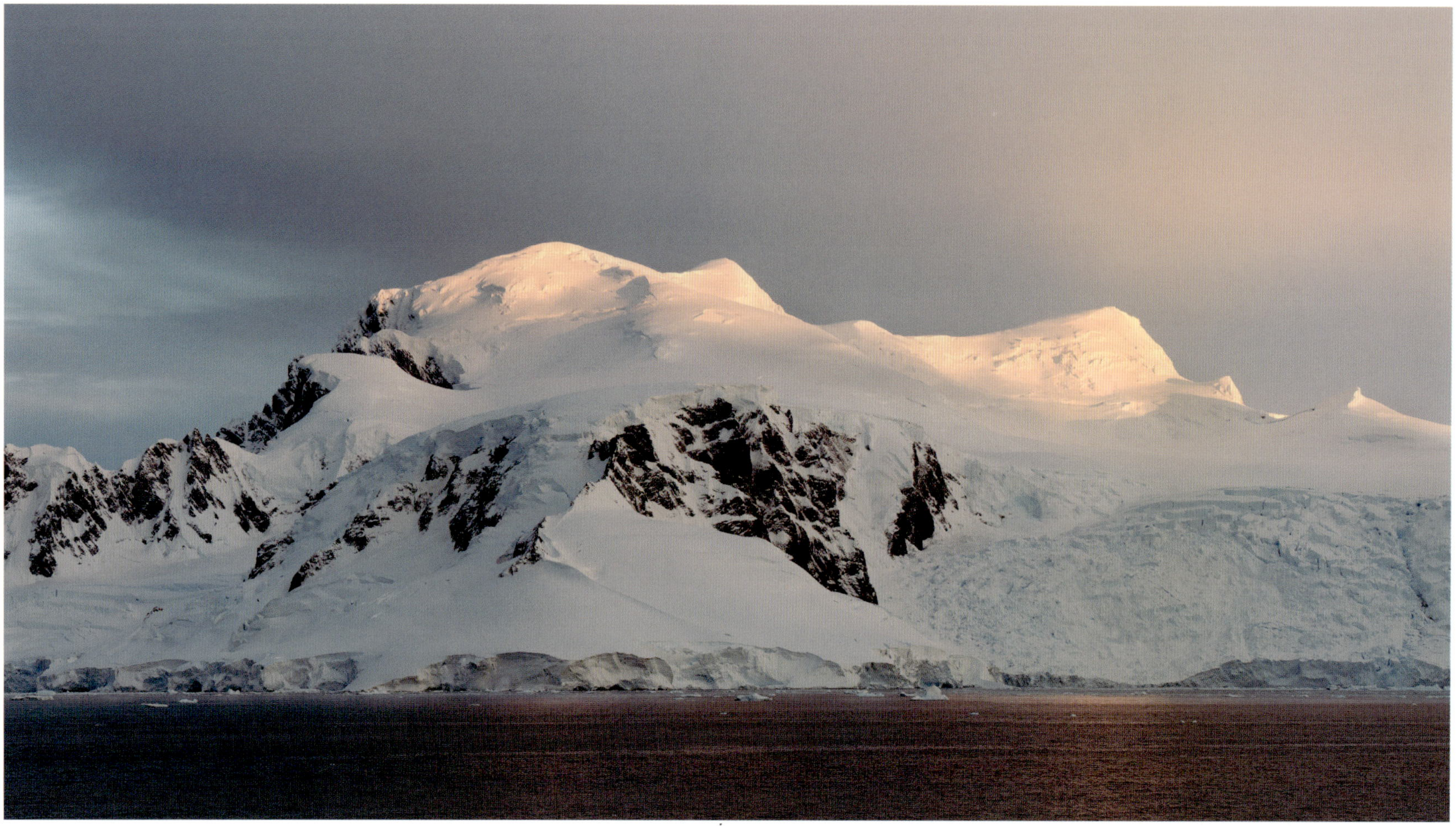

Alpenglow

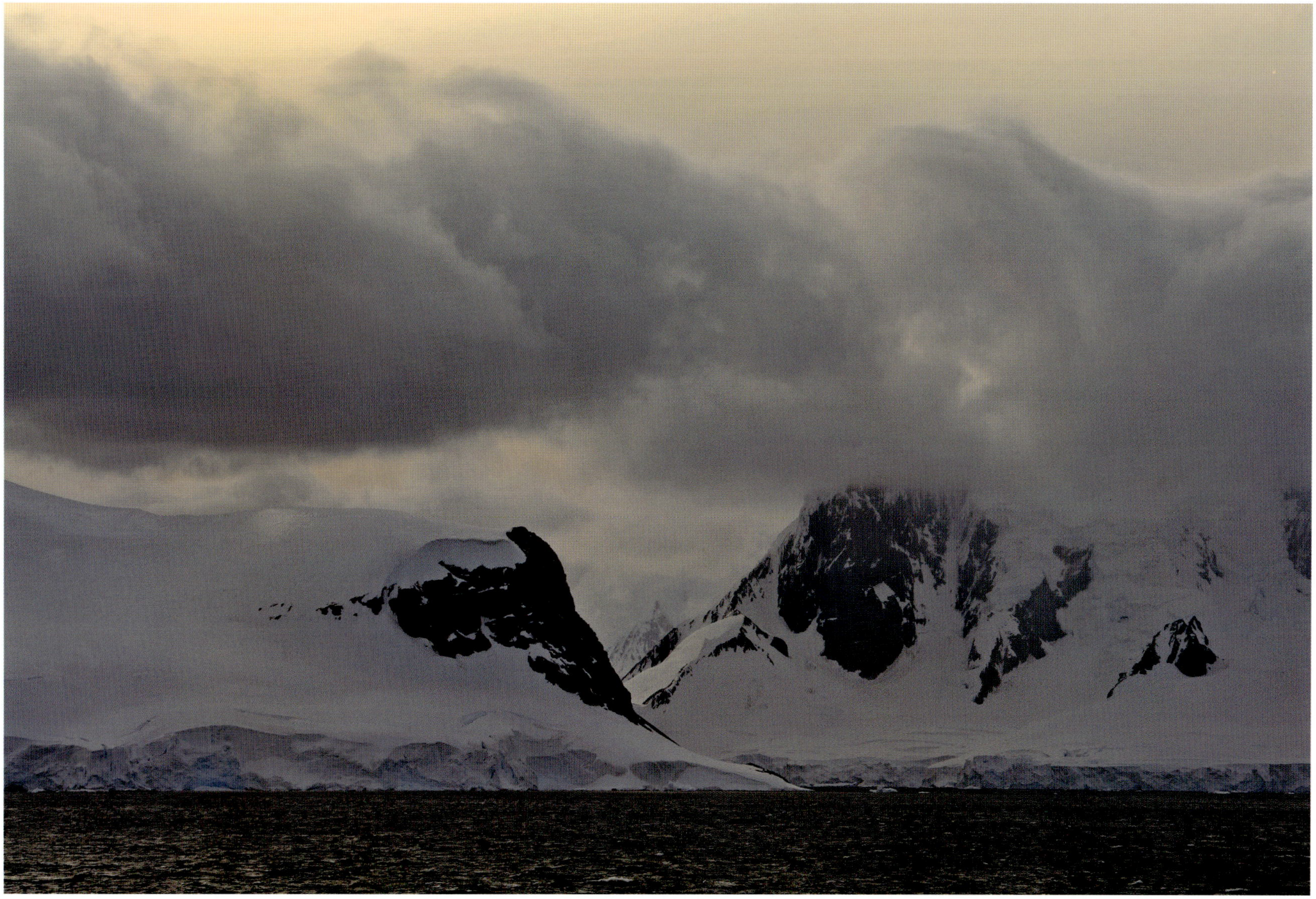

Hidden Gap

SECOND ONLY TO the killer whale the leopard seal is Antarctica's top predator. Weighing in at an average of 540 kilograms or 1200 pounds it is a ferocious hunter that uses its speed and cunning to seize any penguin that lingers or drops its vigilance. When stretched out to a strike its prey, its long head and slender neck give it a daunting reptilian look that exaggerates its ferocious reputation.

It is not known to attack humans yet the few encounters, fatal or not, have been enough to make us wary in its presence. Curious by nature it readily swims aside Zodiacs, raising its sinuous head, nostrils flared, expelling bursts of air. Sometimes gliding by with a raised back it is tempting to reach out only two feet and touch such a captivating animal. With so many tempting buttocks hanging over the Zodiac's edge it is a favourable testament to its benign attitude towards humans. And yet unfortunately the exception exists.

In 2003 a young marine biologist who regularly swam in these waters gathering specimens was grabbed by the head, dragged around for a few moments and seemingly played with. Released and then grabbed again, according to her dive computer, she was then dragged down 230 feet and held there for six minutes. In an incident such as this there is

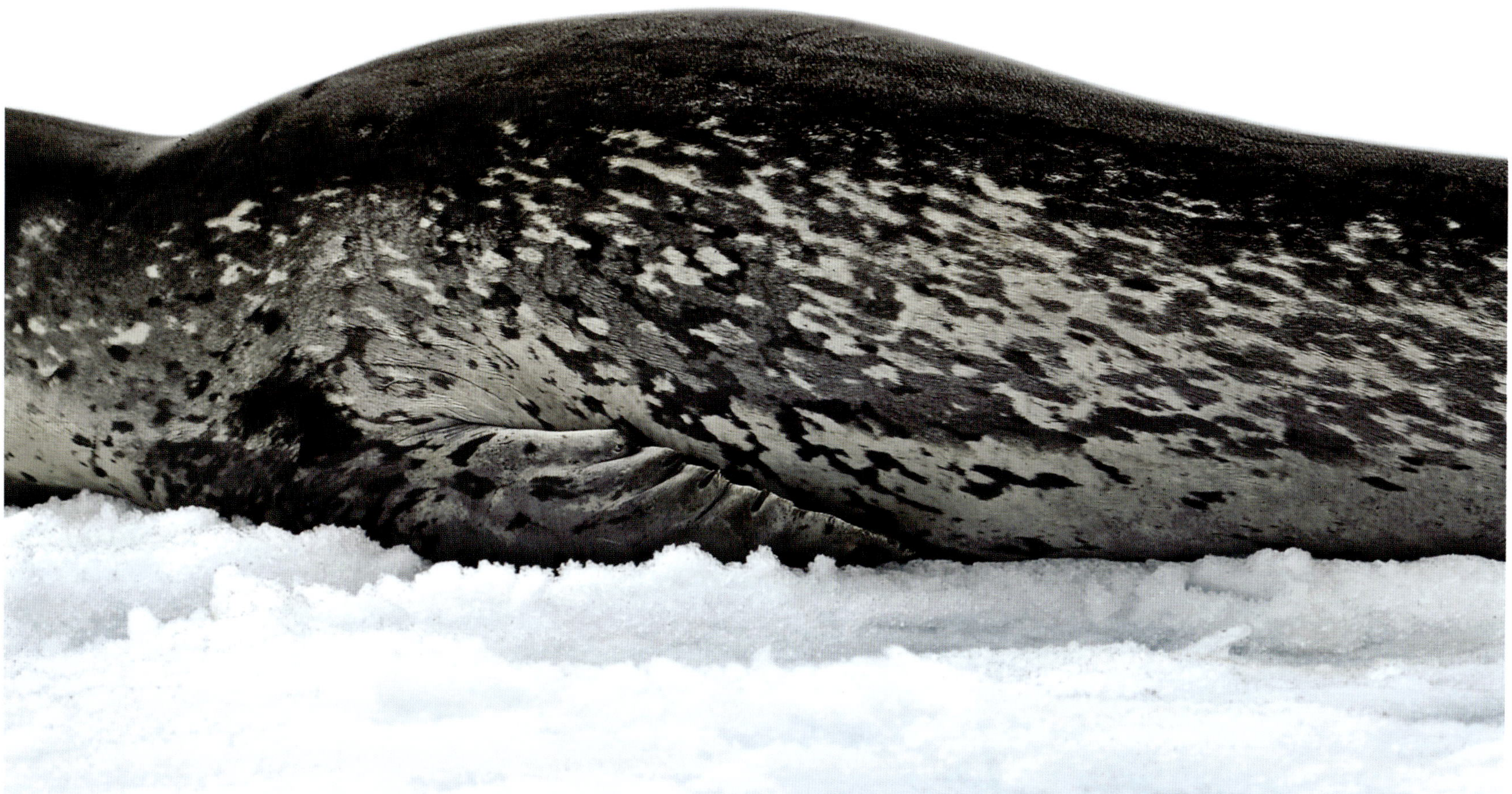

Leopard seal flipper

not much a second person nor people standing watch could do. A rare occurrence but it definitely raises an awareness of what could happen. A memorial plaque dedicated to her is mounted close to the research station where she worked.

One afternoon we had just landed a couple of Zodiacs on a shallow pebbly beach at Neko Harbour. Everyone quickly moved towards the penguin colony but I just stood there three feet from the water's edge setting up my tripod, admiring the penguins that were tobogganing down the hill. A sharp call from Sebastien, a member of the ship's crew, warned me of a leopard seal ten feet behind me. I turned to see the seal's head held high looking me over curious as to what I was. I doubt there was another intention but my confidence was not total.

After the attack on Kirsty Brown our polar plunges were strictly monitored. Twice they were cancelled due to a loitering leopard seal. Eventually some brave souls did get to plunge but only for ten seconds and very close to shore. Responsibility runs high and I suspect the monitors do not particularly enjoy this event.

Sometimes very playfully a leopard seal will slide behind a slow moving Zodiac. With a spinning body it swims in till it is about a foot from the spinning propeller and its face is fully surrounded with the bubbles exiting the engine's exhaust at the base of the outboard motor. I suspect they do this for extended periods to enjoy the tickling effect.

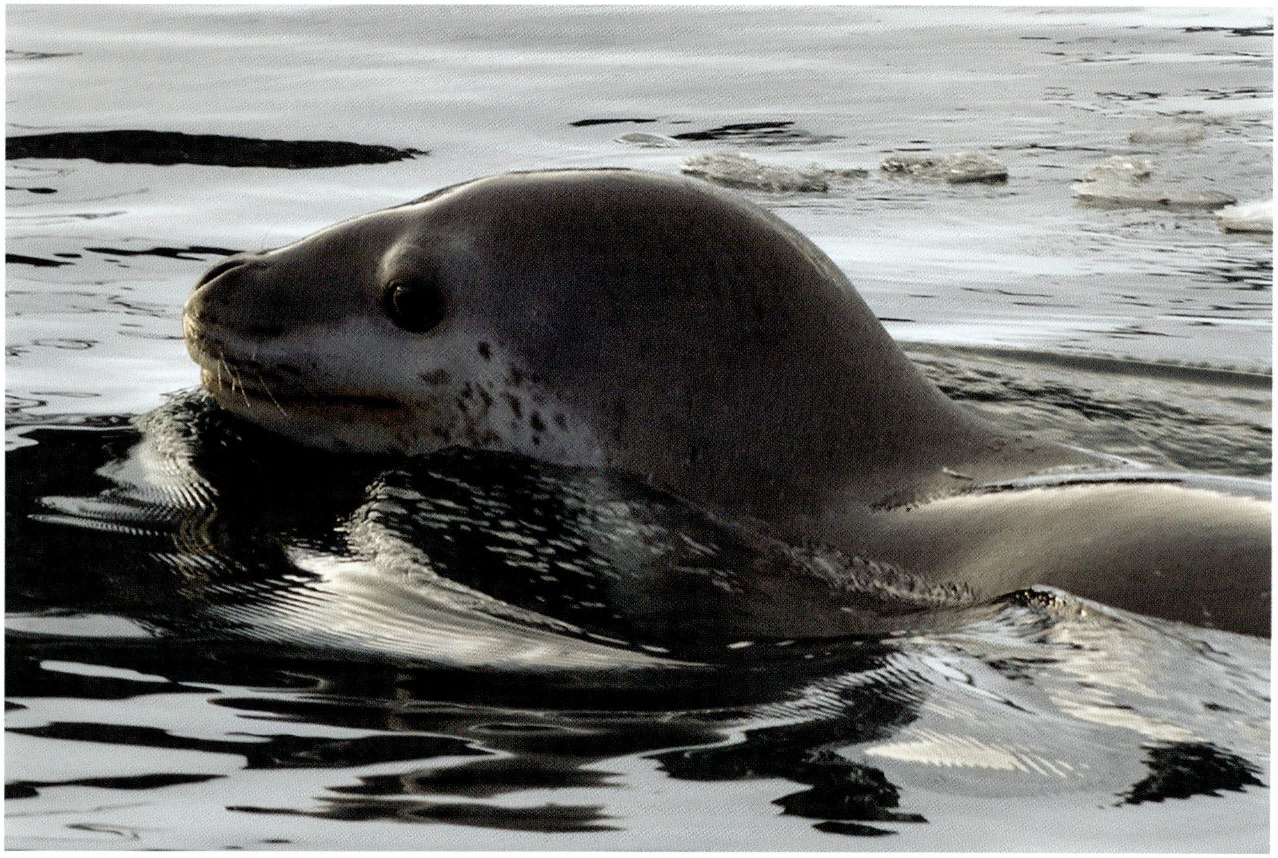

Leopard seal (Hydrurga leptonyx)

THE STRUCTURE YOU SEE here is called Wordie House named after a member of Shackleton's crew that tried to first reach the South Pole. It was constructed on the base of an older British station (1935-36) that was washed away in 1946 by a tsunami. With materials salvaged from a hut at Port Lockroy and also from the whaling station at Deception Island this became a research station to five or six men until its operations were transferred to the Ukrainian station at Vernadsky.

Many of its artifacts such as stove, kettle, furniture and various tinned food supplies remain. Wordie House has been designated an Antarctic Historic Heritage Site as an example of one of the first scientific research stations in Antarctica.

It was through this channel and along this glacial front that the tsunami of 1946 reached White Island and destroyed the British station.

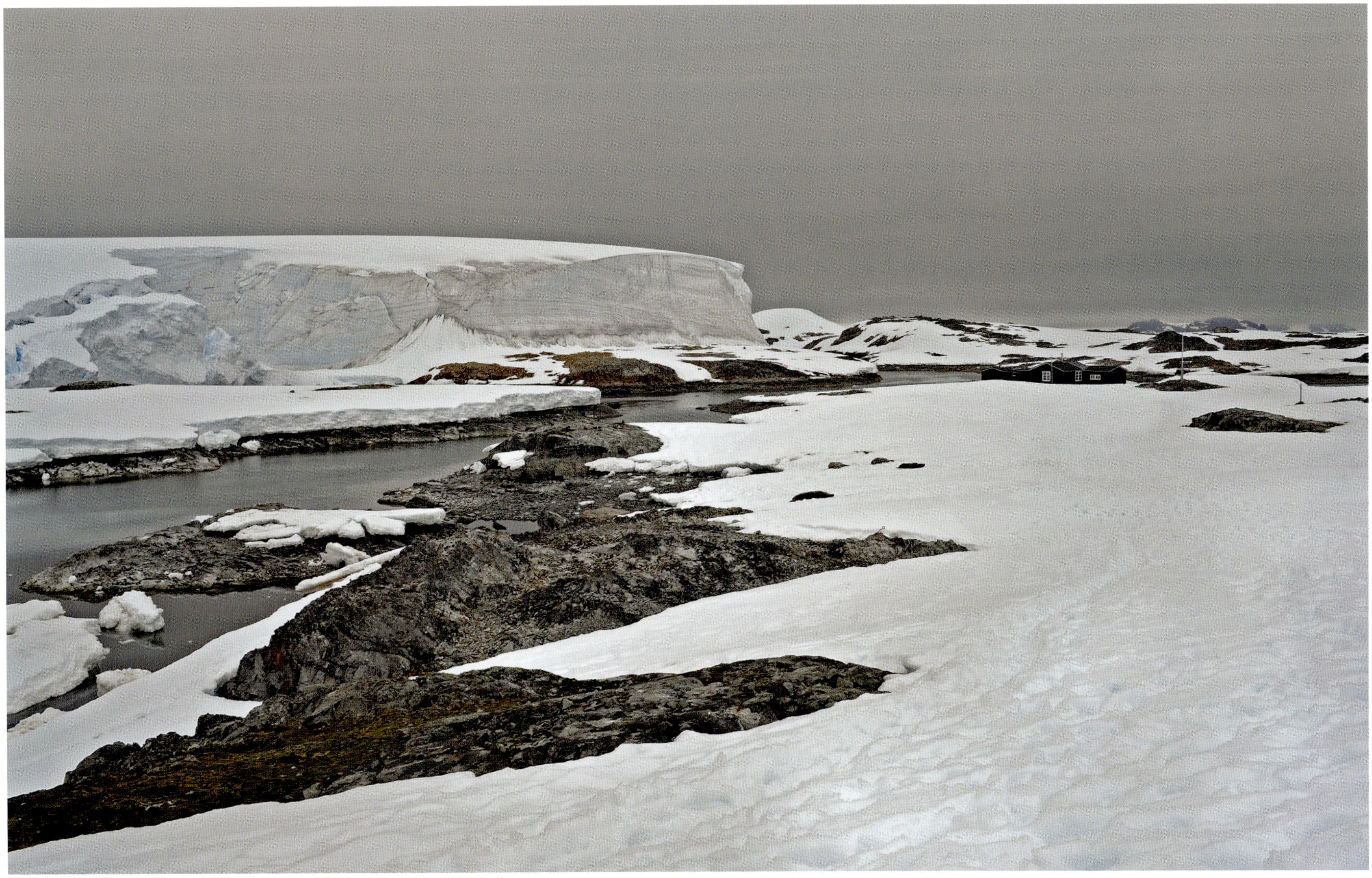

Wordie House

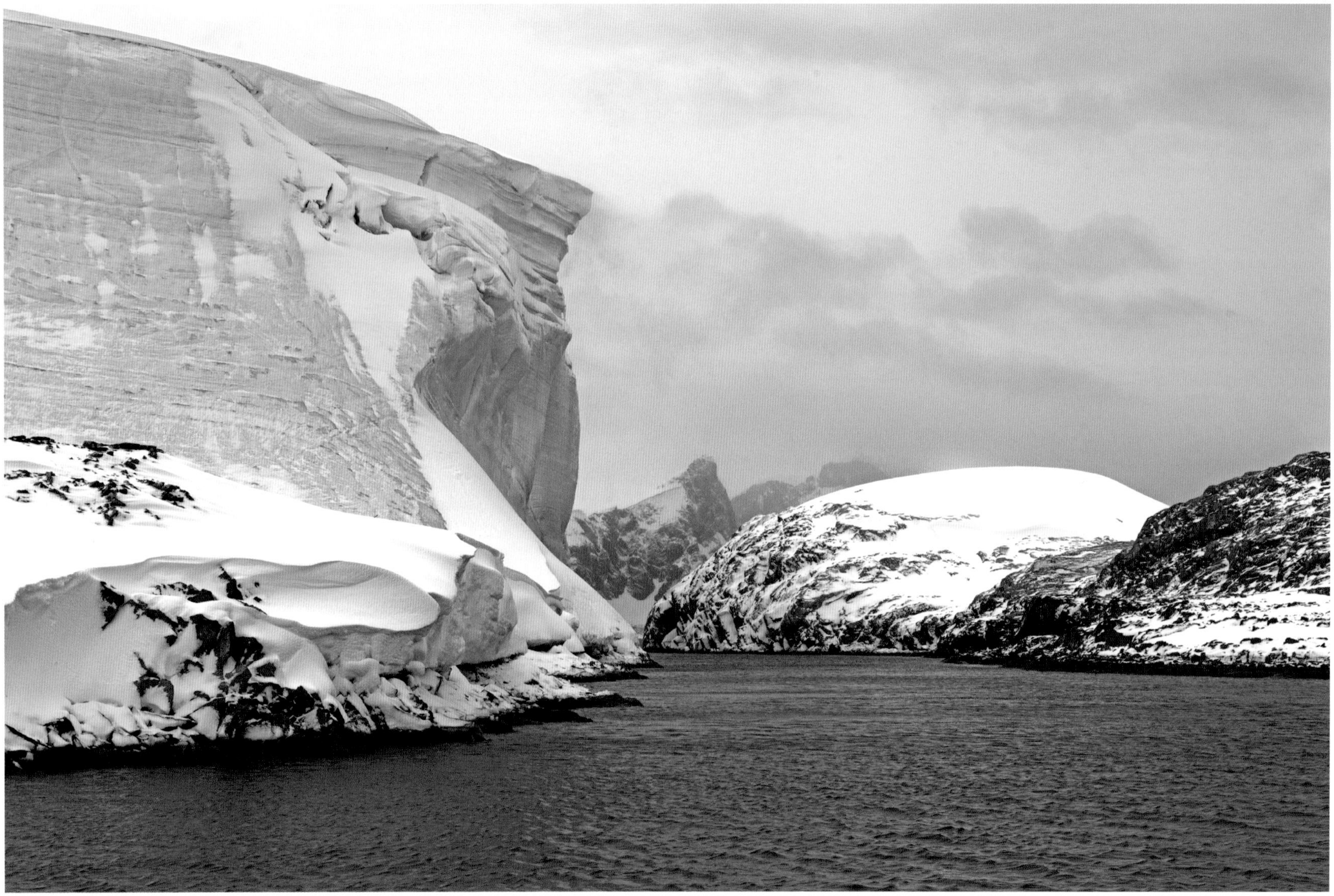

Channel to the open sea

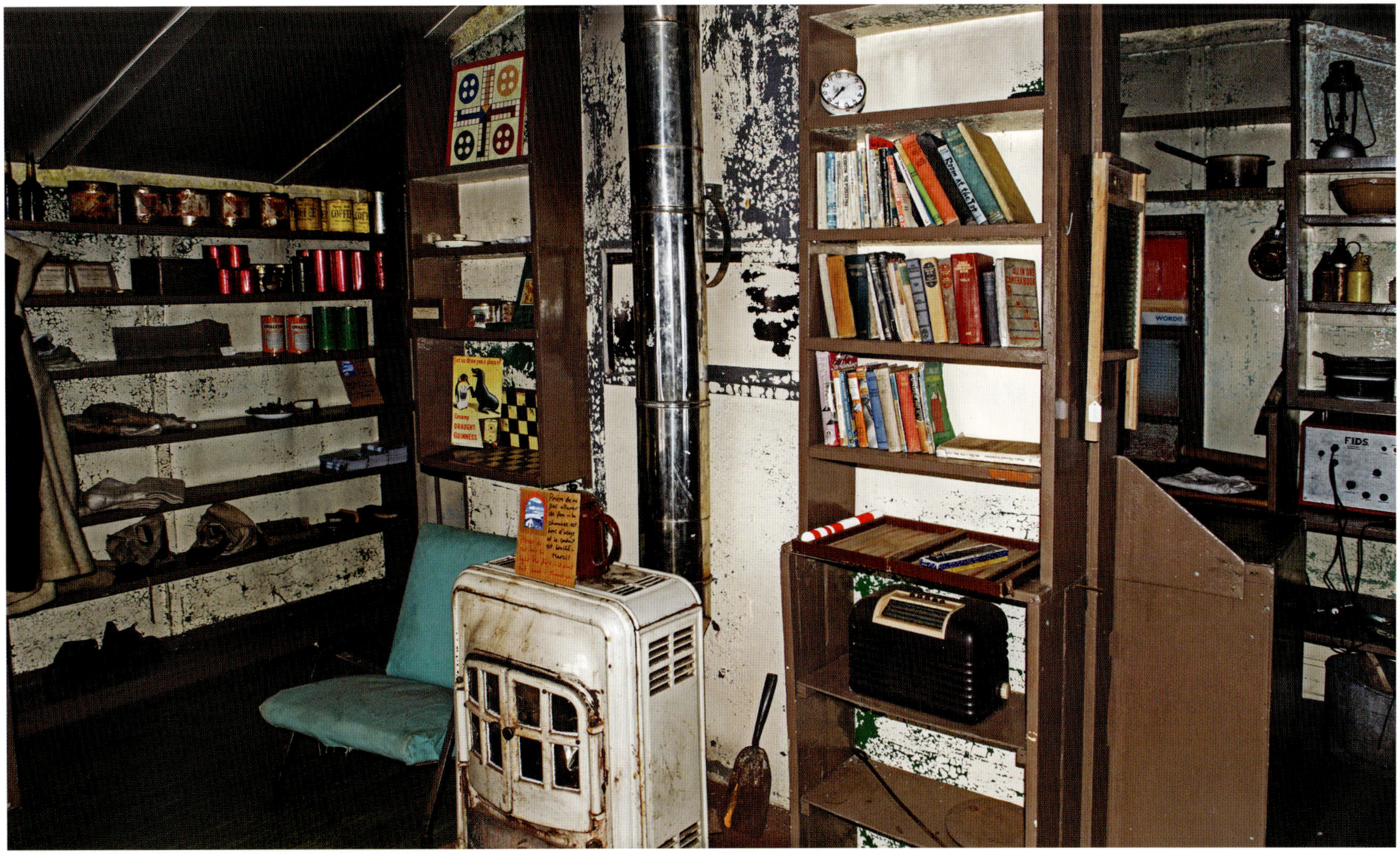

Wordie House interior

I WAS SO busy looking out on the other side of the ship that I nearly missed this image. The landscape I was gazing at had potential and the light was improving. Scanning the sky to locate the sun I looked around the funnel of the ship and noticed that the light in the cloud clinging to a long mountain was changing quickly from a dark grey to a luminous glow as if someone was working a dimmer switch. So dramatic was the change that I grabbed my tripod and ran to the unobstructed side of the ship. I pulled the tripod legs together to work as a makeshift monopod and stuck the end on my soft boot. It was the best I could do to eliminate the engine's vibrations. The brilliant effect in the hovering cloud was fading. I had taken the picture at the right moment and was checking my camera when it suddenly darkened then brightened again for but a moment but it had lost its inner glow and all transformed into a dull shadow.

I had completely forgotten about what a few moments ago had so preoccupied me on the other side. The visual impact was so strong I could not tear myself away and just lingered staring at what was one of the most compelling visual displays of how the momentary infusion of light could transform a landscape. I still stare at the image and expect it to change momentarily.

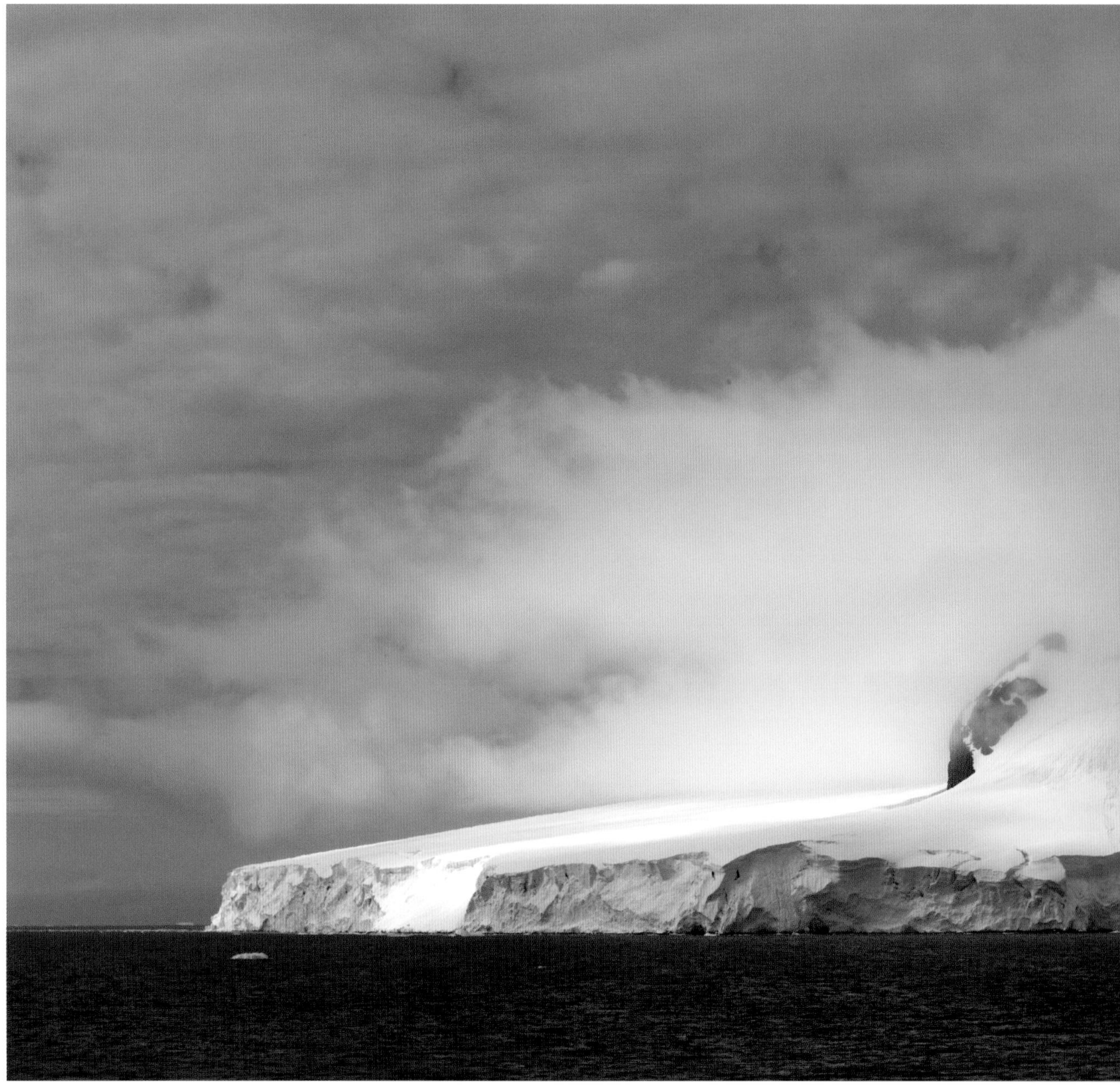

Radiant glacier luminous cloud

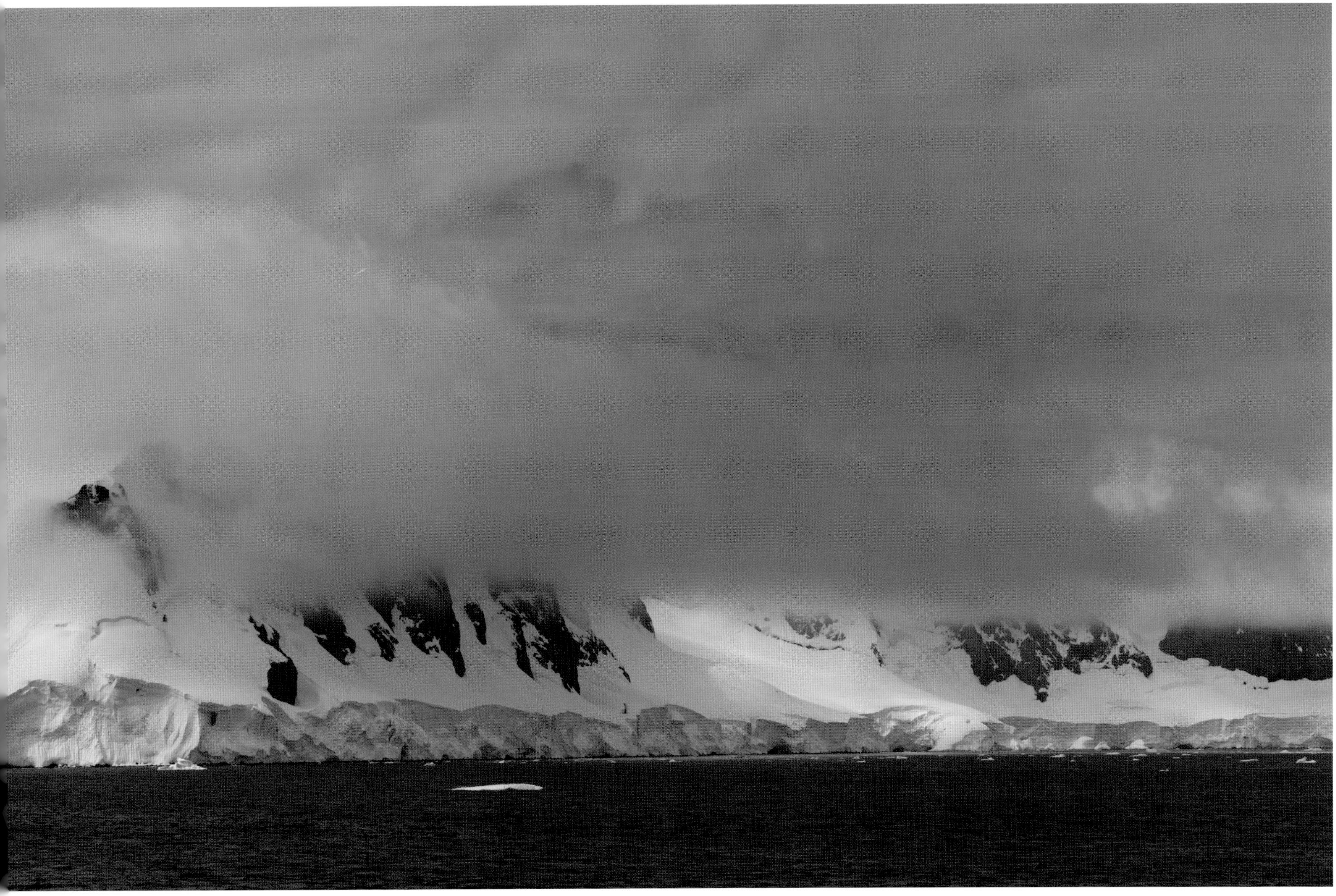

IT IS QUITE an incredible experience to slowly pass by an illuminated glacier front. The colours that emanate from the deep crevices are stunningly vivid and deep with colour. At first your impression is filled with disbelief that in a setting so stark and white that such a bright saturated light can emanate from deep within an icy crevasse.

Hues range from a light bottle green to a glowing cyan to a brilliant neon blue. Each glacier, depending on its age and depth, draws light from deep within projecting the unabsorbed colours. Fresh ice, still filled with air, transmits the whole spectrum as nearly perfect white. Some rare older ice has been so compressed that it can nearly look black by absorbing most of the visible light. Such smaller bergs resemble glass and are easily missed when gliding past in an ocean as black as tar.

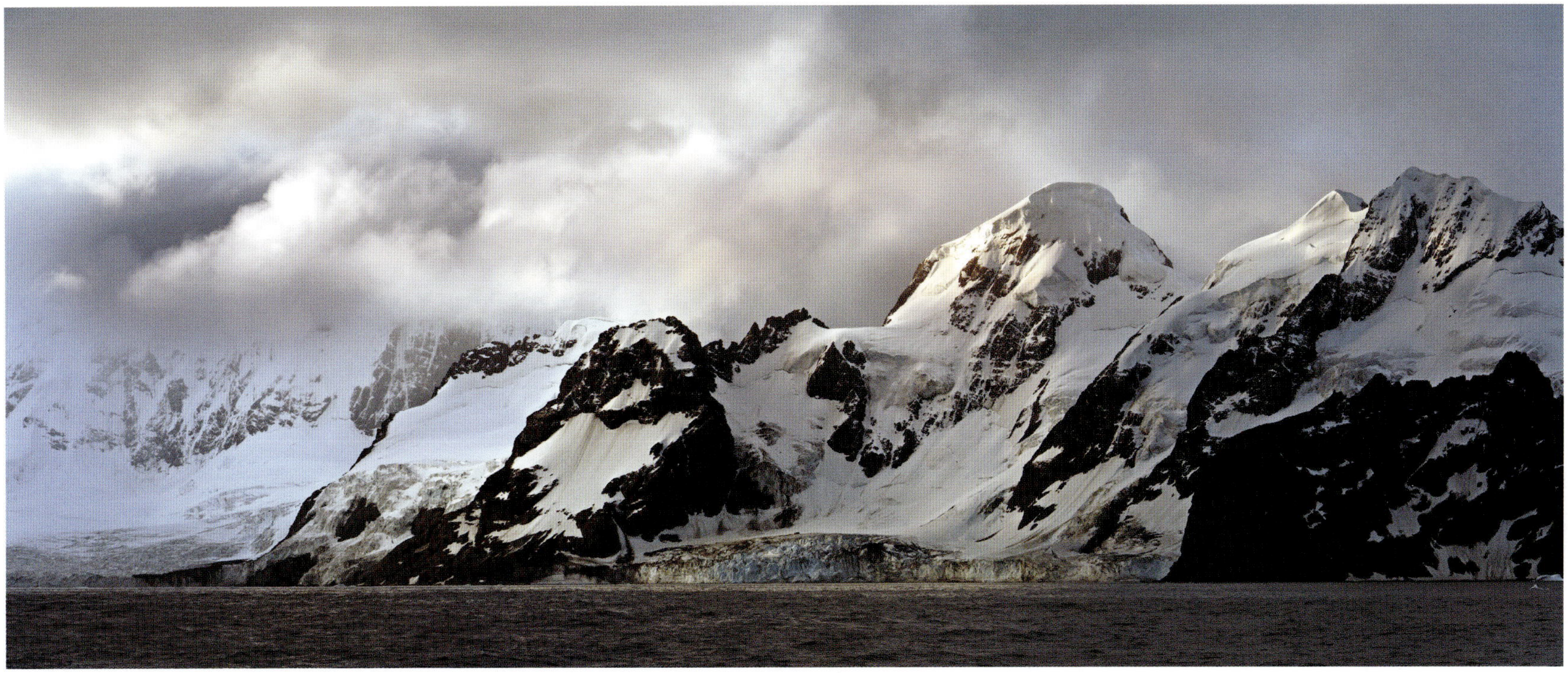

Glacier Elephant Island

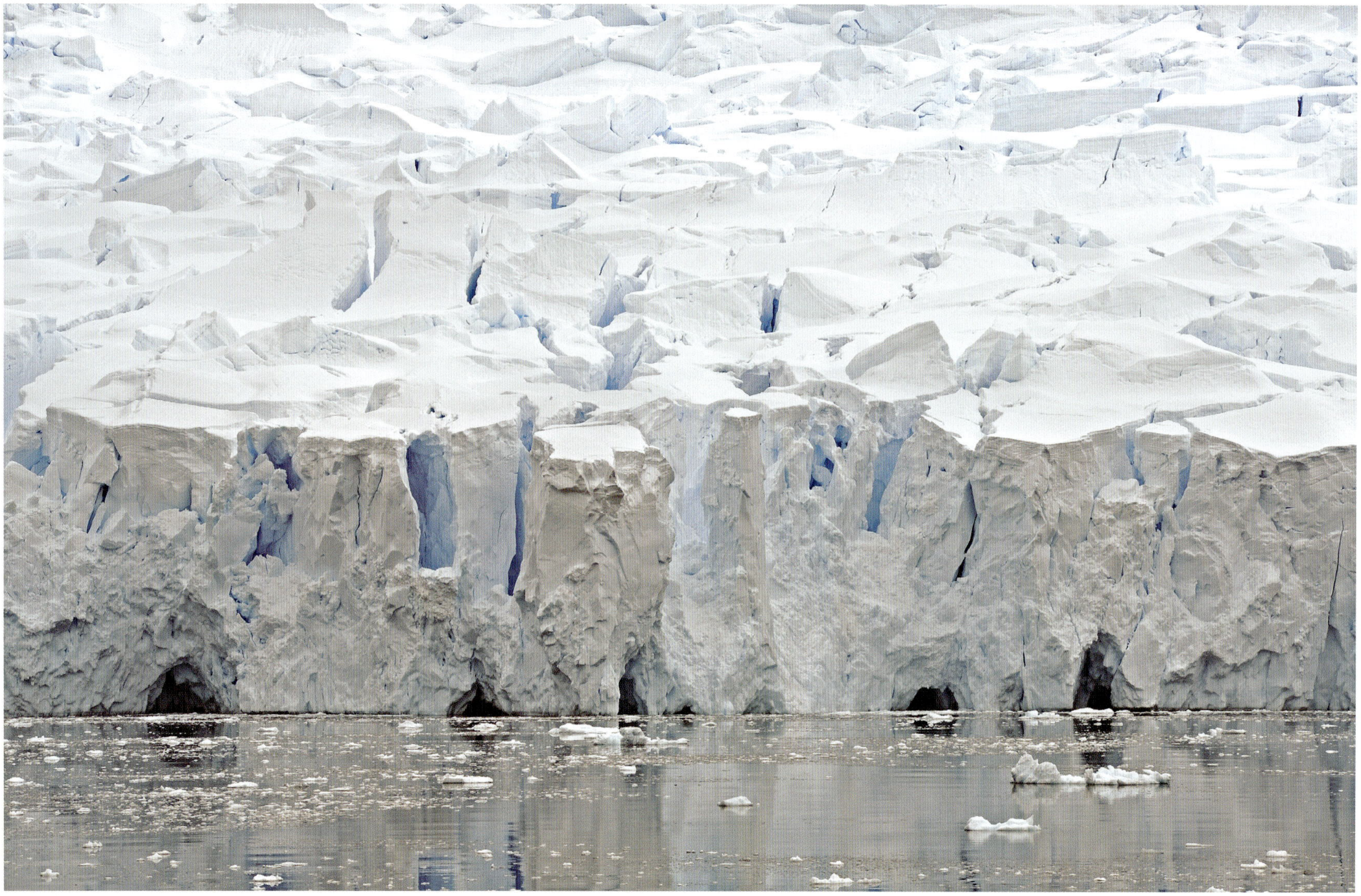

Neon blue

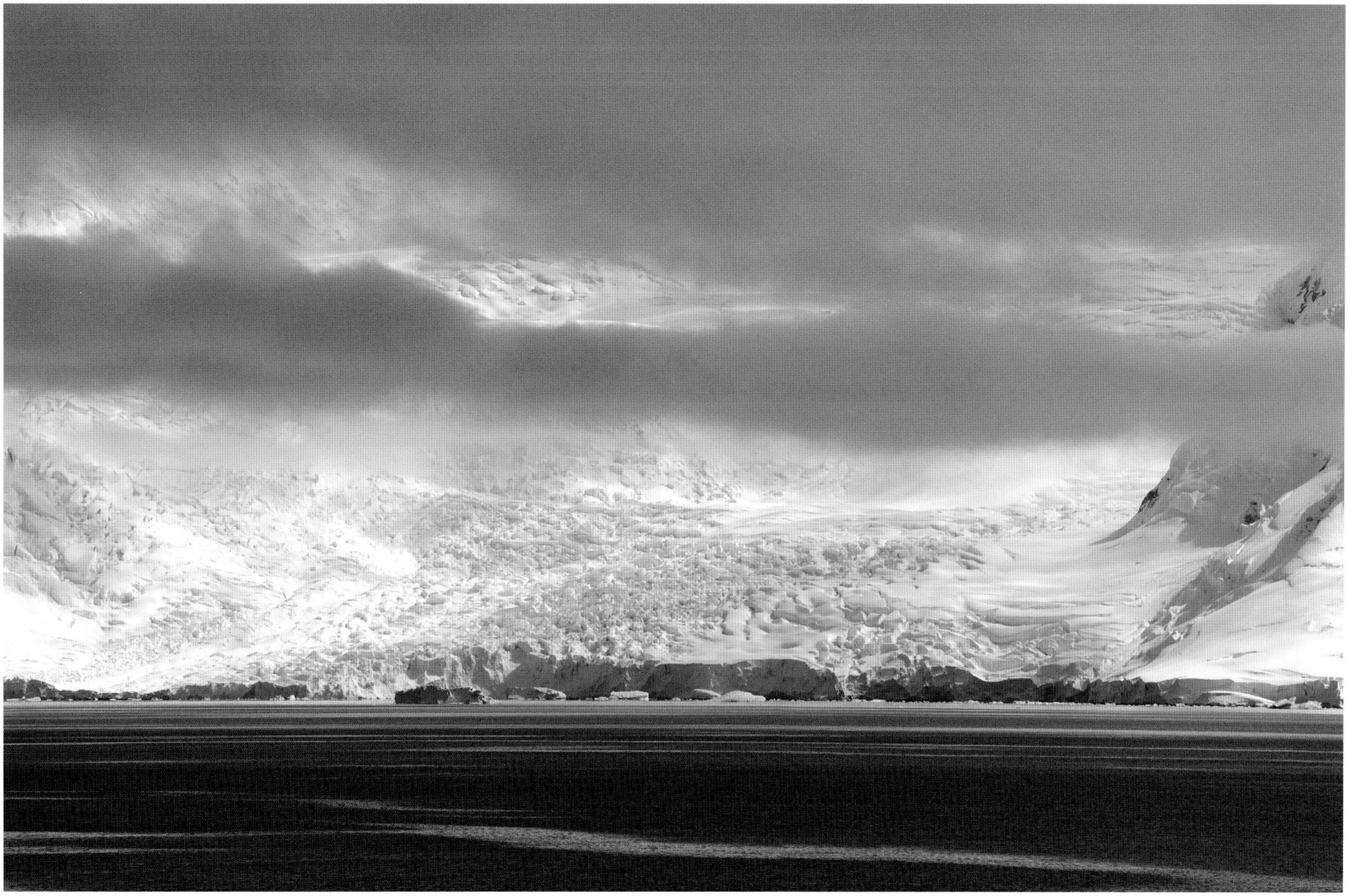

Spilling glacier and dark sea

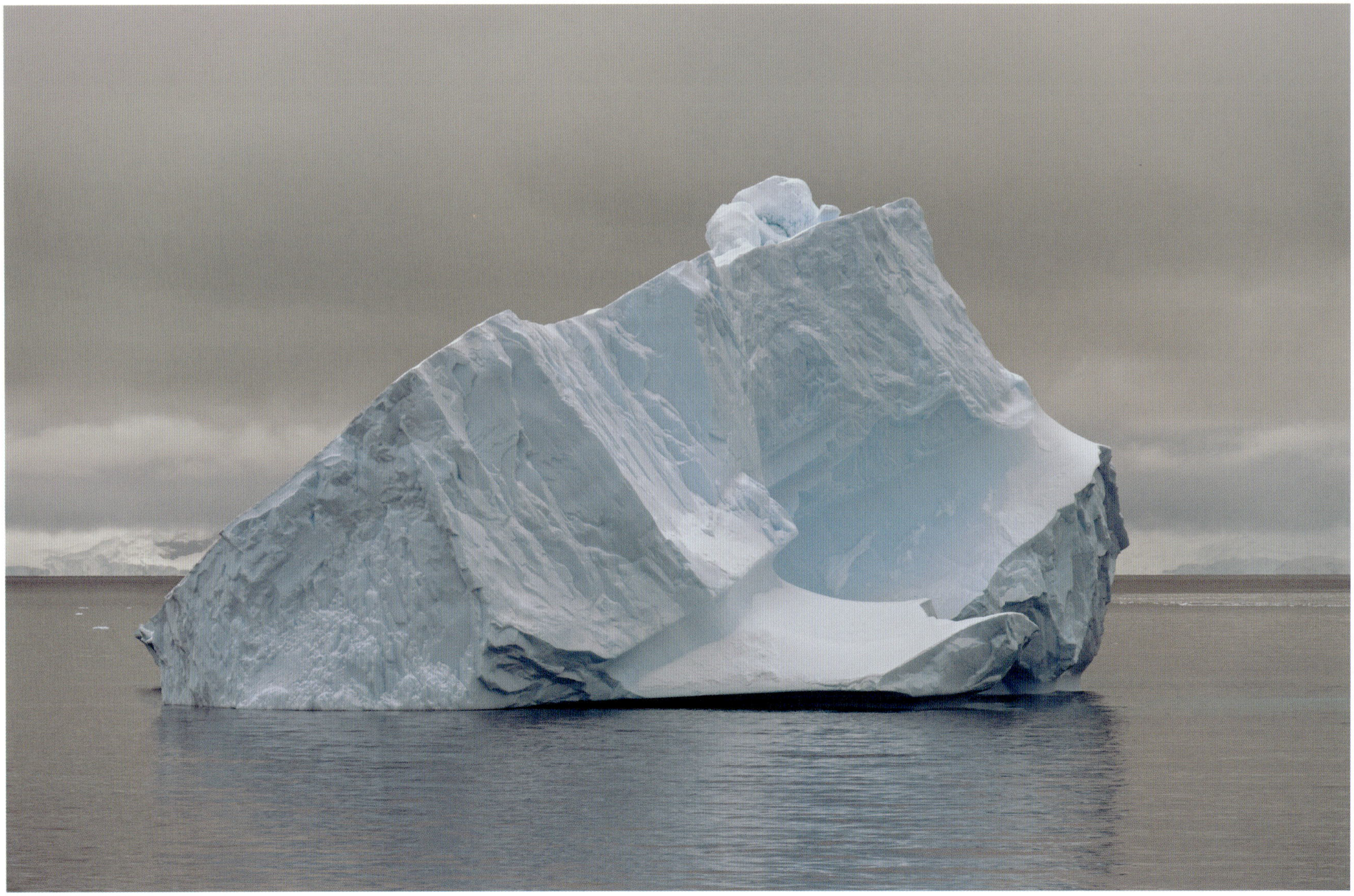

Blue Iceberg

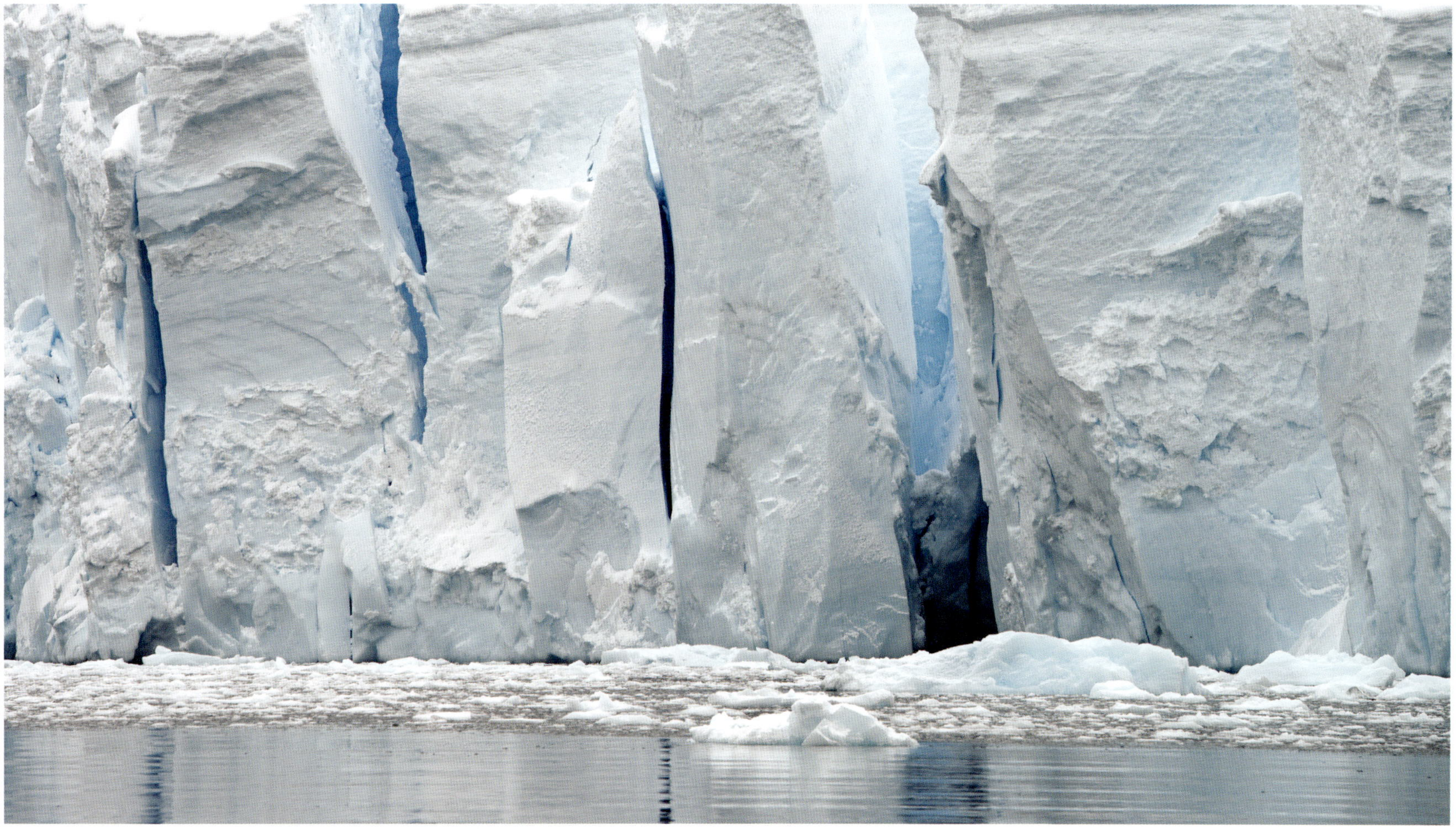

Glacial front

THESE TWO IMAGES were taken on the same day. On this particular morning the sun rose over a dense sky and the light remained even for a long time. I was on deck early hoping to get a picture of the rising sun but the light was so uniform that it was difficult to predict where it would appear. The wind was down and the ship had a soothing motion as it gently rocked through the swells leaving me in a trance-like state. Once in a while the dark horizon would lighten with a touch of yellow but then close up again leaving a somber skyline. It was then that I realized the entire sea and sky appeared colourless. I glanced to the other side of the ship and it was the same all around. The effect on me was immediate. I was now no longer interested in the rising sun but in the effect it had on the colouration of the water. When I arrived on deck it was dark black but now with the brightening sky it was taking on a glossy metallic look. The undulating sea shimmered in all directions and the smoother parts rose and fell as if it was a single sheet of metallic foil. The display was captivating. Later during the day as the sky brightened the sea took on an uninspiring familiar hue.

The morning experience left me wondering about the different colours of the sea and I studied the reflecting coloured surfaces throughout the day. We had now entered

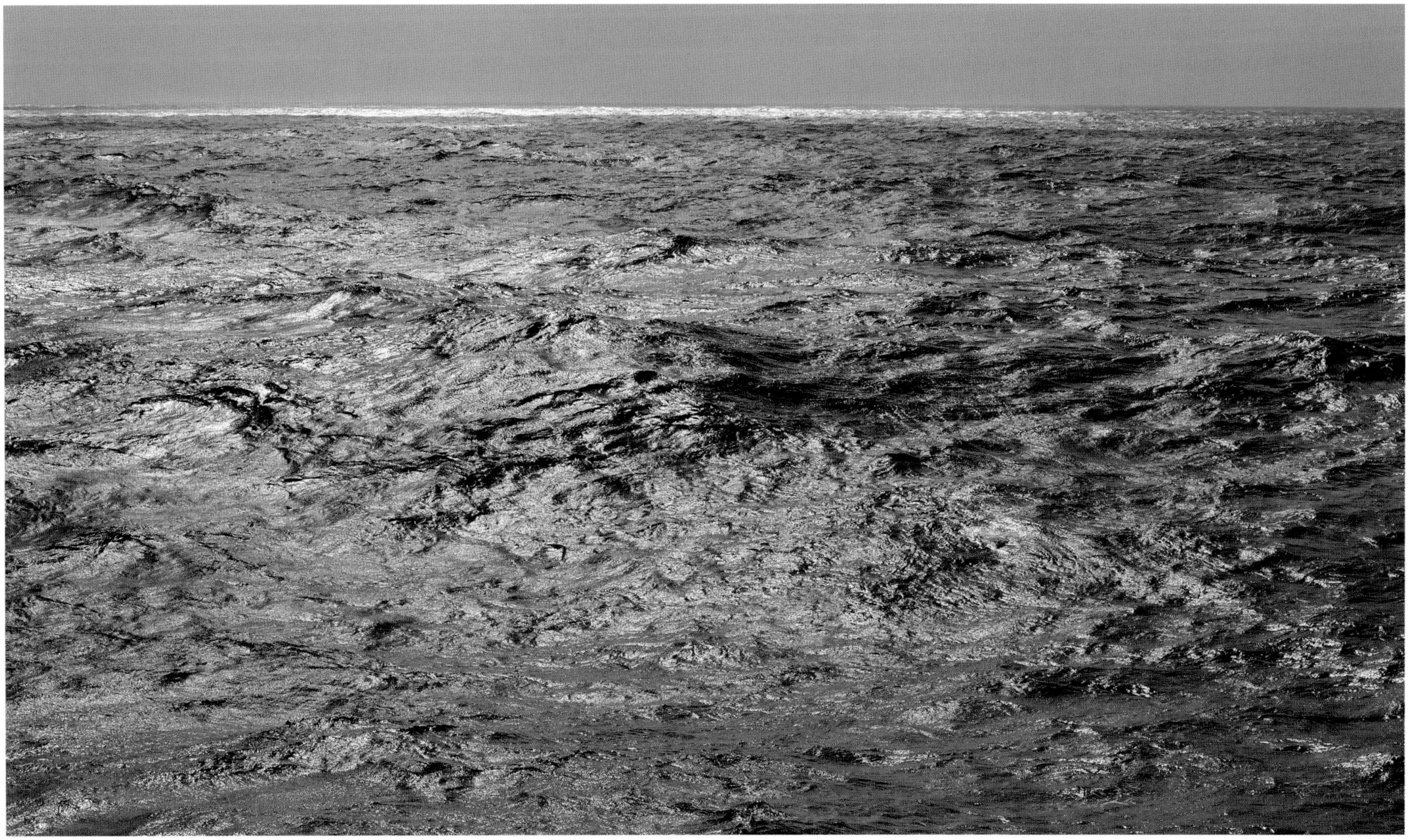

Pewter sea

the Antarctic convergence where the cold water of the south meets the warmer water of the north bringing with it a variety of icebergs and a definitive change in climate. I had spent the day focused on trying to capture the multiple hues reflected and transmitted by the waves. It became frustrating trying to identify the variety of shades and changing colours. I had to contend with a whole new colour palette. Figuring out the source of the colours was also difficult. One moment you were convinced it was the reflected light and in the next you concluded it was transmitted by the water itself. I eventually concluded that in such an unvarying sky it was the quality of the water that changed.

The evening brought on another change. Many layers of cloud formations were at various levels in the atmosphere adding to the diverse changes in light. The lowering sun would break through different layers of cloud lighting up another section of the sky. Where once was a dark grey mass of moisture there now appeared bright warm and richly coloured clouds. In these southern latitudes the spectacle was long lasting with the sun setting on a long shallow oblique angle. I took advantage of the opportunity and remained on deck till midnight. One of the last images of the day was this image of a coloured sky and a wine coloured sea.

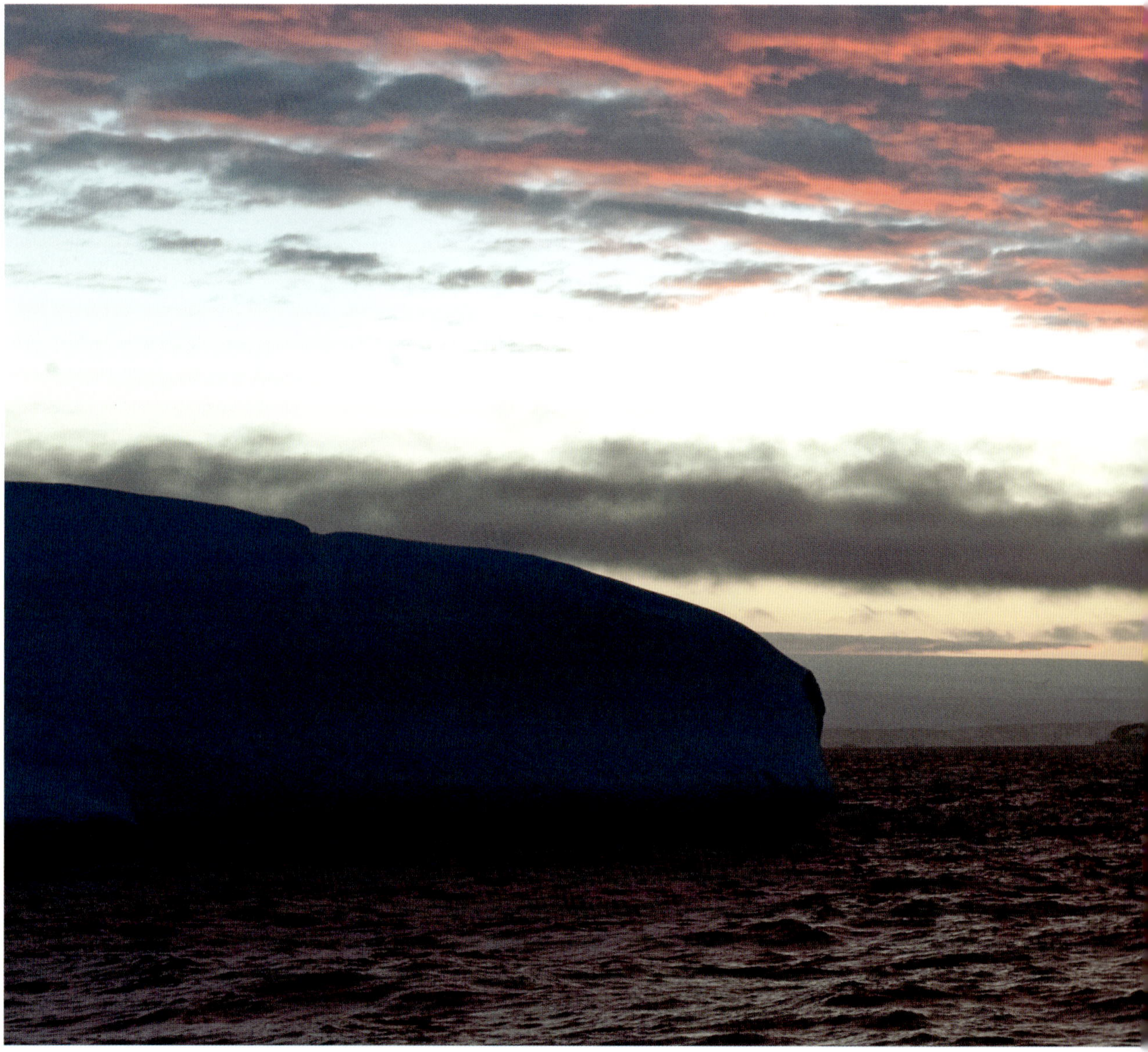

Wine coloured sea

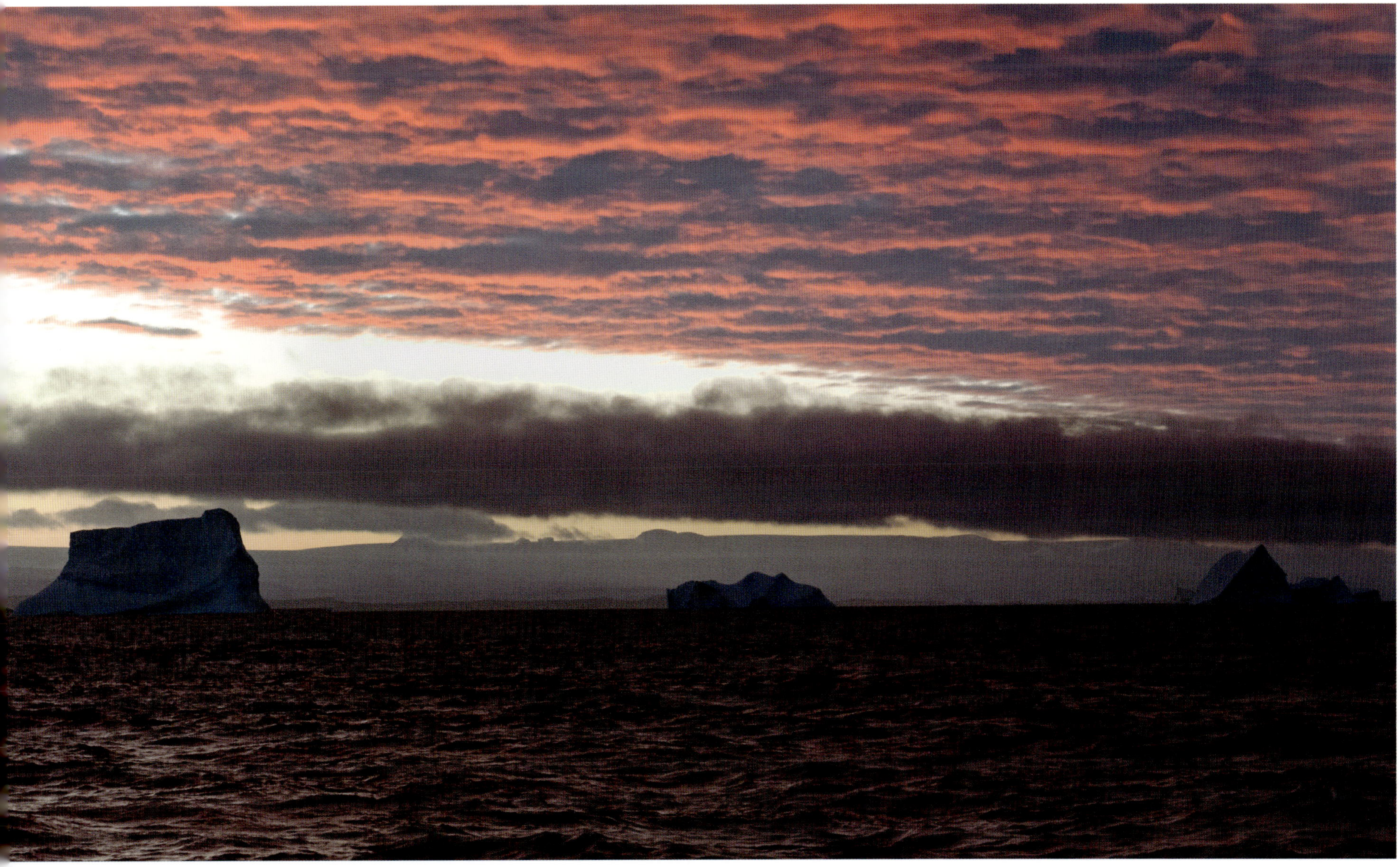

THANKS TO A dustless atmosphere that is more transparent than anywhere else in the world mountains that are miles away still show all their characteristics of shape and colour.

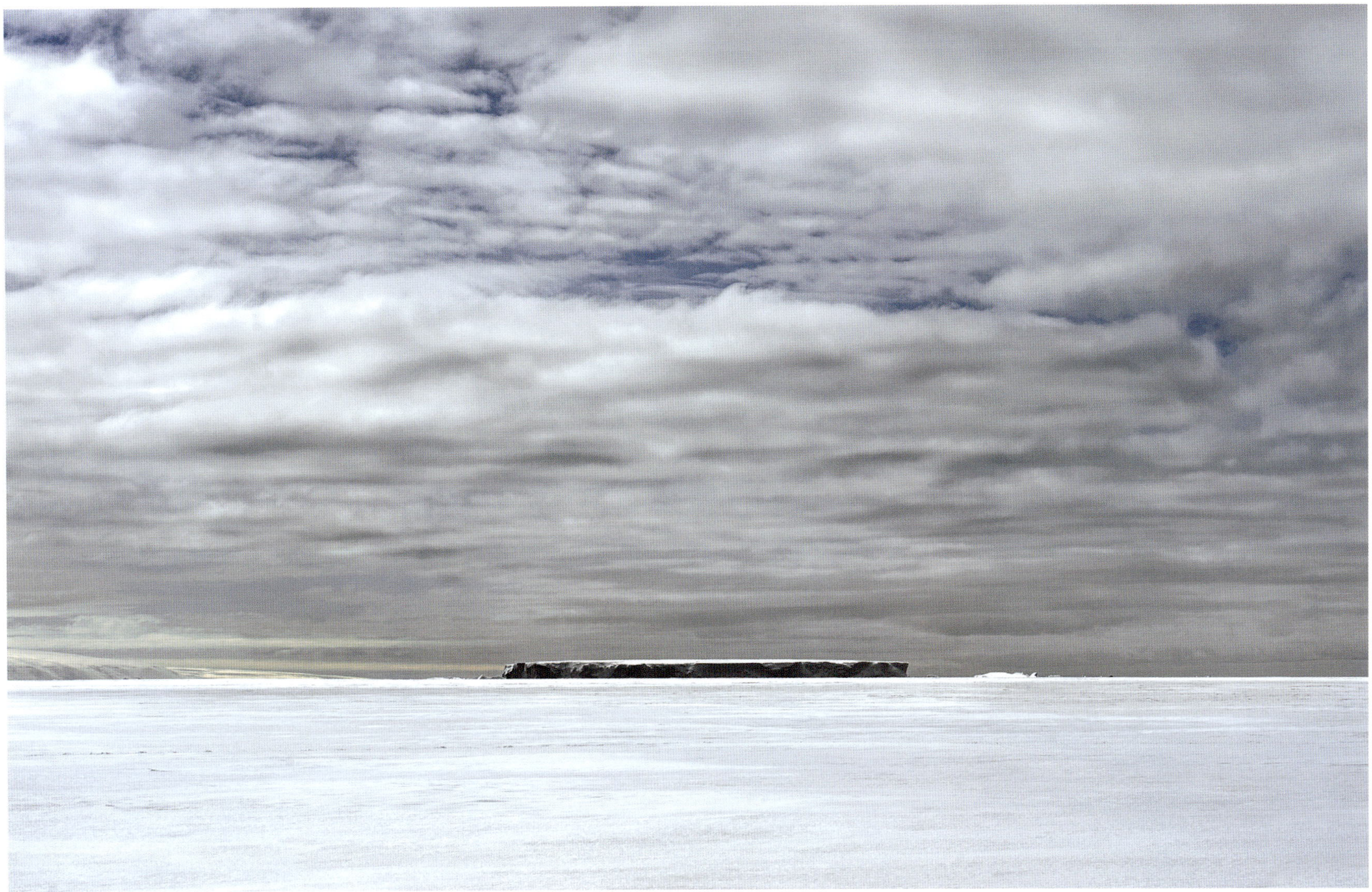

Trapped in an ice pan

LEMAIRE CHANNEL IS but a flooded valley approximately eleven kilometres long and eight hundred metres at its narrowest. On the north side is Booth Island, 985 feet in elevation, and on the south the Antarctic peninsula. I have had the opportunity to pass through it four out of five times. One spring we had arrived too early and the channel was so packed it looked like an ice field and definitely impassable. On another occasion there was plenty of loose ice called berry bits and also larger growlers, the truck-size chunks of ice that would scrape and grind along the whole length of the ship sounding not unlike a heavy door with a rusty hinge. Sometimes you could feel the vibration of the resisting ice rise through the hull to the upper decks. It seemed impassible but the captain skilfully slalomed through. The vessel was ice reinforced so a few solid bumps just added to the thrill and not the danger.

What I appreciated the most was the proximity of the shore and the seeming closeness of the surrounding walls. However it was still with difficulty that you could try and estimate size and distance. There was no reference point. One year a gull flew by a descending glacier and turned to shore. It disappeared and yet you felt sure it would have only been a stone's throw away. I had experienced this phenomenon in the Arctic but never with such intensity.

Repeated visits to any location offers one the luxury of experiencing the place under different conditions.

With multiple exposures to different light and weather comes an understanding that a single visit could not provide. These images of Lemaire were taken over four trips.

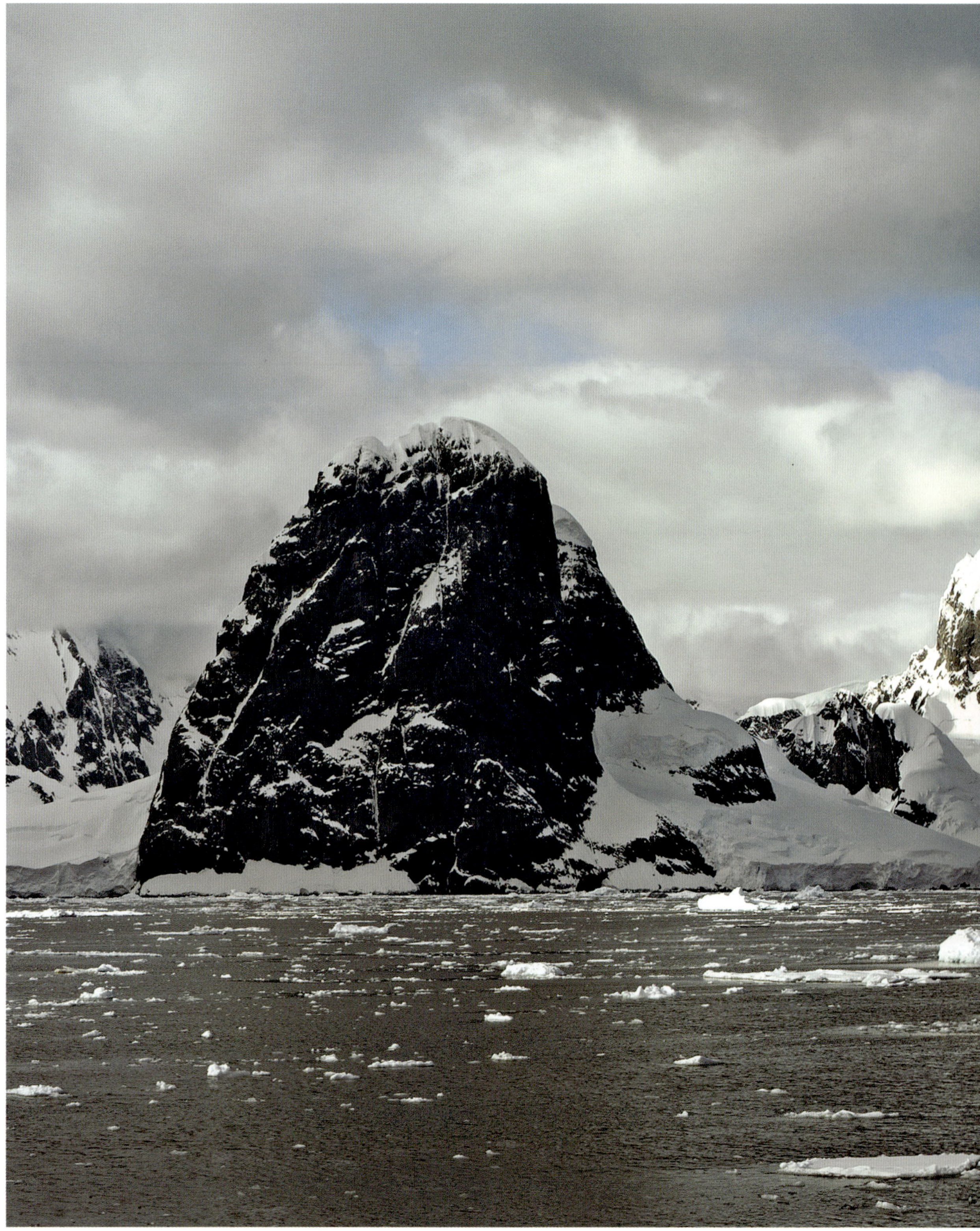

Cape Renard Lemaire Channel

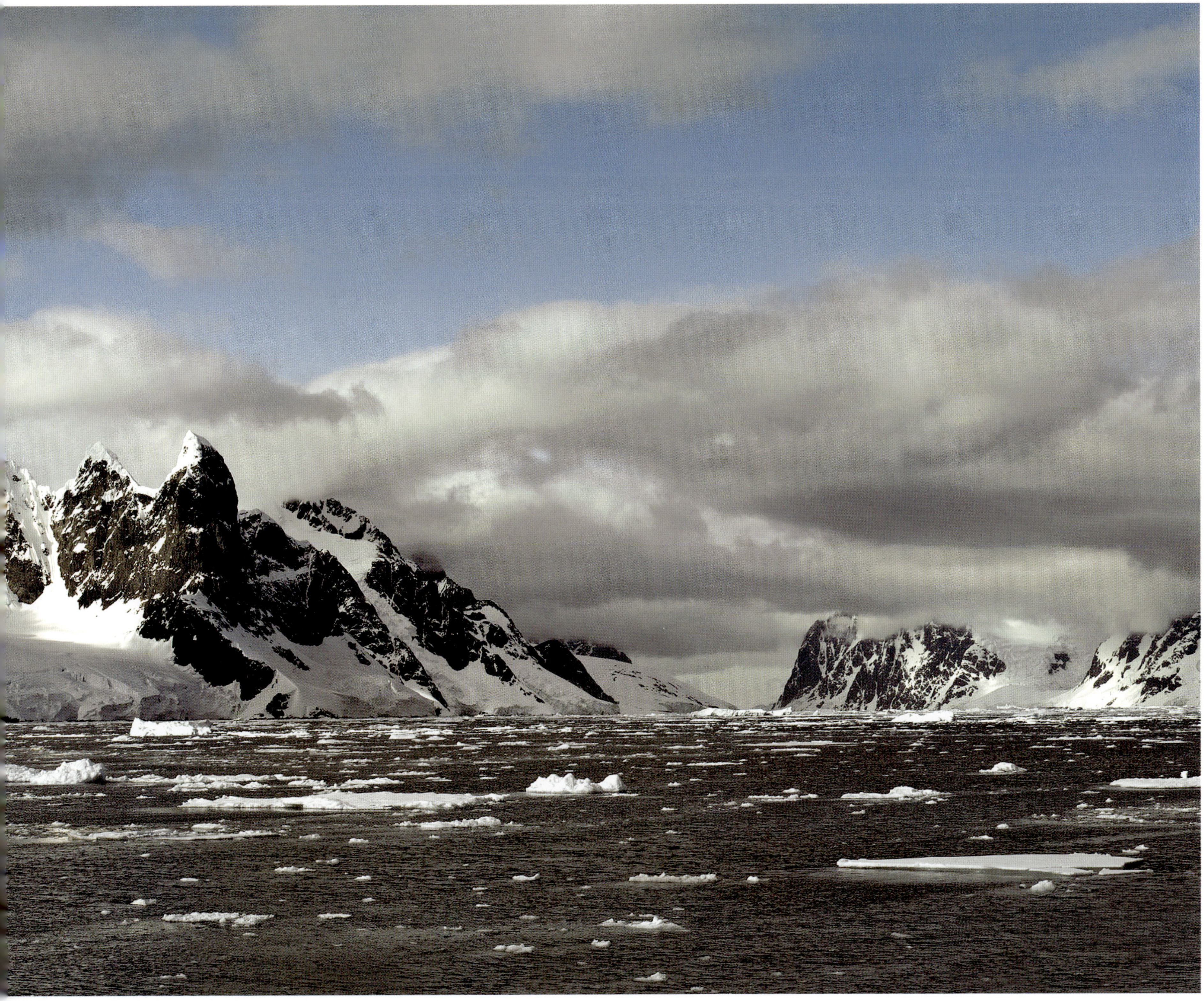

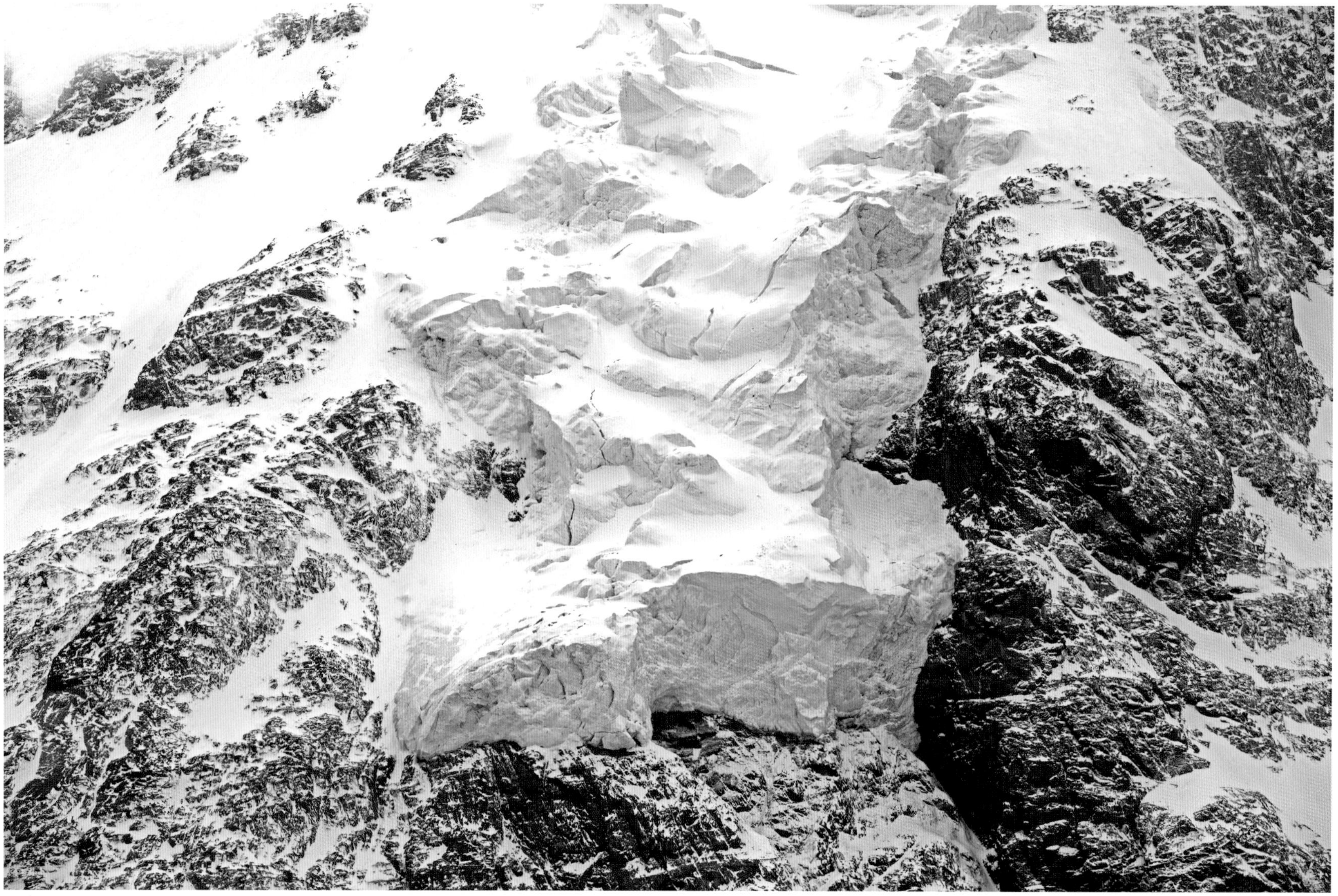

Icefall

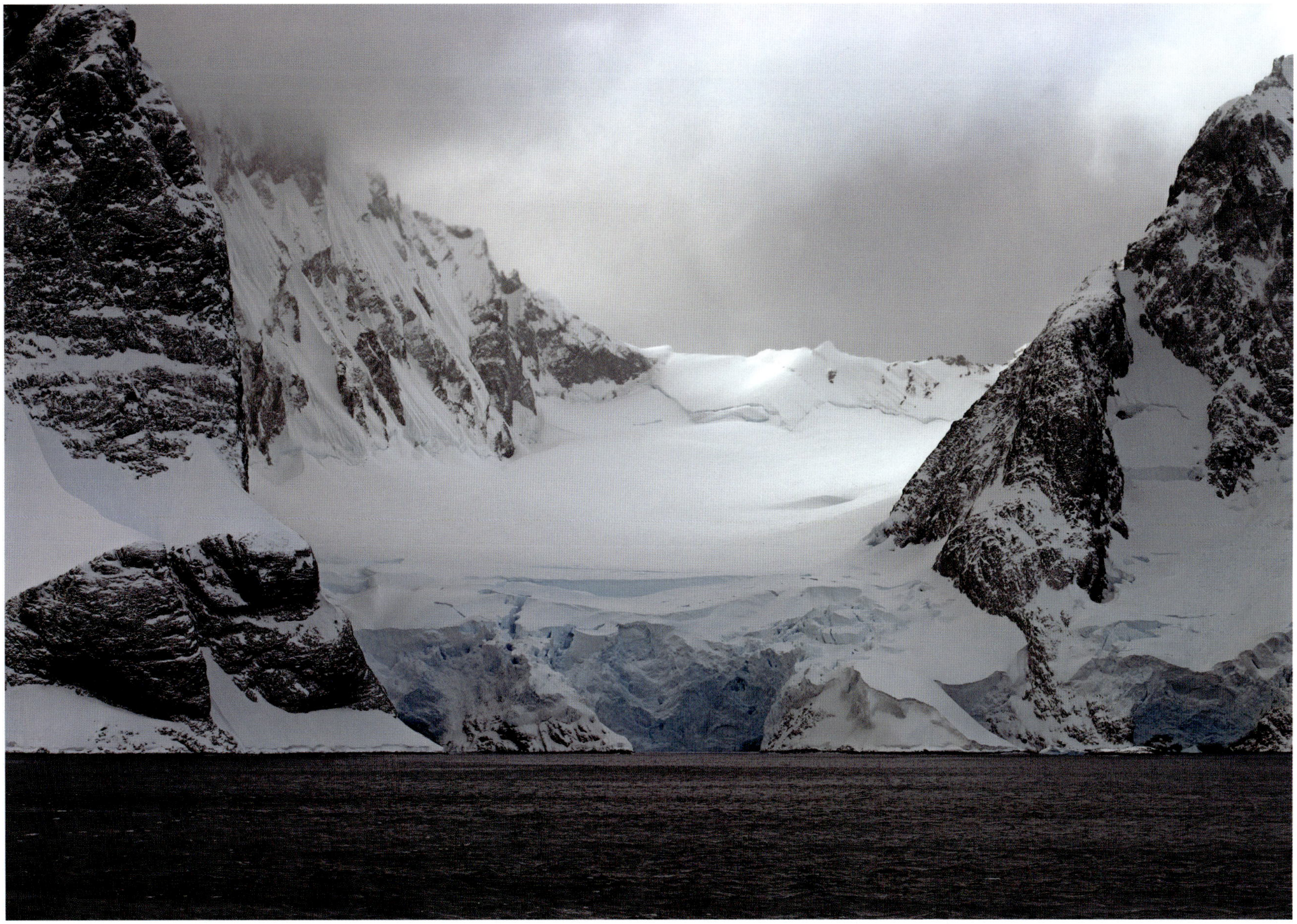

Deloncle Bay and Hotine Glacier, Lemaire Channel

THE ADAGE THAT no one steps into the same river twice can also apply to anyone experiencing Antarctica.

You can return to this turbulent continent many times, stand on the same coordinates, and feel you are there for the first time. The cause of this uncertainty is the unpredictable and unstable weather patterns that continuously blend and mix all the elements that make up weather into an innumerable number of combinations.

Temperature, wind, humidity and precipitation in all their varied forms supply a scope of components that range at the best of times from manageable to conditions that can easily put you at high risk and threaten your life. On rarer days a sailor can experience a calm sea that actually looks and undulates like oil while only a few hours later with a sudden drop in air pressure that person can be struggling to stay upright in the world's stormiest seas. Antarctica's sometimes dry air can betray your sense of distance so much so that mountains rising over the horizon a hundred miles distant seem to be only a dozen miles away. But there are also periods where just the opposite is true. A bright and clear day can suddenly be met with a rapidly forming fog that quickly forms a dense mist feeling

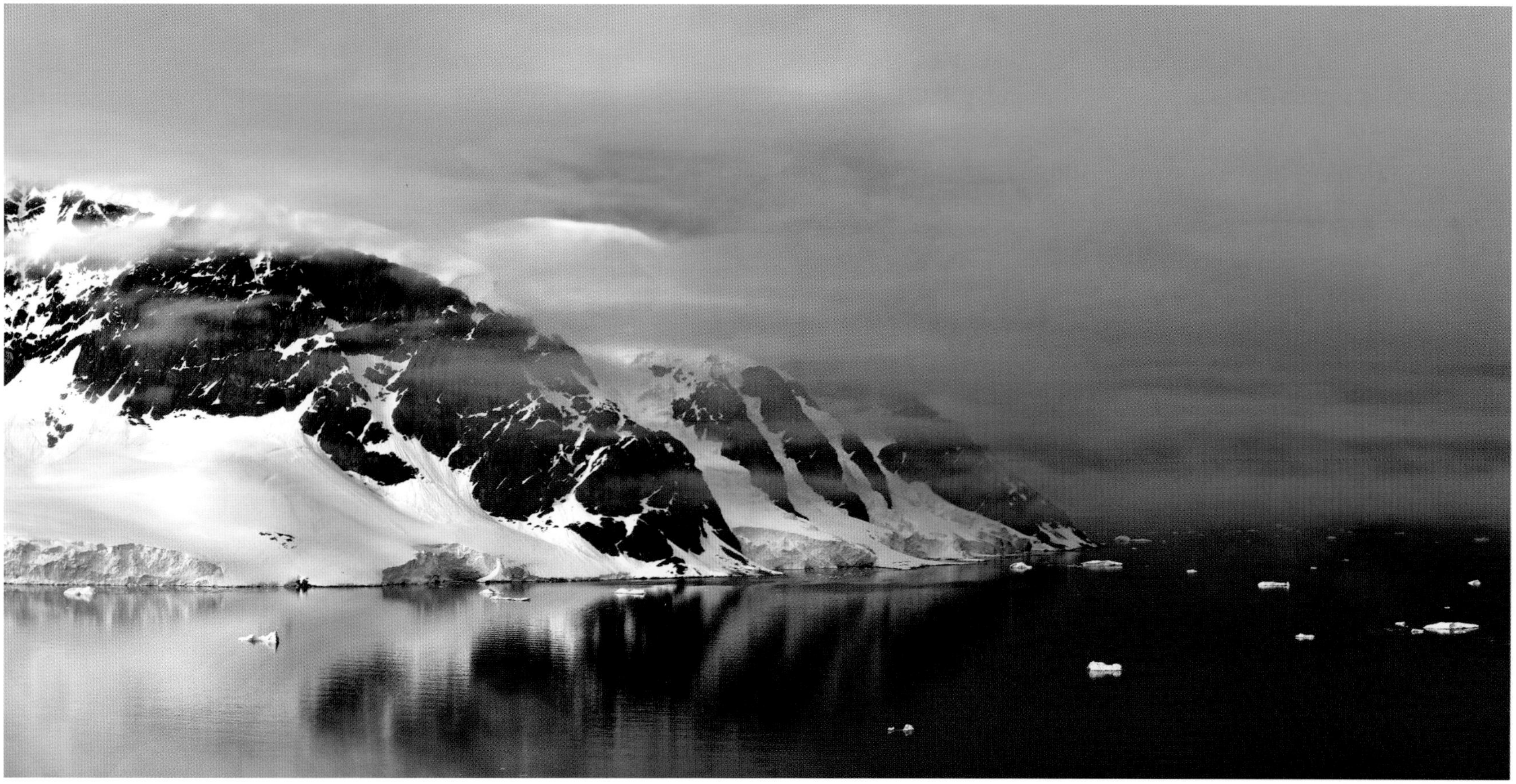

Errera Channel

like suspended rain. The endless horizon has vanished and now a ship without radar is put on hold. In conditions like this even walking over the level snow becomes a challenge. In the absence of visual relief surrounded by a uniform light distance and shape disappear making it easy to misstep. I would compare it to snow blindness not on the brighter side but on the darker.

If you add to this diversified mix of weather the slanted seasonal light of Antarctica you have the most diverse display of colours anywhere. In a land that is known for its brilliant and omnipresent white it always astonishes the visitor with its varying intensities of blues, cyans, yellows, reds, purples and blacks. Some sunsets even give you the impression of a light green; a result of some remarkable melding of yellow skies and the delicate blues transmitted by light escaping from densely compacted ice. Many times on my repeated trips the ordinary has become extraordinary and vice versa so much so that I get fooled into thinking that I have never been here before.

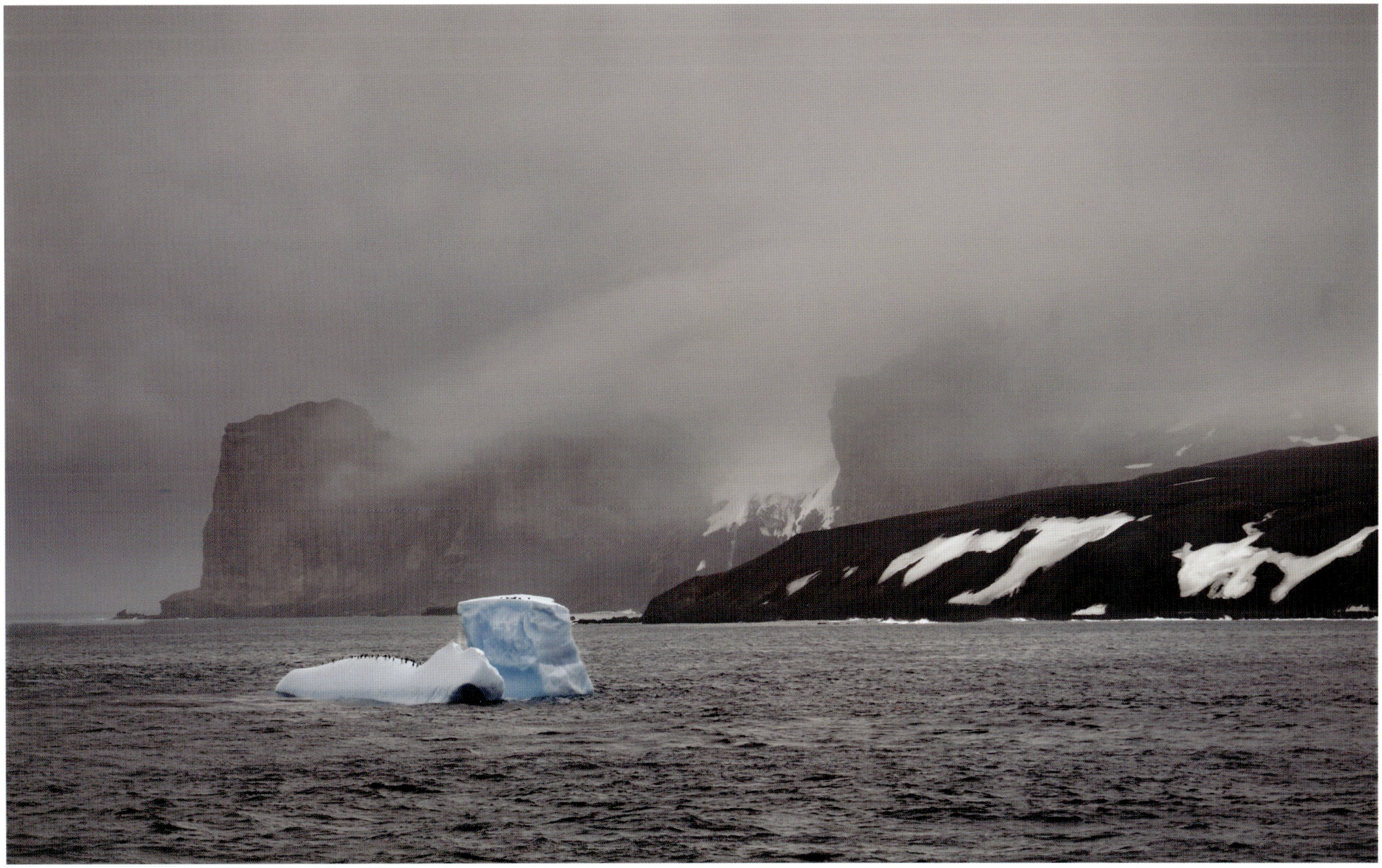

Penguins on blue ice

STANDING ON DECK or floating in a Zodiac one easily feels that you are being slowly drawn into a fantasy. Nothing whether in size, colour or shape seems familiar. The only thing that one is sure of is the vastness that draws you in. All is much larger than expected.

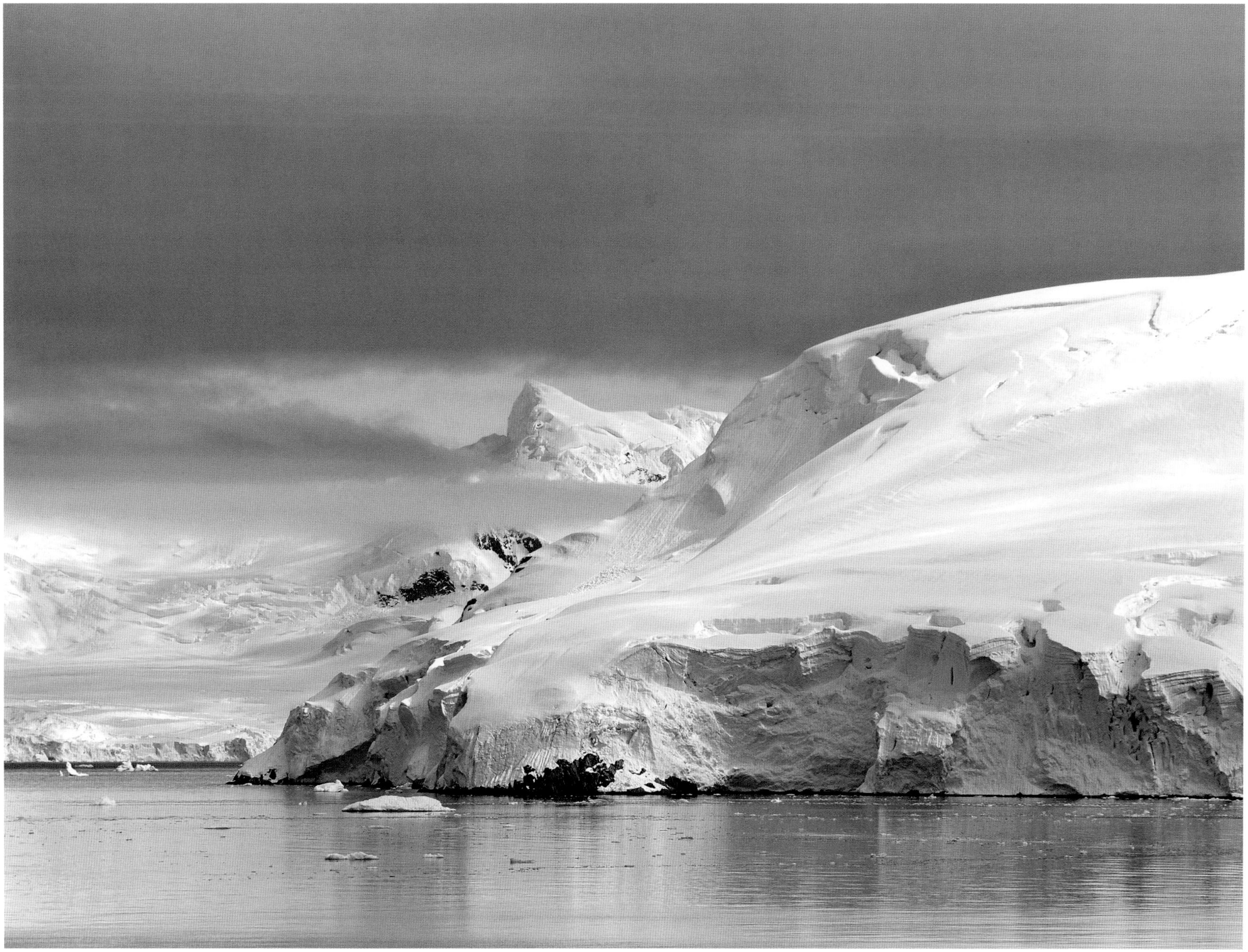

Lost in mystery

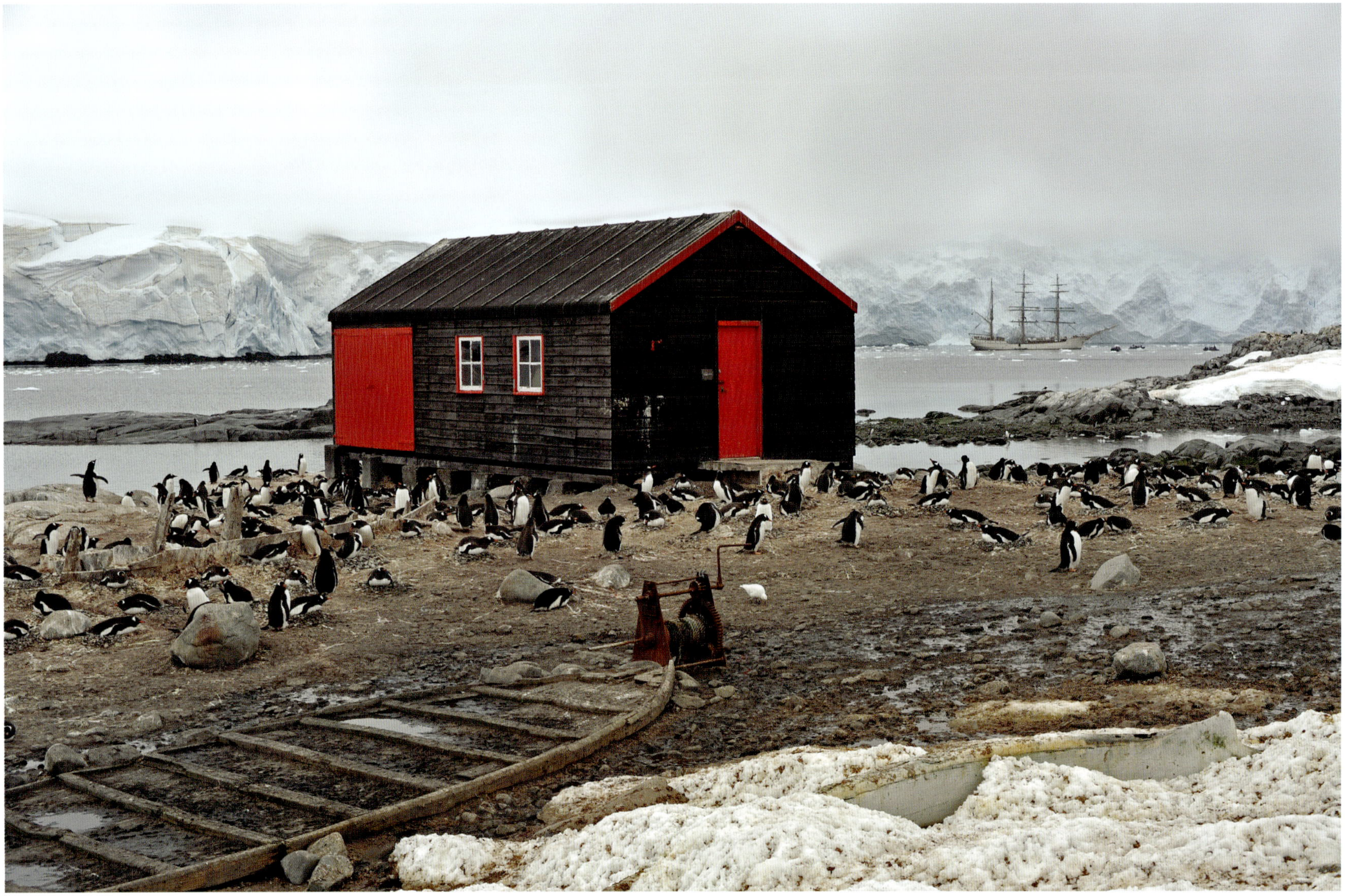

Port Lockroy

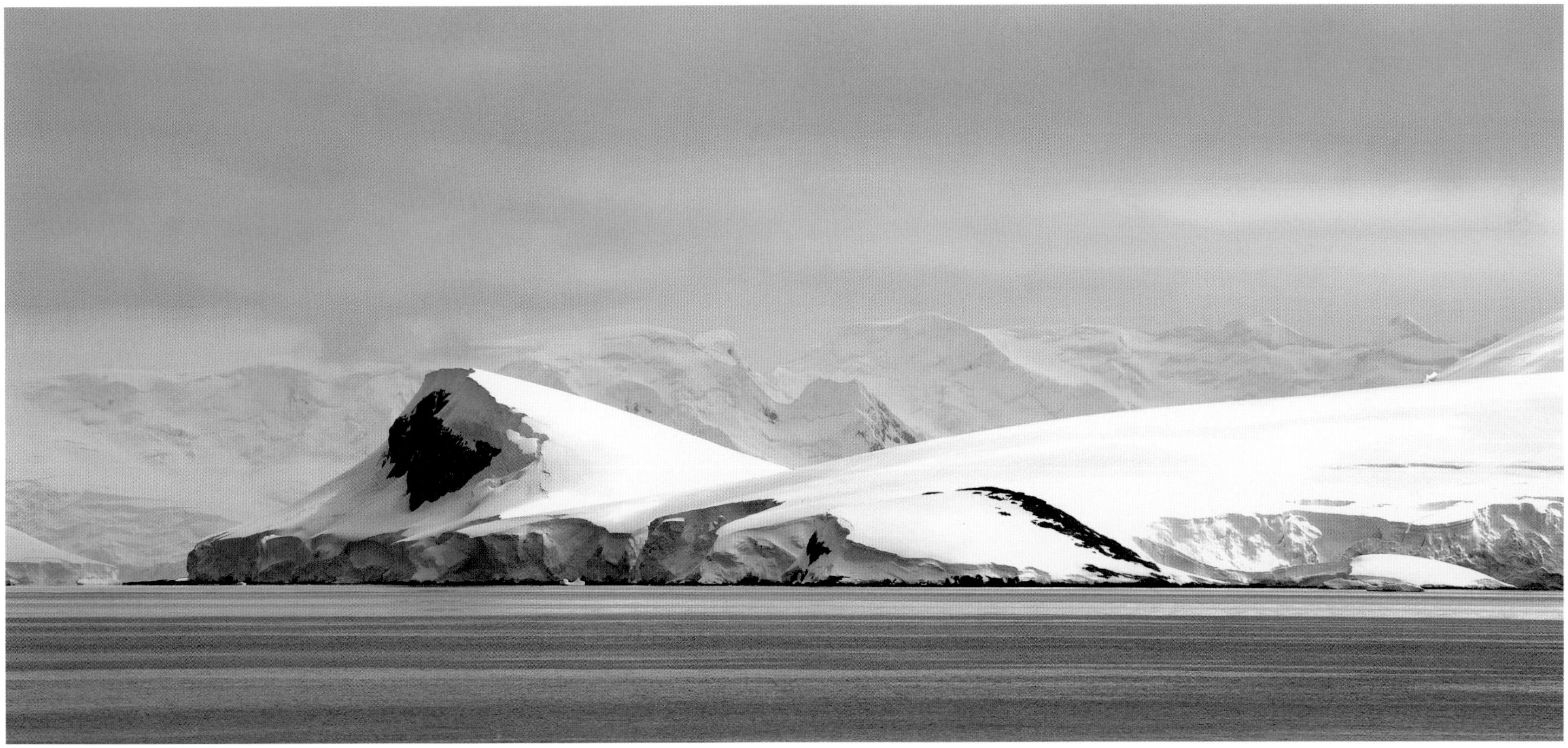

Spit

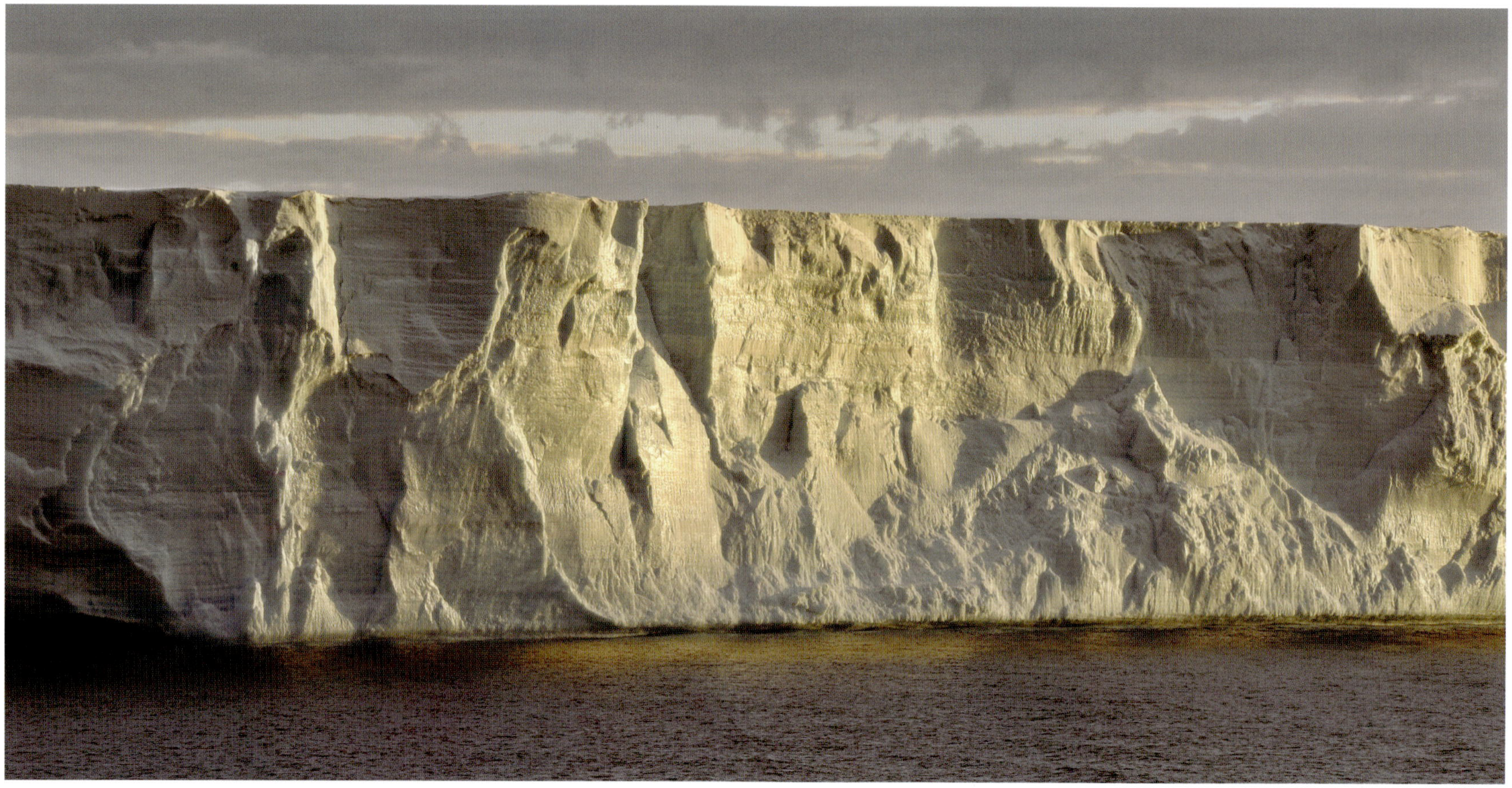

Iceberg detail

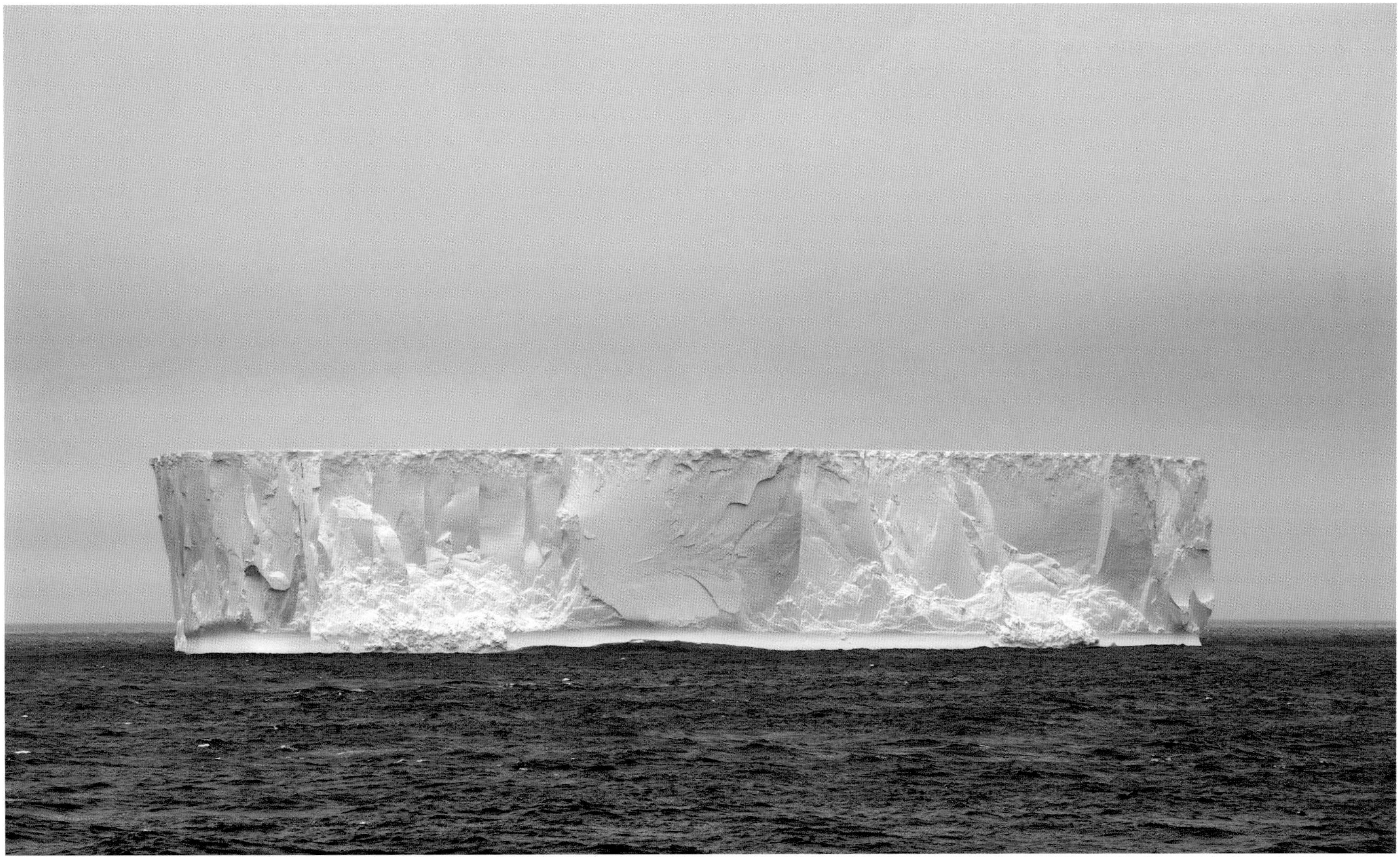

Small tabular iceberg Weddell Sea

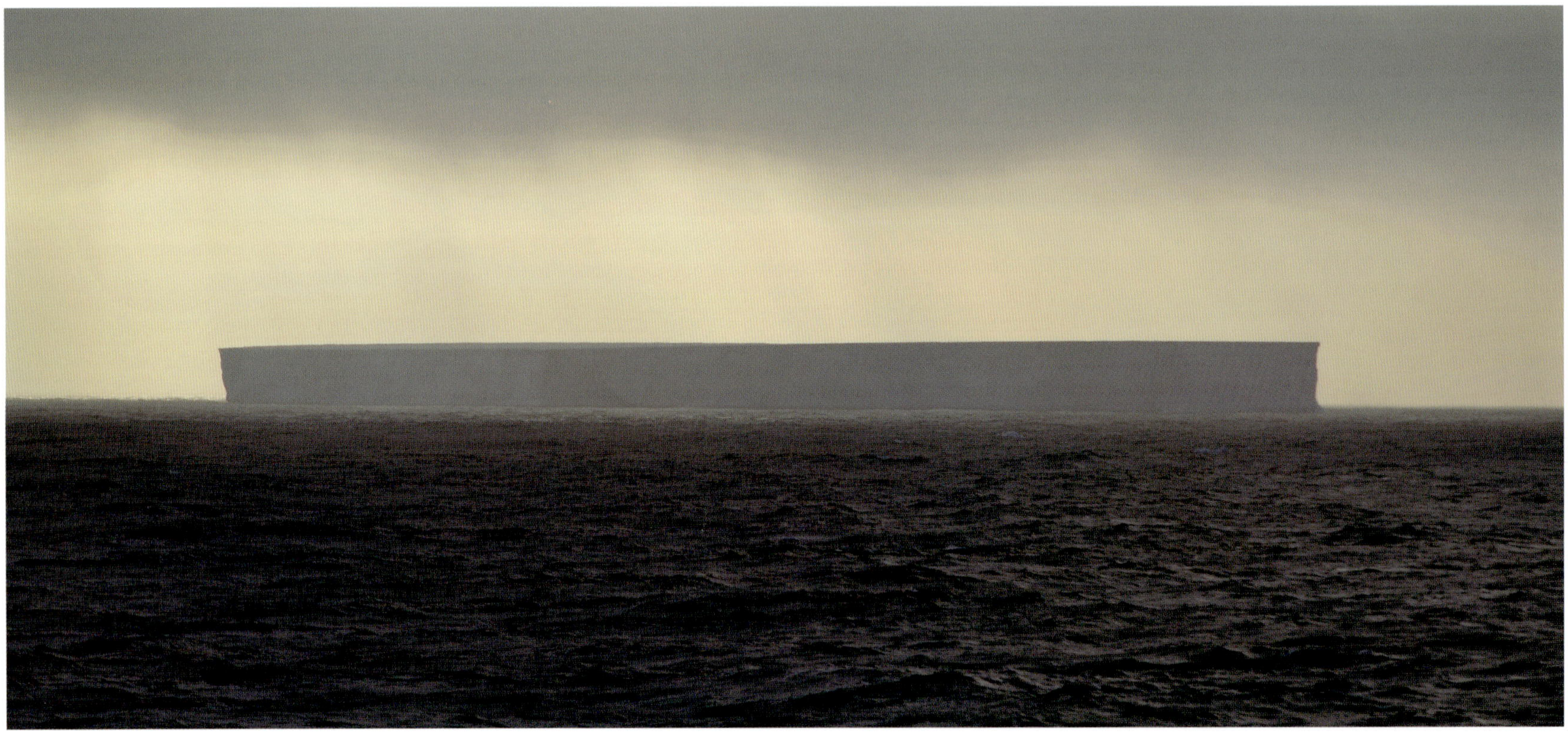

Tabular iceberg

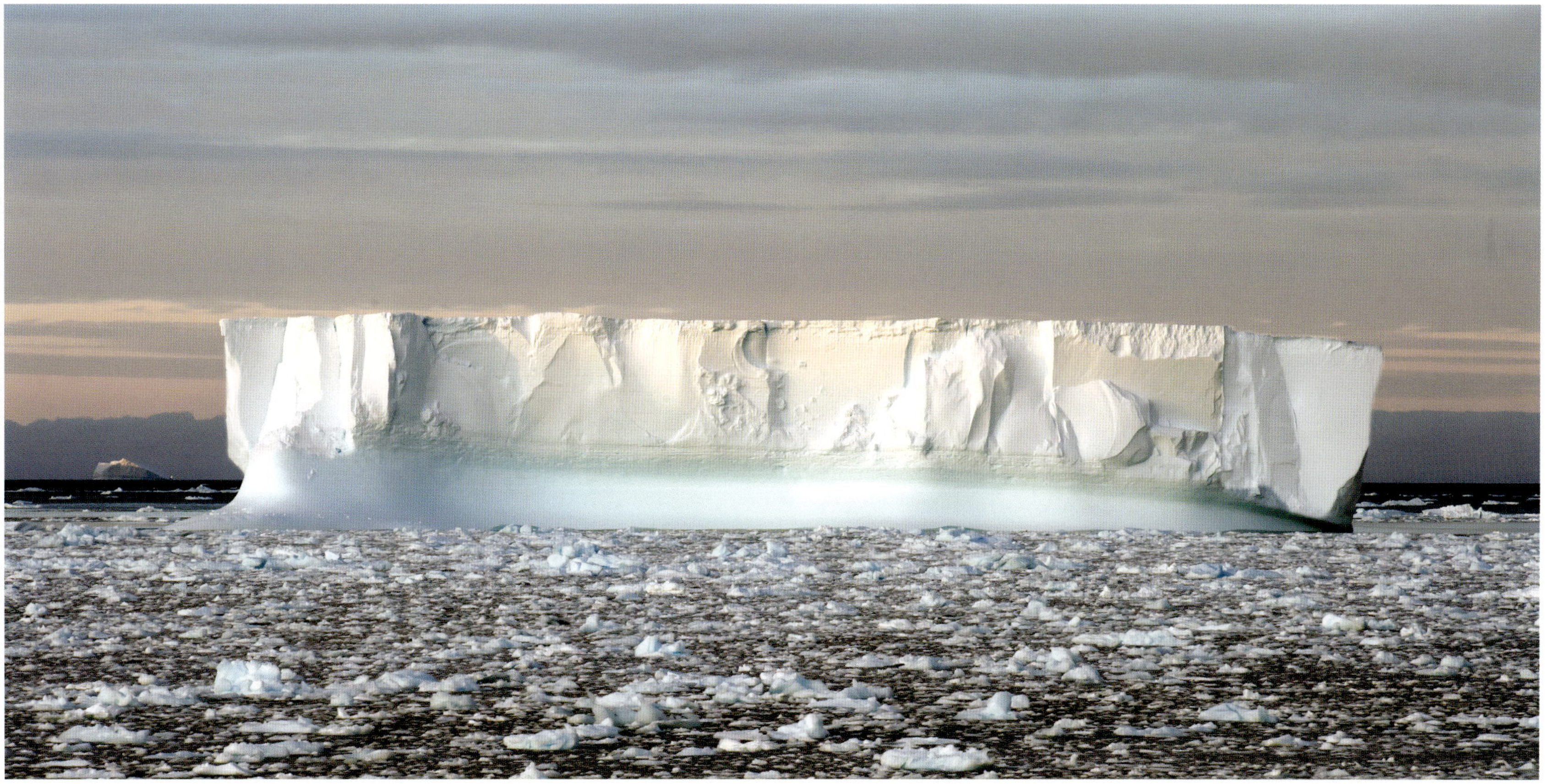

Iceberg and brash ice

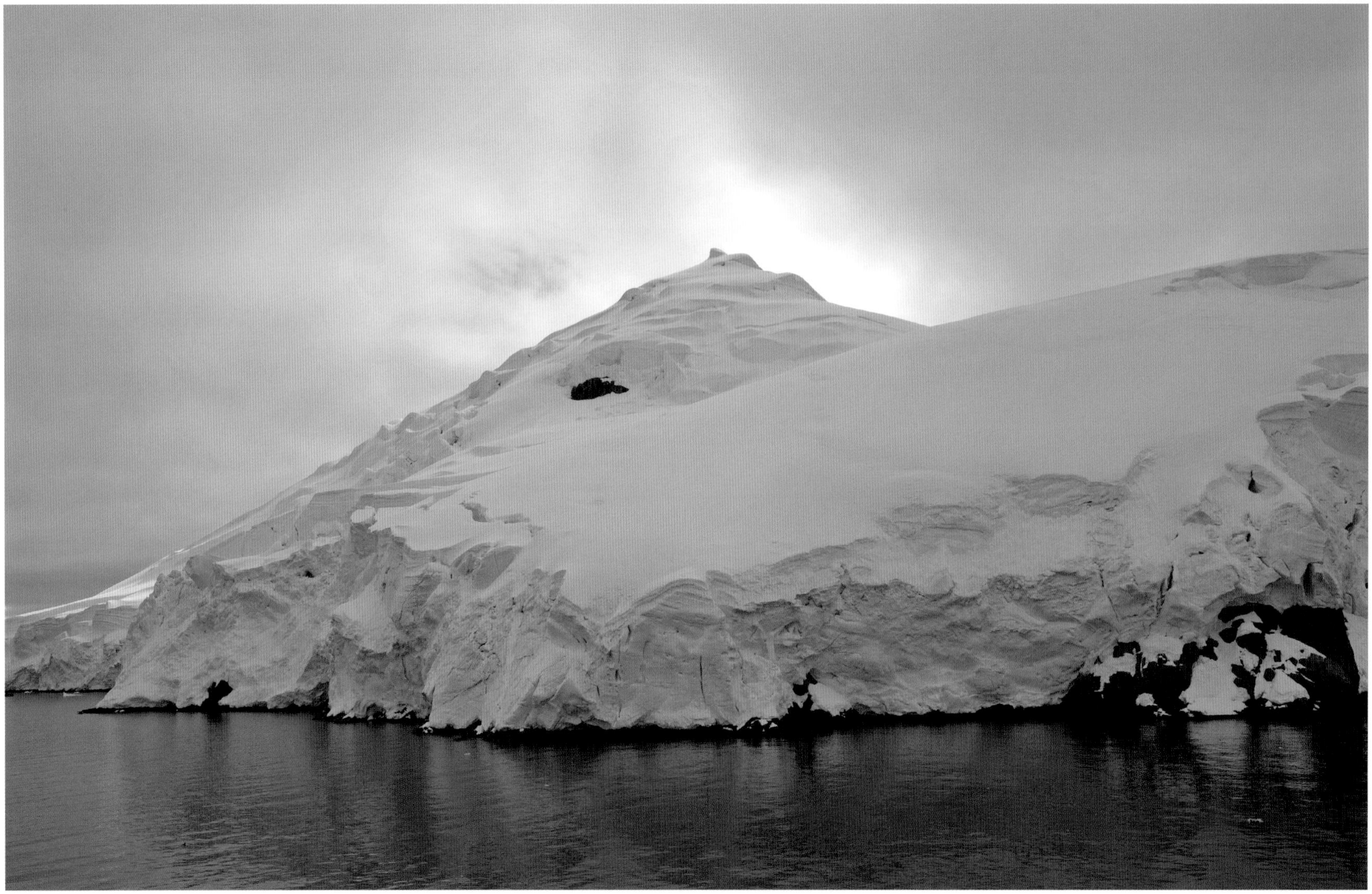

Backlit

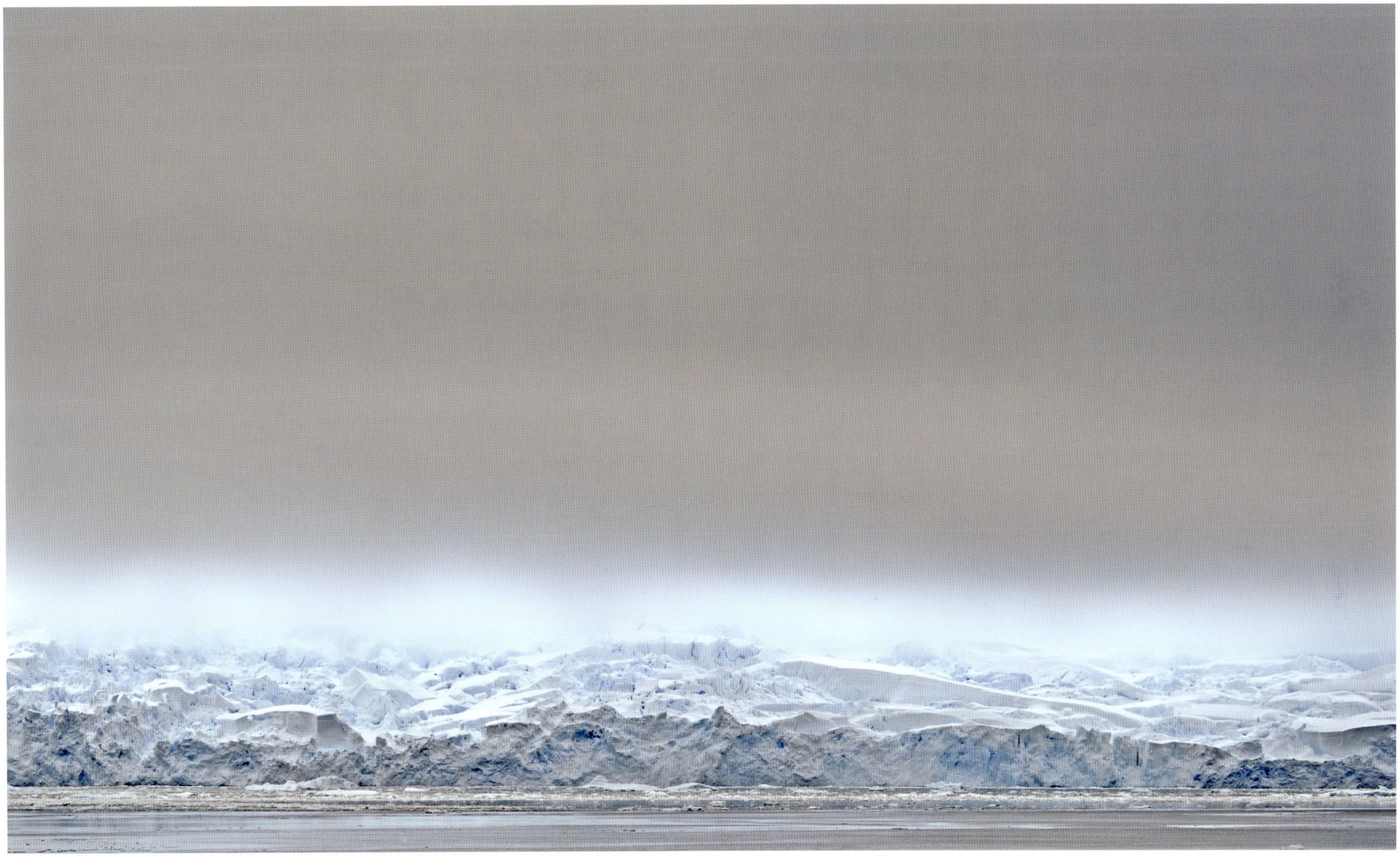

Infused colour, glacier front

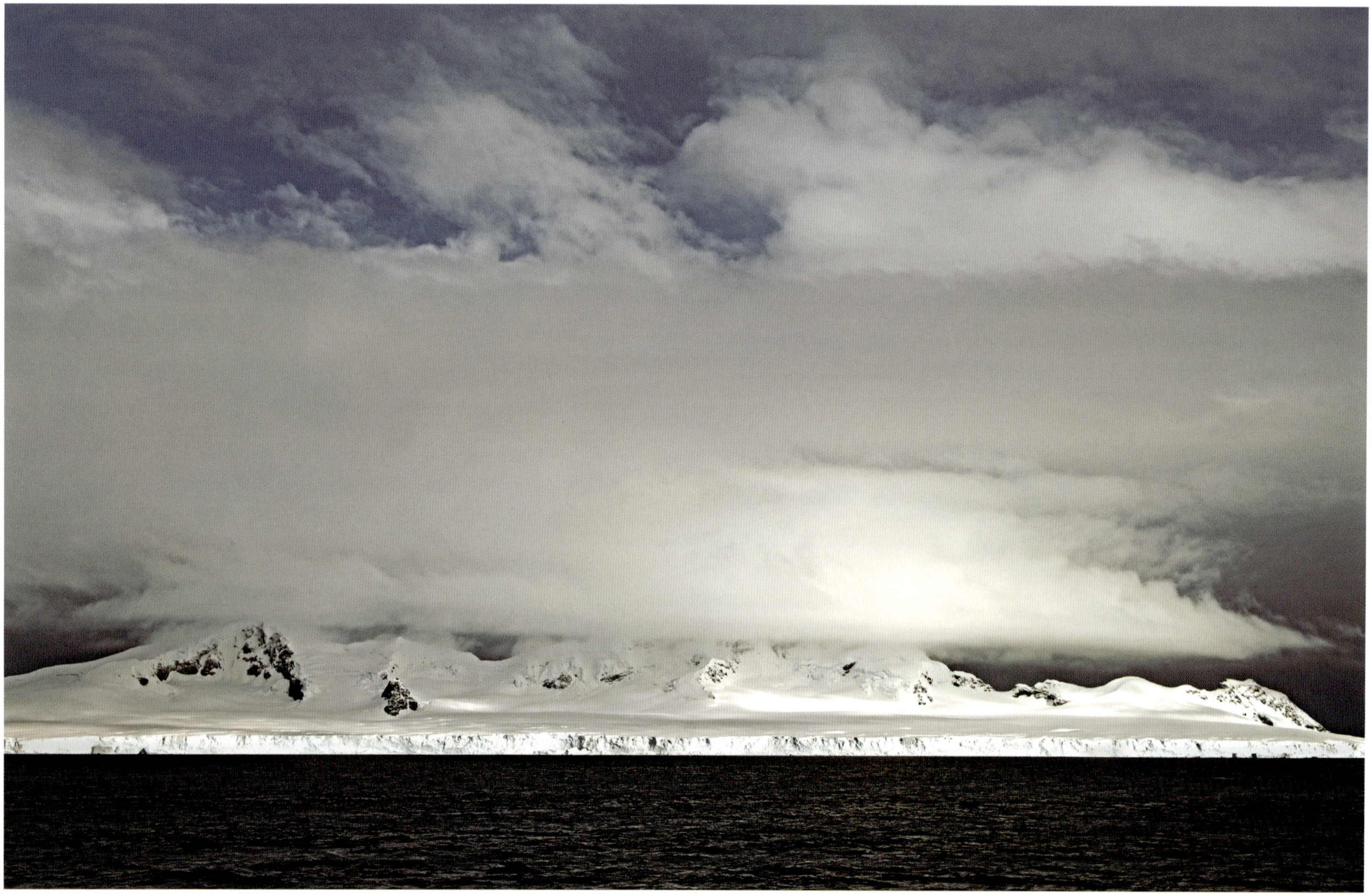

Permeating light

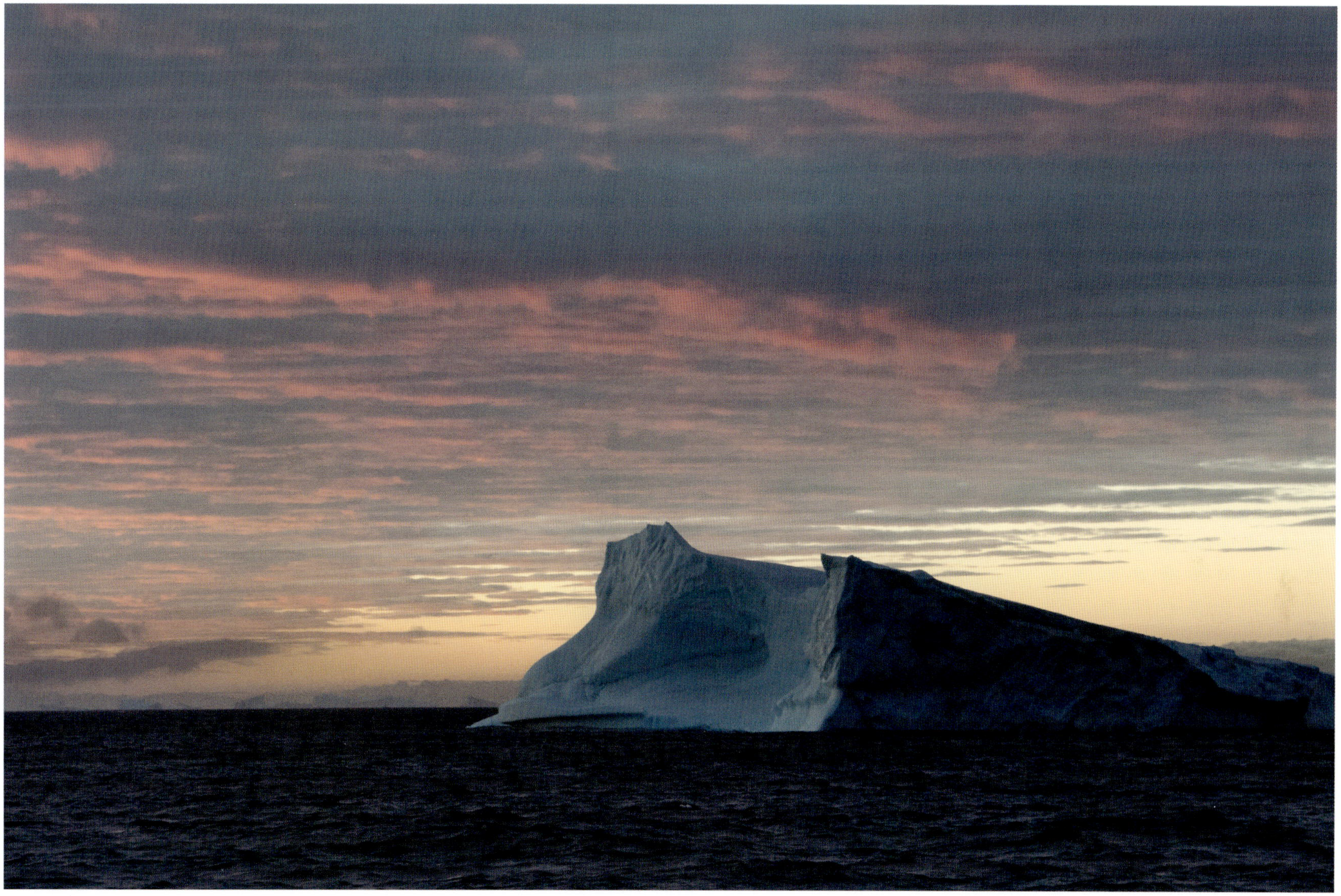

Close of day Antarctica and uncertainty of the night

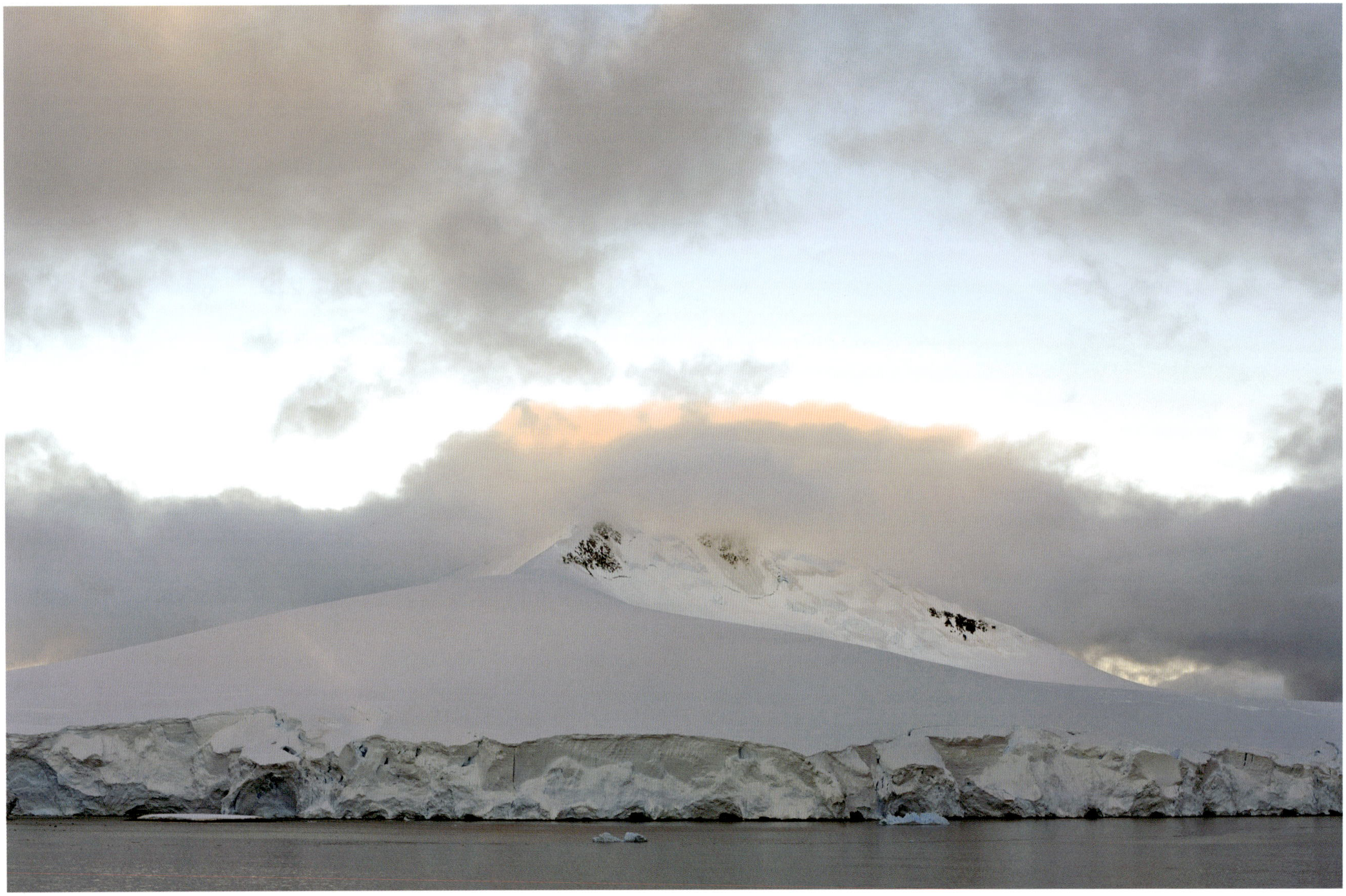

Clouds at dusk

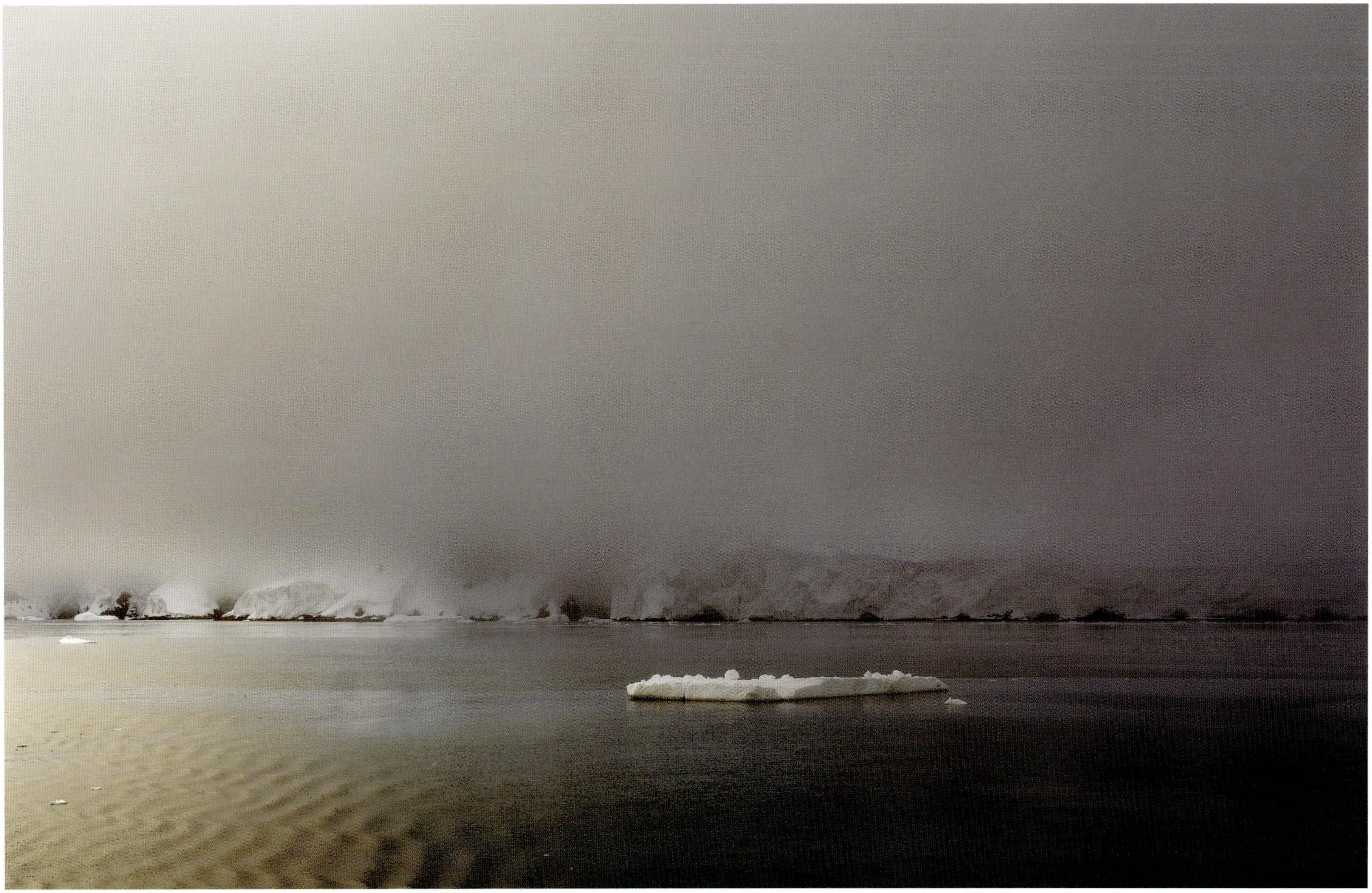

Sunburst

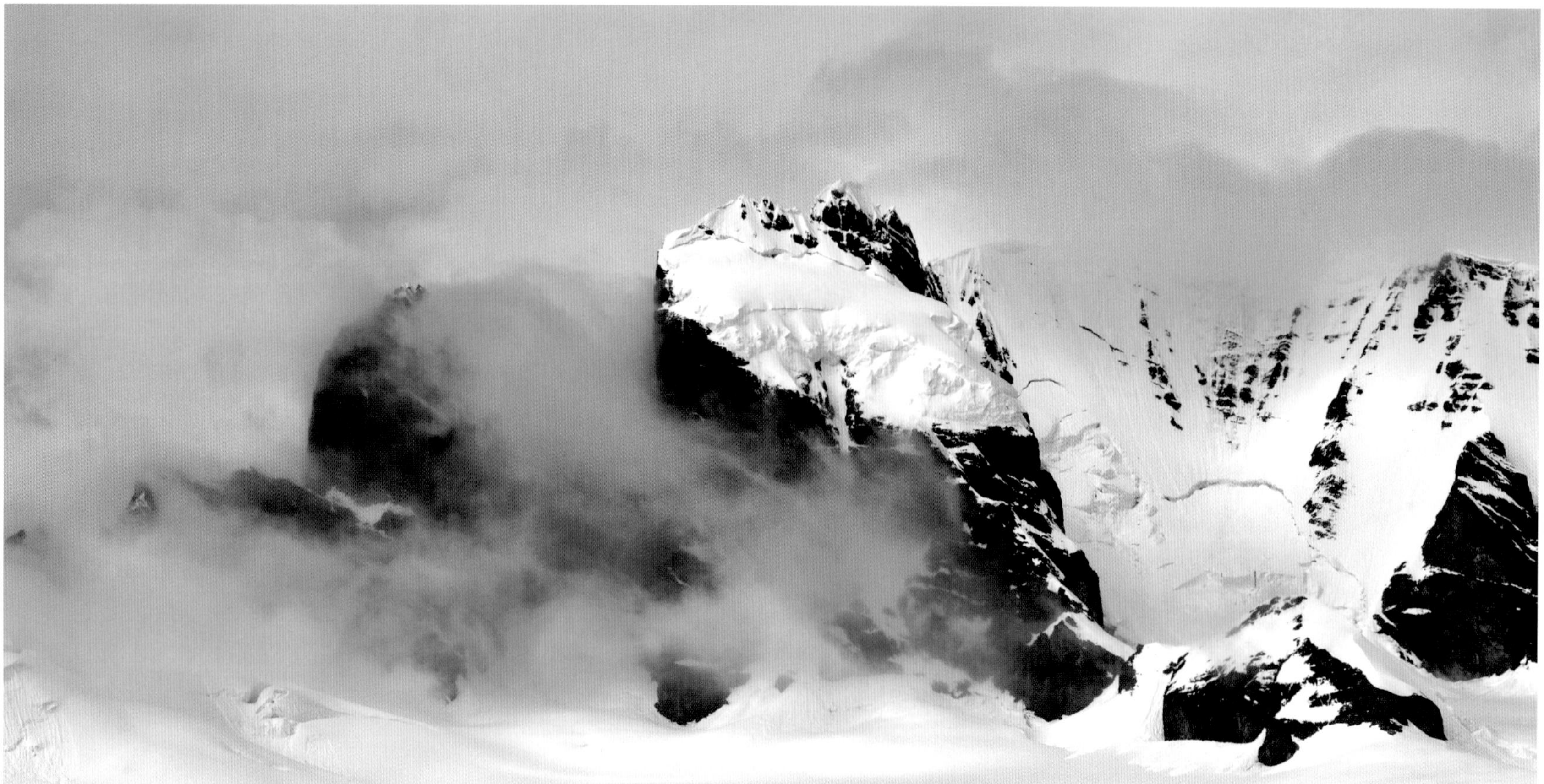

Ice-capped nunatak

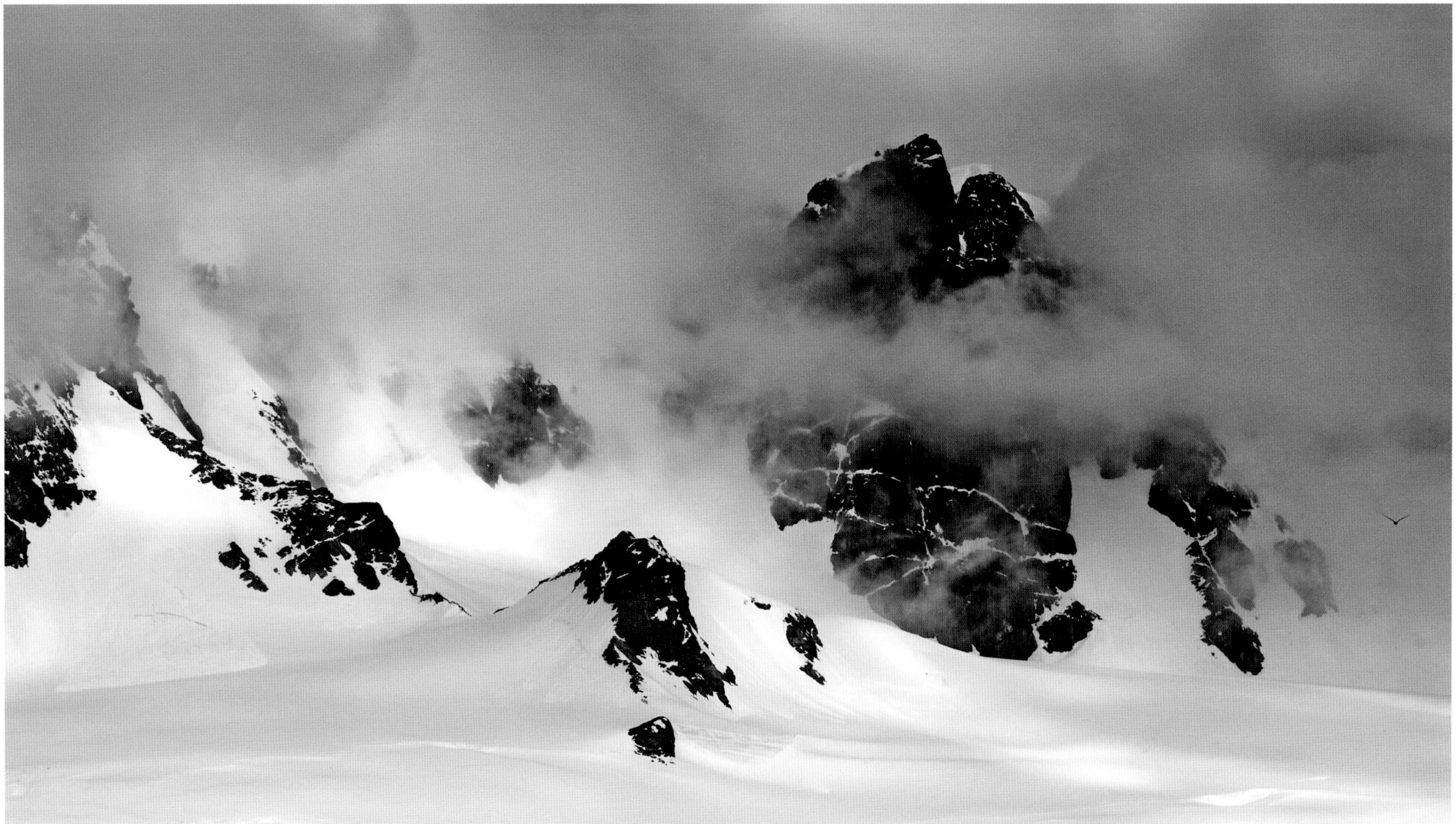

Shrouded peaks

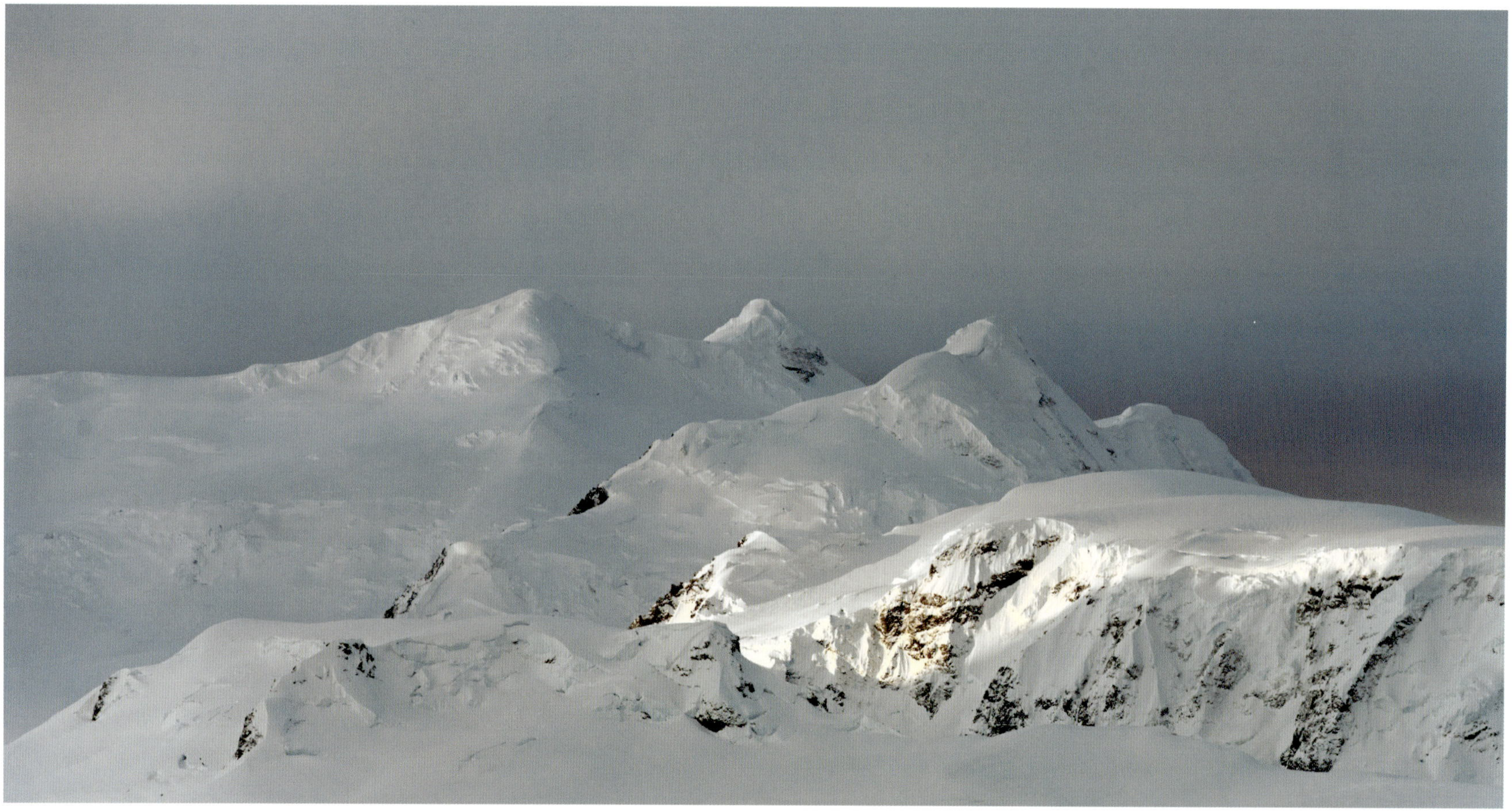

Blue horizon

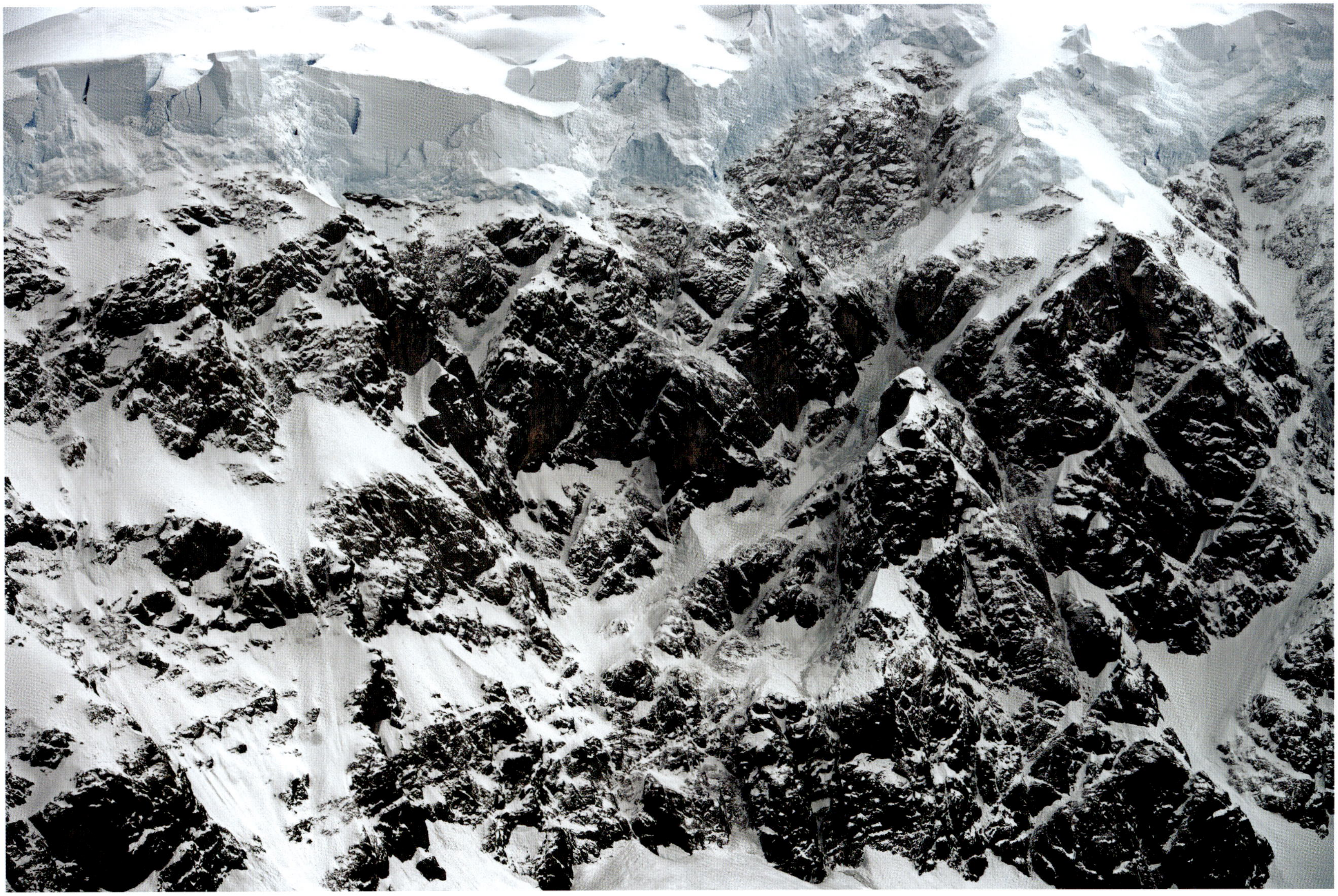

Glacier end

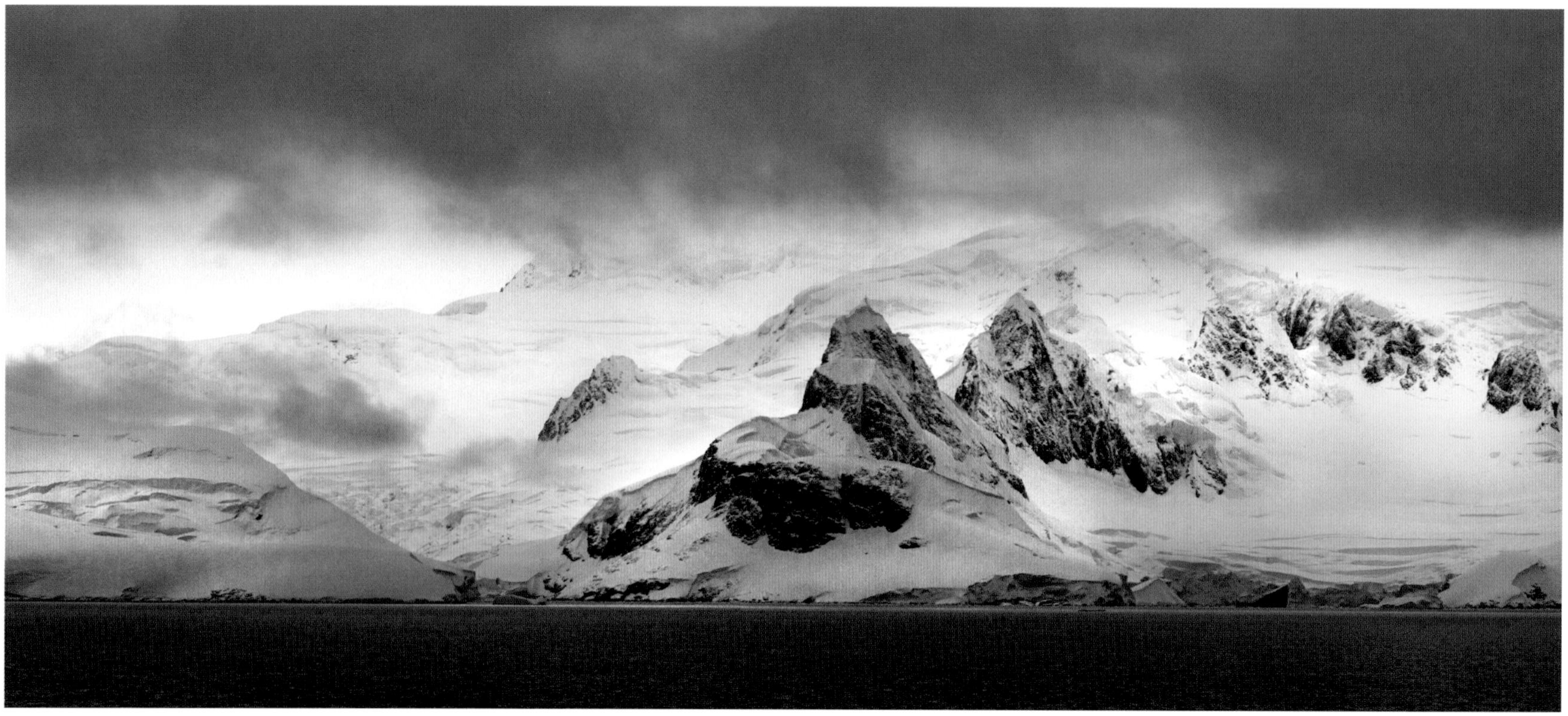

Antarctic coast

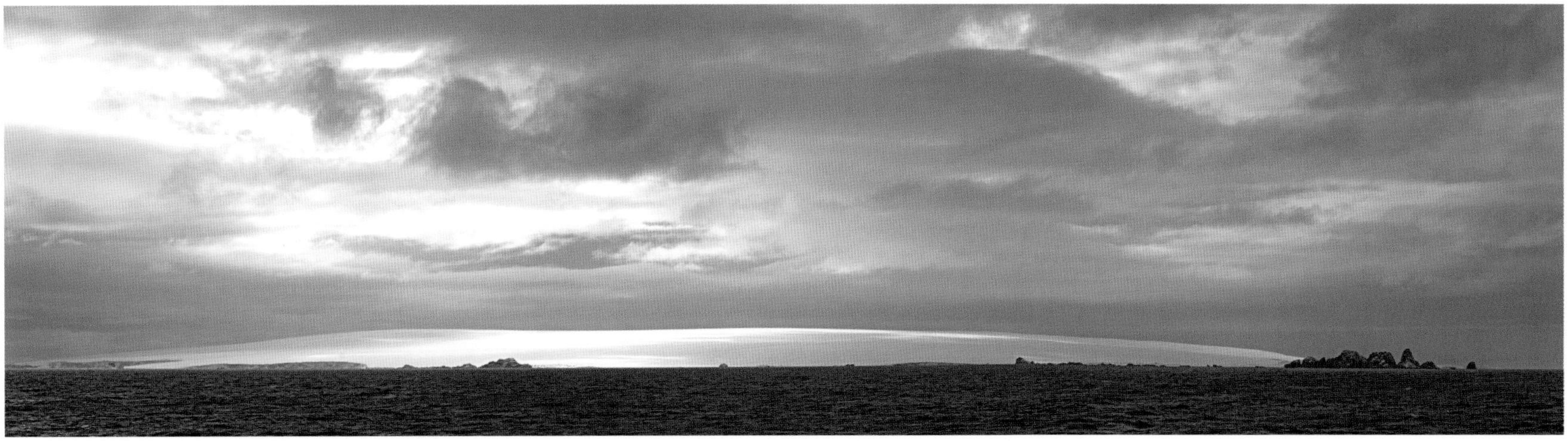

Low Island

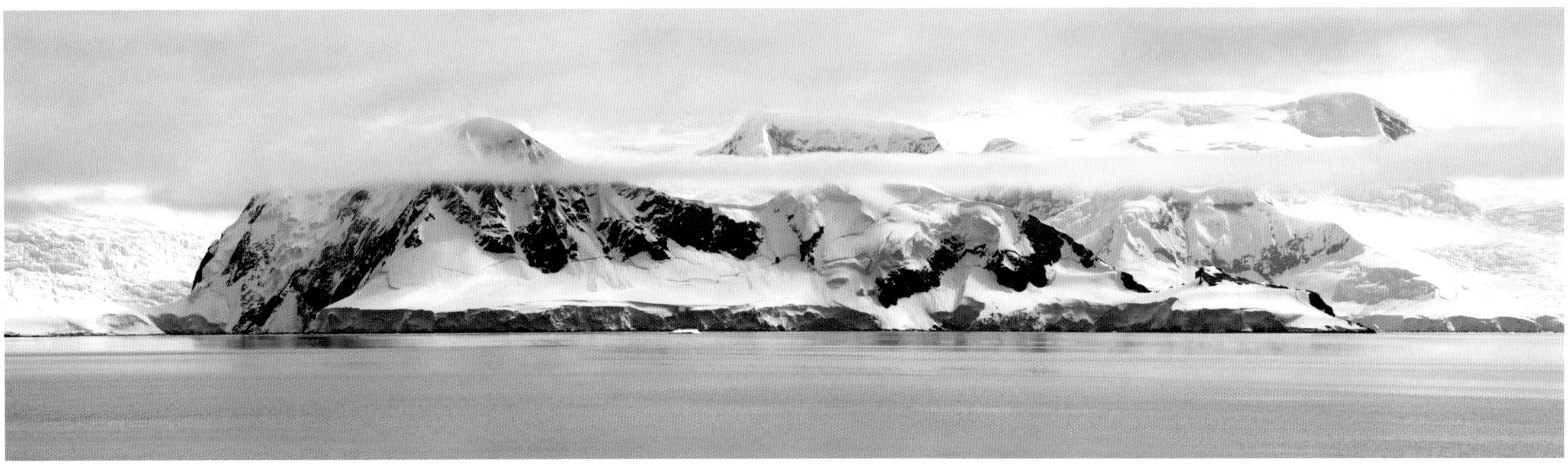

Mountains in the Gerlache Strait

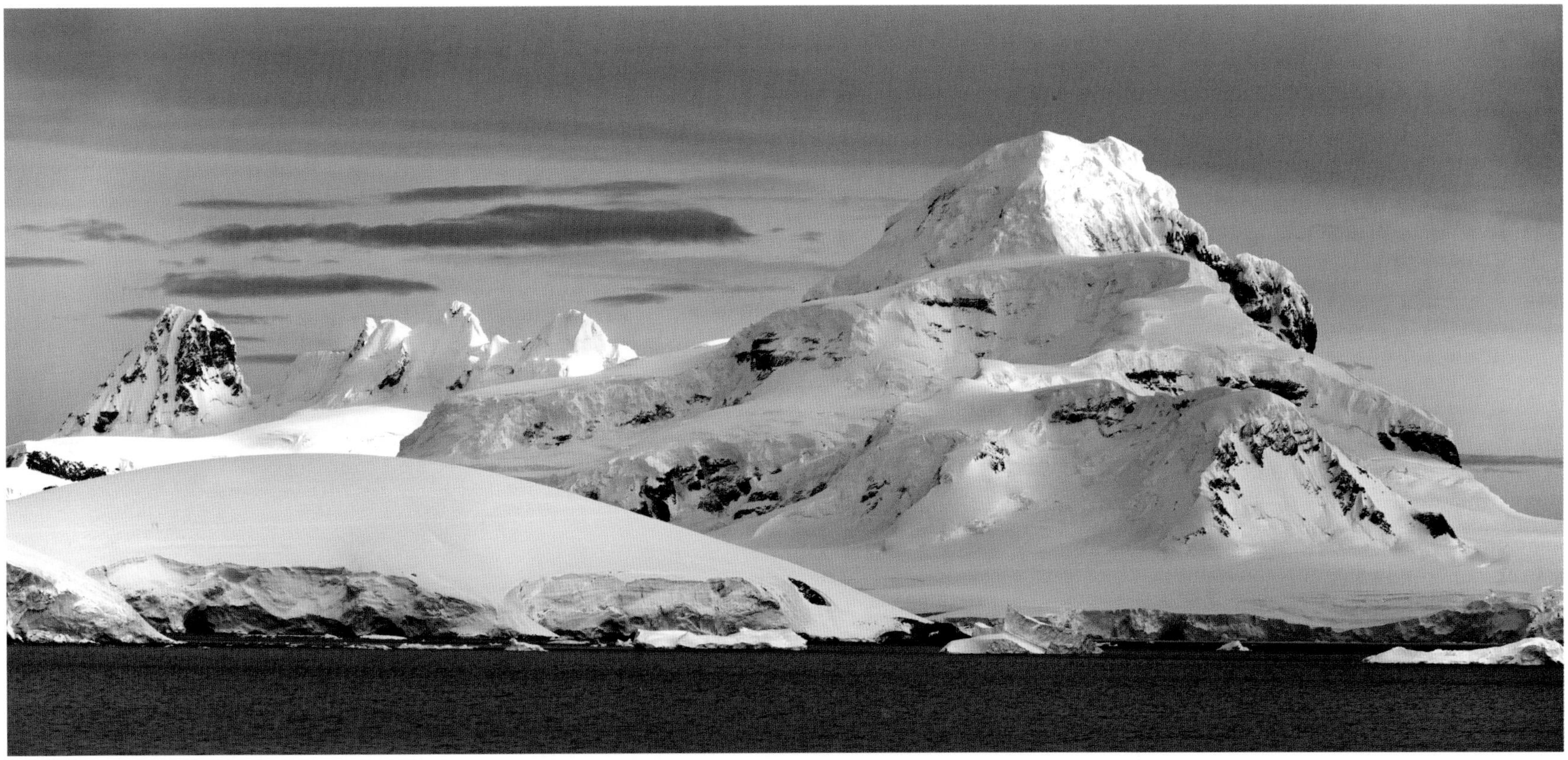

Snow-capped peaks

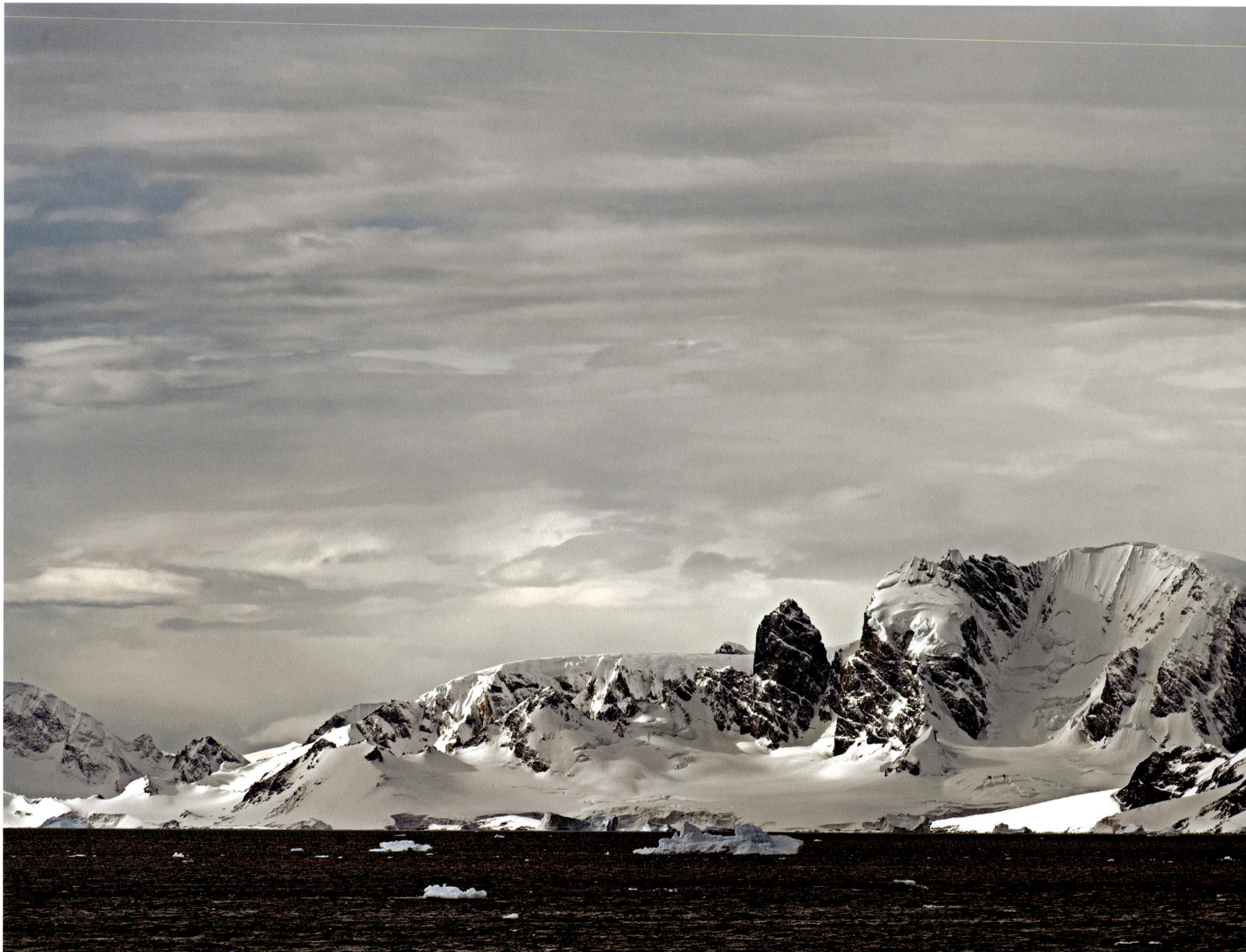

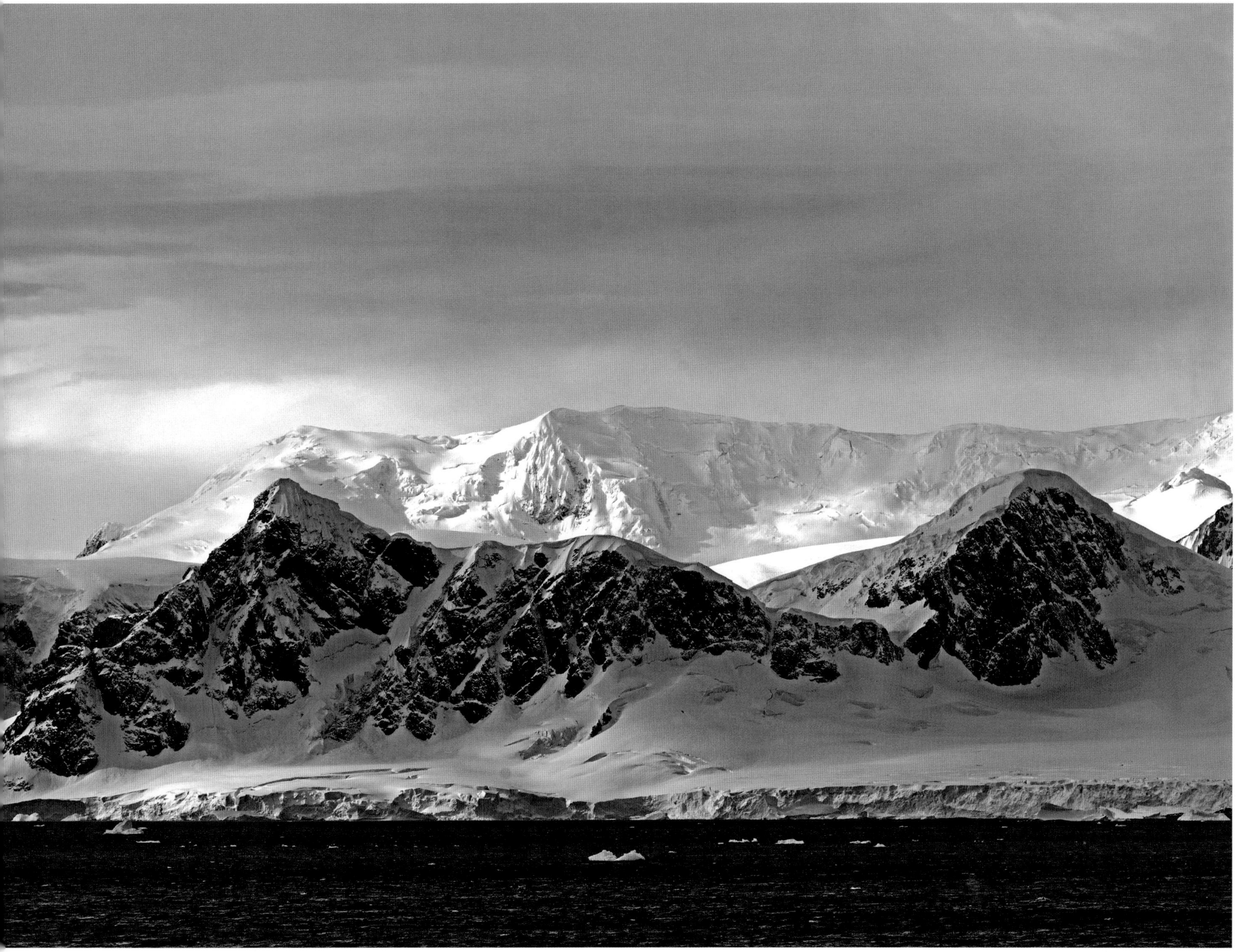

Spigot Peak Errera Channel

IN THIS EXTENSIVE and varied archipelago Smith island has the distinction of having the highest elevation, 2105 metres (6906 feet).

Discovered in 1819 by a notorious sealer and explorer Captain William Smith named the islands New South Britain but that was quickly modified a year later to its present name. Antarctic scholars are slowly warming to the idea that he, an Englishman, and not a Russian nor American, was the true discoverer of the last continent.

This was the last image I took of Antarctica as we entered the Drake Passage for the trip home. Mysterious, imposing, and wrapped in dark clouds, this island massif shines in a brilliant white. A fitting evidence to the beauty and allure of a continent that we are now, after two hundred years, only beginning to appreciate and understand.

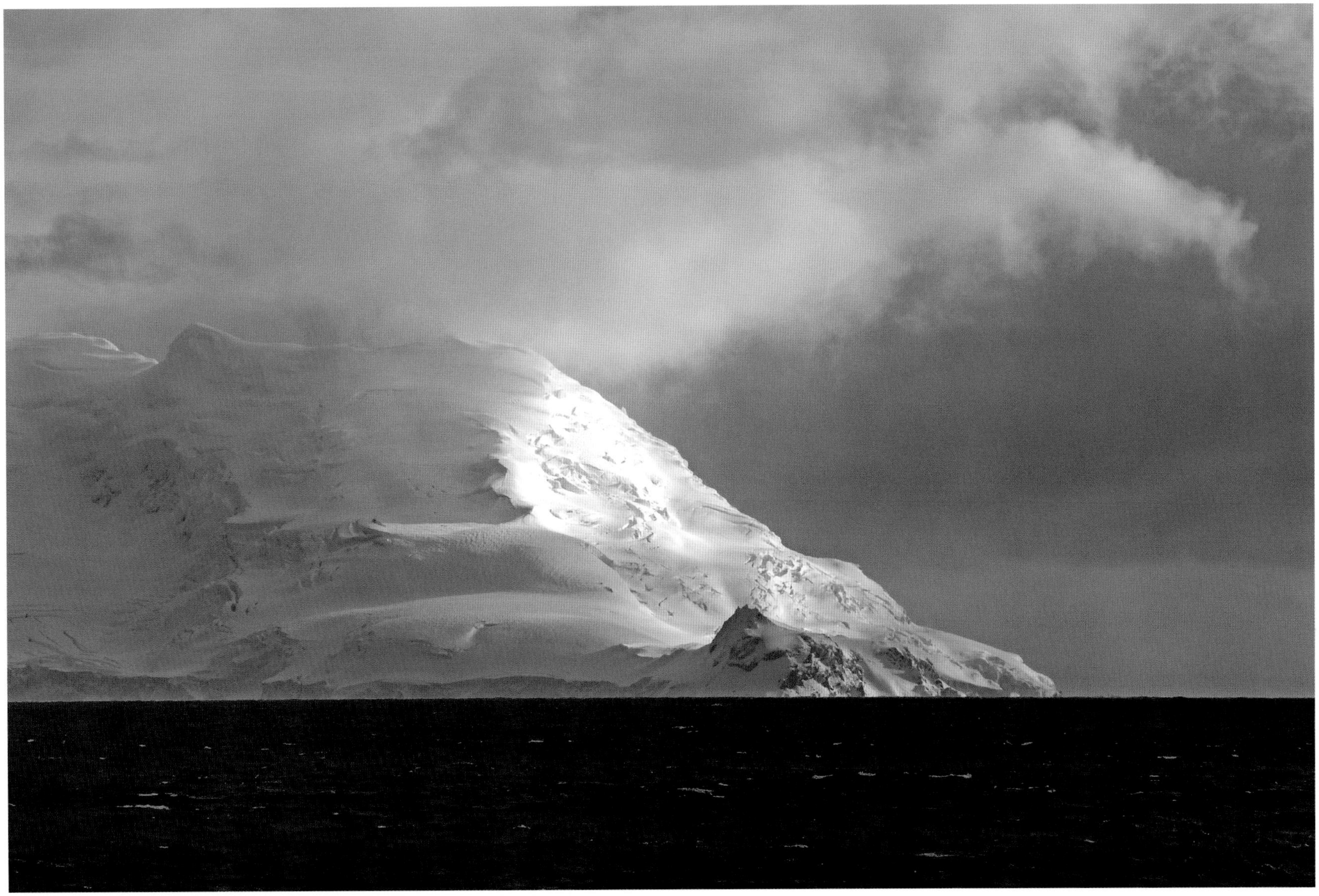

Leaving Antarctica Smith Island South Shetlands